16. JAN 08		
18. MAY		

Reliability Modeling, Analysis and Optimization

SERIES ON QUALITY, RELIABILITY AND ENGINEERING STATISTICS

Series Editors: *M. Xie* (*National University of Singapore*)
T. Bendell (*Nottingham Polytechnic*)
A. P. Basu (*University of Missouri*)

Published

Vol. 1: Software Reliability Modelling
M. Xie

Vol. 2: Recent Advances in Reliability and Quality Engineering
H. Pham

Vol. 3: Contributions to Hardware and Software Reliability
P. K. Kapur, R. B. Garg and S. Kumar

Vol. 4: Frontiers in Reliability
A. P. Basu, S. K. Basu and S. Mukhopadhyay

Vol. 5: System and Bayesian Reliability
Y. Hayakawa, T. Irony and M. Xie

Vol. 6: Multi-State System Reliability
Assessment, Optimization and Applications
A. Lisnianski and G. Levitin

Vol. 7: Mathematical and Statistical Methods in Reliability
B. H. Lindqvist and K. A. Doksum

Vol. 8: Response Modeling Methodology: Empirical Modeling for Engineering and Science
H. Shore

Vol. 9: Reliability Modeling, Analysis and Optimization
H. Pham

Vol. 10: Modern Statistical and Mathematical Methods in Reliability
A. Wilson, S. Keller-McNulty, Y. Armijo and N. Limnios

Vol. 11: Life-Time Data: Statistical Models and Methods
J. V. Deshpande and S. G. Purohit

Series on Quality, Reliability and Engineering Statistics Vol. 9

Reliability Modeling, Analysis and Optimization

Hoang Pham
Rutgers University, USA

World Scientific

NEW JERSEY • LONDON • SINGAPORE • BEIJING • SHANGHAI • HONG KONG • TAIPEI • CHENNAI

Published by

World Scientific Publishing Co. Pte. Ltd.
5 Toh Tuck Link, Singapore 596224
USA office: 27 Warren Street, Suite 401-402, Hackensack, NJ 07601
UK office: 57 Shelton Street, Covent Garden, London WC2H 9HE

British Library Cataloguing-in-Publication Data
A catalogue record for this book is available from the British Library.

RELIABILITY MODELING, ANALYSIS AND OPTIMIZATION
Series on Quality, Reliability and Engineering Statistics — Vol. 9

ISBN 981-256-388-1

Typeset by Stallion Press
Email: enquiries@stallionpress.com

Printed in Singapore by World Scientific Printers (S) Pte Ltd

To

Michelle, Hoang Jr. and David

Preface

In today's technological world, nearly everyone depends upon the continued functioning of a wide array of complex machinery and equipment for our everyday safety, security, mobility and economic welfare. We expect our electric appliances, lights, hospital monitoring control, next-generation aircraft, nuclear power plants, data exchange systems, and aerospace applications, to function whenever we need them. When they fail, the results can be catastrophic, injury or even loss of life. As modern information-age society grows in complexity both in embedded systems and applications, so do the complex problems and challenges in reliability.

This volume presents current research and system modeling and optimization in reliability and its applications by many leading experts in the field. The book comprised of twenty-three chapters, organized in four parts: Reliability Modeling, Software Quality Engineering, Software Reliability Modeling, and Maintenance and Inspection Policies. The subjects covered include system reliability modeling, reliability optimization, software reliability, software quality, maintenance theory, maintenance inspection and policies, engineering reliability analysis, reliability failure analysis, sampling plans and schemes, high performance screening scheme, software development process and improvement, stochastic process modeling, statistical distributions and analysis, fault-tolerant performance, software measurements, software cost effectiveness, queuing theory and applications, system availability, reliability of repairable systems, testing sampling inspection, software capability maturity model, accelerated life modeling, statistical control, and HALT testing.

This volume can serve as a reference for researchers and professors, and also prove useful for practitioners in reliability and maintenance engineering, software and information engineering, and safety and systems engineering. It also can serve as an advanced textbook for graduate and postgraduate students wishing to study and engage in reliability research.

Part I consists of six chapters, focusing on reliability modeling. The numerical scheme proposed in Chapter 1 is devoted to the computing for the marginal distributions of a semi-Markov process with a finite state space in general and the availability of a component with general failure and repair rates in particular. Chapter 2 analytically describes an optimal checkpointing interval that minimizes the mean time to completion of the process for double modular redundant systems with one spare module. This chapter also compares the two numerical schemes where one is the rollback recovery of only two modules and the other is the roll-forward recovery of two and spare modules. Chapter 3 describes some existing statistical control approaches such as cumulative quantity control and cumulative sum control chart which can be used to modeling the time-between-event monitoring events. Some comparisons based on in-control average run length and in-control average time to signal performance are also discussed.

Chapter 4 discusses various stochastic optimal interval policies for certificate revocation list in the public key infrastructure architectures that minimize the expected costs. Chapter 5 describes an unreliable economic manufacturing quality model in a discrete time framework under general discrete failure time and discrete repair time distributions. The optimal production policy is also discussed for geometric and discrete Weibull failure distributions. Chapter 6 investigates accelerated lifetime models using a highly accelerated life testing (HALT) method to predict the product's reliability and lifetime in nominal conditions.

Part II consists of six chapters, focusing on software quality engineering. Chapter 7 introduces the application of Poisson regression analysis to software quality data known to have a Poisson distribution. This chapter develops regression fault models and also compares the predictive quality of the two models to classify the software system into low- and high-risk groups with respect to the number of expected faults. Chapter 8 describes spatial complexity measures of object-oriented software based on the definition and usage of classes and objects. The significance of new spatial complexity measure has been demonstrated based on fifteen object-oriented projects. Chapter 9 conducts software development experiments to classify human factors and their interactions affecting software reliability by considering human factors which consist of inhabitors and inducers. This chapter also introduces a quality engineering approach based on a signal-to-noise ratio to clarify the relationships among human factors and software reliability measured by the number of seeded faults detected by review activities.

Chapter 10 describes an automated and simplified genetic programming based decision tree modeling technique for the software quality classification problems. The model can be used to predict the class membership of software modules depending on the type of quality factors used such as number of faults or code churn. Chapter 11 discusses an approach to quantifying the effort spent performing each process step and correlating that effort to the resulting product's quality. Chapter 12 discusses the software process improvement activities from the two project applications by analyzing on some key process areas from the nonattained requirement items based on the capability maturity model (CMM).

Part III consists of five chapters, focusing on software reliability modeling. Chapter 13 introduces a software reliability growth model which incorporates both imperfect debugging and introduction of defects. The asymptotic properties of the maximum likelihood estimators of the model are also discussed. Chapter 14 introduces a new testing methodology for software systems based on continuous sampling plan. This chapter also describes the performance measures of the proposed plans using Markov approach. In Chapter 15, a software reliability growth model incorporating a fault-detection process during the system test phase of the distributed development environment using a technique of stochastic differential equations of an Ito type is presented. This chapter also discusses optimal software release policies based on the reusable rate of software components which minimizes the expected total software cost.

Chapter 16 discusses an infinite server queuing model considering the time distribution of fault-isolation process based on a concept of a delayed S-shaped software reliability growth model. Chapter 17 discusses the relationship between the number of debugging and the software availability measurement when the system is used intermittently. The time-dependent behaviors of the users and the system alternating between up and down states by a Markov process is also described in this chapter.

Part IV consists of six chapters, focusing on maintenance and inspection policies. Chapter 18 describes the extended inspection model where a system is checked at both periodic times and at the same times as its working times. The total expected cost until the detection of system failure is obtained where the working times of a system are random times. Chapter 19 introduces a decision rule for detecting these failure-prone products under a screening scheme. The screening test can be applied either at the end of the production or after some stress tests, or both, so as to minimize the field failures. This chapter also derives the decision rule based on a modified binomial distribution for fitting real life data from the test.

Chapter 20 discusses the periodic policies for self-diagnosis systems with two types of inspections: Type 1 inspection is done at periodic times jT and Type 2 inspection is done at periodic times knT for some specified n. This chapter also obtains the optimal inspection policies which minimize the expected total cost. Chapter 21 describes an optimal managerial damage level which is below a pre-specified level and derives the expected cost per unit time. This chapter also discusses an optimal policy that minimizes the expected cost.

Chapter 22 introduces a maintenance model for periodically inspected degraded repairable systems subject to degradation and random shocks. This model can be used to determine the optimal preventive maintenance threshold and inspection time that minimizes the average long-run maintenance cost rate. This chapter also presents the optimal solution for the average long-run maintenance cost rate using the Nelder–Mead downhill simplex method. Chapter 23 discusses an age-dependent maintenance model for complex maintenance systems with interaction failures. The optimal policies that minimize the long-run average cost per unit time for various scheduled schemes are also obtained in this chapter.

Hoang Pham

Piscataway, New Jersey

September 2004

List of Contributors

K. K. Aggarwal	GGS Indraprastha University, India
M. Arafuka	Kinjo Gakuin University, Japan
W. Bodhisuwan	King Mongkut's Institute of Technology North Bangkok, Thailand
W.-T. Cheong	National University of Singapore, Singapore
J. K. Chhabra	National Institute of Technology, India
C. Cocozza-Thivent	Université de Marne-la-Vallée, France
T. Dohi	Hiroshima University, Japan
B. Dumon	ISTIA, France
R. Eymard	Université de Marne-la-Vallée, France
T. Fujiwara	Fujitsu Peripherals Limited, Japan
B. C. Giri	Hiroshima University, Japan
T. N. Goh	National University of Singapore, Singapore
F. Guérin	ISTIA, France
C. Hoffman	General Dynamics Decision Systems, USA
S. Hwang	Rutgers University, USA
S. Inoue	Tottori University, Japan
K. Ito	Mitsubishi Heavy Industries, Ltd. Japan
H. Kawai	Tottori University, Japan
T. M. Khoshgoftaar	Florida Atlantic University, USA
H. Kondo	Nanzan University, Japan
P. Lantieri	ISTIA, France
W. Li	Rutgers University, USA

Y. Liu	Florida Atlantic University, USA
R. Matsuda	Tottori University, Japan
S. Mizutani	Aichi Institute of Technology, Japan
S. Nakagawa	Kinjo Gakuin University, Japan
T. Nakagawa	Aichi Institute of Technology, Japan
S. Nakamura	Kinjo Gakuin University, Japan
Y. Okuda	Aichi Institute of Technology, Japan
H. Pham	Rutgers University, USA
Y. Saitoh	Tottori University, Japan
T. Satow	Tottori University, Japan
N. Seliya	Florida Atlantic University, USA
P. R. Sharma	National University of Singapore, Singapore
Y. Singh	GGS Indraprastha University, India
T. Sugiura	Aichi Institute of Technology, Japan
R. M. Szabo	IBM Corporation, USA
Y. Tamura	Tottori University of Environmental Studies, Japan
L.-C. Tang	National University of Singapore, Singapore
K. Tokuno	Tottori University, Japan
G. Twaites	General Dynamics Advanced Information Systems, USA
M. Uchida	Tottori University, Japan
M. Xie	National University of Singapore, Singapore
S. Yamada	Tottori University, Japan
P. Zeephongsekul	RMIT University, Australia
Q. Zhao	Tottori University, Japan

Contents

Preface vii

List of Contributors xiii

I. RELIABILITY MODELING

1. **Numerical Computation of the Marginal Distributions of a Semi-Markov Process** 1
   C. Cocozza-Thivent and R. Eymard

2. **Optimal Checkpointing Interval for Task Duplication with Spare Processing** 29
   S. Nakagawa, Y. Okuda and S. Yamada

3. **Monitoring Inter-Arrival Times with Statistical Control Charts** 43
   P. R. Sharma, M. Xie and T. N. Goh

4. **Optimal Interval of CRL Issue in PKI Architecture** 67
   M. Arafuka, S. Nakamura, T. Nakagawa and H. Kondo

5. Discrete-Time Economic Manufacturing Quantity Model with Stochastic Machine Breakdown and Repair 81

B. C. Giri and T. Dohi

6. Applying Accelerated Life Models to HALT Testing 107

F. Guérin, P. Lantieri and B. Dumon

II. SOFTWARE QUALITY ENGINEERING

7. A Poisson Regression Model of Software Quality: A Comparative Study 131

T. M. Khoshgoftaar and R. M. Szabo

8. Measurement of Object-Oriented Software Understandability Using Spatial Complexity 155

J. K. Chhabra, K. K. Aggarwal and Y. Singh

9. A Quality Engineering Approach to Human Factors in Design-Review Process for Software Reliability Improvement 183

S. Yamada and R. Matsuda

10. Tree-Based Software Quality Classification Using Genetic Programming 201

T. M. Khoshgoftaar, Y. Liu and N. Seliya

11. An Approach to Quantifying Process Cost and Quality 225

G. Twaites and C. Hoffman

12. **Software Process Improvement Activities Based on CMM** 273
T. Fujiwara and S. Yamada

III. SOFTWARE RELIABILITY MODELING

13. **Asymptotic Properties of a Software Reliability Growth Model with Imperfect Debugging: A Martingale Approach** 289
W. Bodhisuwan and P. Zeephongsekul

14. **A Two-Level Continuous Sampling Plan for Software Systems** 315
S. Hwang and H. Pham

15. **Software Reliability Analysis and Optimal Release Problem Based on a Flexible Stochastic Differential Equation Model in Distributed Development Environment** 339
M. Uchida, Y. Tamura and S. Yamada

16. **An Extended Delayed *S*-Shaped Software Reliability Growth Model Based on Infinite Server Queuing Theory** 357
S. Inoue and S. Yamada

17. **Disappointment Probability Based on the Number of Debuggings for Operational Software Availability Measurement** 373
Y. Saitoh, K. Tokuno and S. Yamada

IV. MAINTENANCE AND INSPECTION POLICIES

18. Optimal Random and Periodic Inspection Policies 393
T. Sugiura, S. Mizutani and T. Nakagawa

19. Screening Scheme for High Performance Products 405
W.-T. Cheong and L.-C. Tang

20. Optimal Inspection Policies for a Self-Diagnosis System with Two Types of Inspections 417
S. Mizutani, T. Nakagawa and K. Ito

21. Maintenance of a Cumulative Damage Model and Its Application to Gas Turbine Engine of Co-Generation System 429
K. Ito and T. Nakagawa

22. An Inspection-Maintenance Model for Degraded Repairable Systems 439
W. Li and H. Pham

23. Age-Dependent Failure Interaction 459
Q. Zhao, T. Satow and H. Kawai

Index 485

CHAPTER 1

Numerical Computation of the Marginal Distributions of a Semi-Markov Process

C. Cocozza-Thivent* and R. Eymard†

Laboratoire d'Analyse et de Mathématiques Appliquées (CNRS-UMR 8050), Université de Marne-la-Vallée, Cité Descartes, 5 boulevard Descartes, Champs sur Marne, 77454 Marne-La-Vallée Cedex 2, France

**christiane.cocozza@univ-mlv.fr*

†robert.eymard@univ-mlv.fr

1. Introduction

Industrial devices must be designed to prevent possible severe consequences from the failure of a device component. A way to ameliorate this design is to use probabilistic models of the device, from which technical and economical expectations can be drawn. Let us first give a mathematical background of such models. We consider a semi-Markov process $(\eta_t)_{t\geq 0}$ taking its values in a finite space E. Let $T_0 = 0$ and T_n $(n \geq 1)$ be the successive jump times of this process. We assume that the semi-Markov kernel of the process has a density q with respect to the Lebesgue measure. This means that, for all $i_0, i_1, \ldots, i_{n-1}, i, j \in E$, all $0 < s_1 < \cdots < s_n$, and all bounded measurable function f defined on $\mathbb{R}_+$, we have

$$\begin{aligned}
&\mathbb{E}(1_{\{\eta_{T_{n+1}}=j\}} f(T_{n+1} - T_n)/\eta_0 = i_0, \eta_{T_1} = i_1, T_1 = s_1, \ldots, \\
&\quad \eta_{T_{n-1}} = i_{n-1}, T_{n-1} = s_{n-1}, \eta_{T_n} = i, T_n = s_n) \\
&\quad = \mathbb{E}(1_{\{\eta_{T_{n+1}}=j\}} f(T_{n+1} - T_n)/\eta_{T_n} = i) \\
&\quad = \int_{\mathbb{R}_+} f(t)\, q(i, j, t) dt\,.
\end{aligned}$$

Let us define the transition rates $a(i, j, t)$ between states i and j at time t by:

$$a(i, j, t) = \frac{q(i, j, t)}{\mathbb{P}(T_1 > t/\eta_0 = i)} = \frac{q(i, j, t)}{\int_t^{+\infty} \sum_{k \in E} q(i, k, u) du}\,. \tag{1}$$

Let us note that, since the values T_n, $n \in \mathbb{N}_*$, are jump times, the relation $0 = q(i, i, t) = a(i, i, t)$ must hold for all $t \in \mathbb{R}_+$.

Remark 1. An important case is the study of some component with a general failure rate, denoted by $\lambda(t)$, and a general repair rate, denoted by $\mu(t)$. The state of this component is then described by an alternating renewal process, i.e., a semi-Markov process taking its values in the set $\mathbb{E} = \{0, 1\}$, the values 1 and 0, respectively, representing the up-state and the down-state. The transition rates are then given by:

$$a(1, 0, t) = \lambda(t)\,, \quad a(0, 1, t) = \mu(t)\,.$$

The following properties help to understand the meaning of the transition rates a. It can be shown[1] that:

$$\begin{aligned}
&\mathbb{P}(\eta_{T_1} = j,\ T_1 \le t/\eta_0 = i) \\
&\quad = \int_0^t a(i, j, s) \exp\Big(- \int_0^s \sum_{k \in E} a(i, k, u) du \Big) ds\,.
\end{aligned}$$

Therefore we get

$$\mathbb{P}(T_1 > t/\eta_0 = i) = \exp\Big(- \int_0^t \sum_{k \in E} a(i, k, u) du \Big), \tag{2}$$

meaning that the hazard rate of T_1 knowing $\{\eta_0 = i\}$ is given by:

$$b(i, t) = \sum_{j \in E} a(i, j, t), \quad \forall t \in \mathbb{R}_+ .$$

Thus, giving q or a are equivalent, since a is defined from q by Eq. (1) and q is computed from the values of a, using the relation,

$$q(i, j, t) = a(i, j, t) \exp\left(- \int_0^t \sum_{k \in E} a(i, k, u) du \right).$$

Returning to the example given in Remark 1, the availability of the component is then defined by $A(t) = \mathbb{P}(\eta_t = 1)$, that is, one marginal distribution of the process $(\eta_t)_{t \geq 0}$.

This chapter presents a new method to approximate the marginal distributions of a general semi-Markov process in the case of any initial distribution (i.e., any distribution of the process at time 0). In Sec. 2, we obtain the equations satisfied by these marginal distributions. Indeed, introducing the variable X_t, defined as the elapsed time without a jump, the equations fulfilled by the marginal distributions of the Markov process (η_t, X_t) are shown to be the solutions of some transport equations. Since the boundary conditions of these equations are expressed under an integral formulation, no analytical solution can be obtained in the general case. However, in the particular framework of Remark 1, three numerical methods derived from the renewal theory[2] allow a direct computation of the availability (i.e., $\mathbb{P}(\eta_t = 1)$), using discretization schemes for the resolution of Volterra integral equations. Unfortunately, the third method, which appears to present the best efficiency in most cases, fails in some realistic situations (for example, in the case where the failure rate of a component is much smaller than the repair rate). Moreover, the adaptation of the methods given in Ref. 2 to the general case of semi-Markov processes remains to be carefully studied, since no straightforward implementation seems to hold.

Therefore, in Sec. 3, a new numerical algorithm is shown to deliver a convergent approximation of the marginal distributions in the general case of semi-Markov processes. Since the equations satisfied by these distributions are transport equations, a finite volume method is used. An advantage of this method is that it gives the possibility to handle the case of any initial distribution, whatever regularity is considered.

Finally, in Sec. 4, numerical examples in comparison with the methods given in Ref. 2, show an admissible but imprecise solution when other methods fail.

2. Equations for the Marginal Distributions

2.1. *The general case*

We denote by $C_b(E \times \mathbb{R}_+)$ the Banach space of all real bounded functions, which are continuous with respect to the real argument and by $C_b^1(E \times \mathbb{R}_+)$ the class of functions belonging to $C_b(E \times \mathbb{R}_+)$, which are continuously differentiable with respect to the real argument and whose derivatives belong to $C_b(E \times \mathbb{R}_+)$.

Let X_t be the elapsed time without a jump at time t:

$$X_t = t - T_n \quad \text{if } T_n \leq t < T_{n+1} \,.$$

Then the process $(\eta_t, X_t)_{t \geq 0}$ is a Markov process, taking its values in $E \times \mathbb{R}_+$. In all the following, we assume that a is continuous. The following proposition can be proven in a more general case.[1,3]

Proposition 1. *For all* $h \in C_b^1(E \times \mathbb{R}_+)$, *let us define*:

$$Lh(i, x) = \sum_{j \in E} a(i, j, x)(h(j, 0) - h(i, x)) + \frac{\partial h}{\partial x}(i, x) \,.$$

Then, the following equation holds:

$$\mathbb{E}(h(\eta_t, X_t)) = \mathbb{E}(h(\eta_0, X_0)) + \int_0^t \mathbb{E}(Lh(\eta_s, X_s))ds \,. \tag{3}$$

Let ρ_t be the probability distribution of (η_t, X_t). It can be written $\rho_t(dx) = (\rho_t(i, dx))_{i \in E}$. Heuristically, we have

$$\rho_t(i, dx) = \mathbb{P}(\eta_t = i, X_t \in [x, x + dx]) ,$$

and more precisely, for all $h \in C_b(E \times \mathbb{R}_+)$

$$\int h \, d\rho_t = \sum_{i \in E} \int_0^{+\infty} h(i, x) \rho_t(i, dx) .$$

Thus, Eq. (3) can also be written as:

$$\begin{aligned}
&\sum_{i \in E} \int_0^{+\infty} h(i, x) \rho_t(i, dx) \\
&\quad = \sum_{i \in E} \int_0^{+\infty} h(i, x) \rho_0(i, dx) \\
&\qquad + \sum_{i \in E} h(i, 0) \int_0^t \sum_{j \in E} \int_0^{+\infty} a(j, i, x) \rho_s(j, dx) ds \\
&\qquad - \sum_{i \in E} \int_0^t \int_0^{+\infty} b(i, x) h(i, x) \rho_s(i, dx) ds \\
&\qquad + \sum_{i \in E} \int_0^t \int_0^{+\infty} \frac{\partial h}{\partial x}(i, x) \rho_s(i, dx) ds . \qquad (4)
\end{aligned}$$

Our purpose is to find a numerical approximation of ρ_t, viewed as the solution of Eq. (4).

2.2. *Case of an initial distribution given by a density*

Let us suppose that the initial distribution has a density with respect to the Lebesgue measure, i.e., for all $i \in E$, $\rho_0(i, dx)$ can be written:

$$\rho_0(i, dx) = p_0(i, x) dx .$$

It can then be shown that, for all $i \in E$, there exists a function $p_t(i, x)$ such that:

$$\rho_t(i, dx) = p_t(i, x) dx .$$

From Eq. (4), we get that this function verifies:

$$\begin{aligned}\int_0^{+\infty} g(x)p_t(i,x)dx &= \int_0^{+\infty} g(x)p_0(i,x)dx \\ &\quad + g(0)\int_0^t \sum_j a(j,i,x)p_s(j,x)dx\,ds \\ &\quad - \int_0^t \int_0^{+\infty} b(i,x)g(x)p_s(i,x)dx\,ds \\ &\quad + \int_0^t \int_0^{+\infty} g'(x)p_s(i,x)dx\,ds\end{aligned}$$

for all $i \in E$ and $g \in C_b^1(\mathbb{R}_+)$. Assuming that for all $j \in E$, the function $(s,x) \to p_s(j,x)$ is continuously differentiable, we can integrate by parts the last term of the above equation. We then get, by identification, that the function $p_t(i,x)$ is, for all $i \in E$ and $x \in \mathbb{R}_+$, the solution of the following system of linear hyperbolic equations:

$$\frac{\partial}{\partial t}p_t(i,x) + \frac{\partial}{\partial x}p_t(i,x) = -b(i,x)p_t(i,x)\,, \tag{5}$$

with the coupled boundary condition

$$\begin{aligned}p_t(i,0) &= \sum_{j\in E}\int_0^{+\infty} a(j,i,x)p_t(j,x)dx \\ &= \sum_{j\in E}\int_0^{+\infty} a(j,i,x)\rho_t(j,dx)\,.\end{aligned} \tag{6}$$

2.3. *Case of a Dirac initial distribution*

Let us now suppose that there exists $(i_0, x_0) \in E \times \mathbb{R}_+$ such that:

$$\mathbb{P}(\eta_0 = i_0, X_0 = x_0) = 1\,.$$

In such a case, the measure $\rho_t(i_0, dx)$ is no more absolutely continuous with respect to the Lebesgue measure. Indeed, the probability that $X_t = x_0 + t$, which means that no jump occurs before time t, is given by:

$$\alpha(t) = \mathbb{P}(T_1 > t/\eta_0 = i_0, X_0 = x_0) = \exp\left(-\int_0^t b(i_0, x_0+u)du\right).$$

Then, the marginal distributions are given by:

$$\rho_t(i, dx) = 1_{\{i=i_0\}}\alpha(t)\delta_{x_0+t}(dx) + p_t(i, x)dx \,. \tag{7}$$

Suppose that the functions $p_t(j, x)$ are continuously differentiable with respect to x. Following the same steps as in Sec. 2.2, we obtain that Eq. (5) is satisfied, with the initial condition:

$$p_0(i, x) = 0$$

and the boundary condition:

$$p_t(i, 0) = a(i_0, i, x_0 + t)\alpha(t) + \sum_{j\in E}\int_0^{+\infty} a(j, i, x)p_t(j, x)dx \tag{8}$$

$$= \sum_{j\in E}\int_0^{+\infty} a(j, i, x)\rho_t(j, dx) \,. \tag{9}$$

2.4. *Resolution of particular cases using convolution tools*

The solution $p_t(i, x)$ of Eq. (5) satisfies:

$$p_t(i, x) = p_0(i, x - t)\exp\left(-\int_0^t b(i, x - t + u)du\right) \quad \text{if } t < x \,, \tag{10}$$

$$p_t(i, x) = p_{t-x}(i, 0)\exp\left(-\int_0^x b(i, u)du\right) \quad \text{if } x \le t \,. \tag{11}$$

Thanks to Eqs. (6) and (8), for both above particular cases, there exist some functions $h_i(t)$ such that

$$p_t(i, 0) = h_i(t) + \sum_{j\in E}\int_0^{+\infty} a(j, i, x)p_t(j, x)dx \,.$$

Using Eqs. (10) and (11), we deduce

$$
\begin{aligned}
p_t(i,0) &= h_i(t) + \sum_{j\in E} \int_t^{+\infty} a(j,i,x) p_0(j, x-t) \\
&\quad \times \exp\left(- \int_0^t b(j, x-t+u) du \right) dx \\
&\quad + \sum_{j\in E} \int_0^t a(j,i,x) p_{t-x}(j,0) \exp\left(- \int_0^x b(j,u) du \right) dx \\
&= k_i(t) + \sum_{j\in E} \int_0^t p_{t-x}(j,0) q(j,i,x) dx \,, \qquad (12)
\end{aligned}
$$

where the functions k_i are known. The numerical approximation of the solution of Eq. (12) requires some quite complex additional work in the general case.

Let us again consider the framework of the reliability theory such as presented in Remark 1, i.e., the case where $E = \{0, 1\}$. Let us suppose that the component is available at time $t = 0$, which means the initial distribution is a Dirac mass:

$$\mathbb{P}(\eta_0 = 1, X_0 = 0) = 1 \,.$$

Let us denote by f (respectively g) the probability density function of the duration of working periods (respectively failure periods). Using the above notations, we can write

$$x_0 = 0 \,, \quad p_0 = 0 \,, \quad k_0 = h_0 = \lambda\alpha = f \,, \quad k_1 = h_1 = 0 \,.$$

Let us denote $u_i(t) = p_t(i, 0)$. Equation (12) delivers

$$u_1 = u_0 * g \,, \quad u_0 = f + u_1 * f \,,$$

and consequently we get

$$u_1 = f * g + u_1 * f * g \,, \quad u_0 = f + u_0 * f * g \,. \qquad (13)$$

These equations show that u_1 (respectively u_0) is the renewal density associated with the renewal process corresponding to the end of the repair periods (respectively associated with the component breaking up) (see for example Ref. 4, paragraph 4.4, formula (5) or Ref. 5, formula (6.5)). Let us define $\bar{F}(t) = \int_t^{+\infty} f(u)du$. From Eqs. (7) and (11), we deduce that the availability $A(t) = \mathbb{P}(\eta_t = 1)$ of the component is given by:

$$\begin{aligned} A(t) &= \int_0^{+\infty} \rho_t(1, dx) \\ &= \bar{F}(t) + \int_0^{+\infty} p_t(1, x)dx \\ &= \bar{F}(t) + \int_0^t u_1(t-x) \exp\left(- \int_0^x b(1, u)\, du \right) dx\,. \end{aligned}$$

Thanks to Eq. (2), the above equation can be written as:

$$A = \bar{F} + u_1 * \bar{F}\,. \tag{14}$$

Using Eq. (13), we also have

$$A = \bar{F} + f * g * \bar{F} + u_1 * \bar{F} * f * g = \bar{F} + A * f * g\,, \tag{15}$$

which is the usual renewal equation for the availability of a component (see Ref. 6, paragraph 4.2.1 or Ref. 5, example 6.45). The implementation of Eqs. (14) and (15) (which respectively corresponds to methods I and II of Ref. 2) seems to fail on some relevant numerical examples, and will not be considered in this paper.

The following section is devoted to a new approach, also based on Eq. (4), which does not require additional regularity assumptions and whose implementation does not depend on the initial conditions.

3. An Approximation Using a Finite Volume Method

In this section, we consider a numerical method aimed to globally approximate the measures ρ_t, solution of the transport equation (4), for all $t \in [0, T[$, and for all type of initial condition. The interval

$[0, T[$ is divided into N_h intervals of length $h = T/N_h$, and we have

$$\mathbb{R}_+ \times [0, T[= \bigcup_{m \geq 0} \bigcup_{n=0}^{N_h - 1} [mh, (m+1)h[\times [nh, (n+1)h[.$$

For all $i \in E$, we approximate the measure $\rho_t(i, dx)$ by the measure $\bar{u}_t^h(i, x)dx$, where the function $\bar{u}$ is equal to a constant, denoted by $u_n^h(i, m)$, on each square $[mh, (m+1)h[\times [nh, (n+1)h[$:

$$\bar{u}_t^h(i, x) = u_n^h(i, m) \quad \text{if } (x, t) \in [mh, (m+1)h[\times [nh, (n+1)h[.$$

The algorithm is initialized by a discretization of the initial distribution ρ_0:

$$u_0^h(i, m) = \frac{1}{h} \int_{[mh, (m+1)h[} \rho_0(i, dx) .$$

Although our scheme approximates the measures ρ_t for all type of initial data, it can be seen as a numerical approximation of Eq. (5). We thus approximate the quantities $\partial\rho/\partial t$ and $\partial\rho/\partial x$ by:

$$\frac{u_{n+1}^h(i, m) - u_n^h(i, m)}{h} \quad \text{and} \quad \frac{u_n^h(i, m) - u_n^h(i, m-1)}{h} ,$$

which produces the following numerical scheme:

$$u_{n+1}^h(i, m) = \frac{u_n^h(i, m-1)}{1 + hb(i, mh)} \quad \text{for } m \geq 1 . \tag{16}$$

Similarly, the boundary conditions are inspired by Eq. (6) or Eq. (9):

$$u_{n+1}^h(i, 0) = \sum_{j \in E} \sum_{m \geq 1} ha(j, i, mh) u_{n+1}^h(j, m) . \tag{17}$$

Using the discrete values given by the scheme, we approximate the values $\mathbb{P}(\eta_t = i) = \int_0^{+\infty} \rho_t(i, dx)$, for all $t \in [nh, (n+1)h[$, by

$$P_n^h(i) = \int_0^{+\infty} \bar{u}_t^h(i, x)dx = h \sum_{m \geq 0} u_n^h(i, m) .$$

The weak convergence of the measures $\bar{u}_t^h(i,x)dx$ to the measures $\rho_t(i,dx)$ is proven in Ref. 7, using a uniqueness result stated in Ref. 3.

4. Numerical Examples

4.1. *Case of an alternating renewal process*

We again consider the case described in Remark 1. We then set $E = \{0, 1\}$, and the numerical scheme can be written for $n \geq 0$ and $i \in \{0, 1\}$:

$$u_{n+1}^h(i,m) = \frac{u_n^h(i,m-1)}{1+hb(i,mh)} \quad \text{for } m \geq 1\,,$$
$$u_{n+1}^h(i,0) = \sum_{m\geq 1} \left(u_n^h(1-i,m-1) - u_{n+1}^h(1-i,m)\right).$$

We are interested in the availability $A(t) = \mathbb{P}(\eta_t = 1)$ of the component. It is known (see for example [1] Formula (4.2) or [2] Proposition 6.51) that the asymptotic availability is equal to:

$$A(\infty) = \frac{m_1}{m_1+m_2}\,, \tag{18}$$

where m_1 (respectively m_2) is the mean duration of a working period (respectively failure period). We thus compare in Tables 1–3 the values $A(T)$ provided by the finite volume method, by the third method of Ref. 2, and the value given by Eq. (18). We also compare the

Table 1. Asymptotic availability of Example 1.

	Finite volume method		Method III of Ref. 2	
	$A(10\,000)$	comp. time	$A(10\,000)$	comp. time
$N_h = 2000, h = 5$	0.62444	2.6	0.62465	1.2
$N_h = 10\,000, h = 1$	0.62461	84	0.62465	25

comp. time: computation time (CPU).

Table 2. Asymptotic availability for Example 2.

	Finite volume method		Method III of Ref. 2	
	$A(10\,000)$	comp. time	$A(10\,000)$	comp. time
$N_h = 2000, h = 5$	0.99259	2.8	>1 (n.s.)	1
$N_h = 10\,000, h = 1$	0.99375	85	0.99440 (n.s.)	24
$N_h = 60\,000, h = 0.17$	0.99399	$4 \cdot 10^3$	0.99404	$2.5 \cdot 10^3$

comp. time: computation time (CPU), (n.s.): not stabilized.

Table 3. Asymptotic availability for Example 3.

	Finite volume method		Method III of Ref. 2	
	$A(3000)$	comp. time	$A(3000)$	comp. time
$N_h = 2000, h = 1.5$	0.99775	2.8	0.12763 (n.s.)	1
$N_h = 10\,000, h = 0.3$	0.99880	85	0.85355 (n.s.)	25
$N_h = 60\,000, h = 0.05$	0.99896	$4.1 \cdot 10^3$	0.98851 (n.s.)	$2.3 \cdot 10^3$
$N_h = 120\,000, h = 0.025$	0.99897	$1.7 \cdot 10^4$	0.99527 (n.s.)	10^4

comp. time: computation time (CPU), (n.s.): not stabilized.

computing times. When the computed value of the availability does not seem to reach any asymptotic value at large t, the value $A(T)$ is followed by "n.s." for "not stabilized". In the following figures, the horizontal lines represent this asymptotic availability. In Fig. 12, the value of the availability computed by the finite volume method is plotted in Fig. 12(a) whereas the one computed by Method III of Ref. 2 is plotted in Fig. 12(b).

Example 1.

The probability distribution of the working periods is a Weibull distribution with a shape parameter $\beta = 3$ and a mean value equal to

1000. The probability distribution of the failure periods is a Weibull distribution with a shape parameter $\beta = 3.5$ and a mean value equal to 600. The results are plotted in Figs. 1 and 2. In this case, the application of Eq. (18) gives: $A(\infty) = 0.62498$.

In this example, Method III of Ref. 2 is faster and slightly better than the finite volume method.

Figure 3 shows the approximation of the measures ρ_{1900} by $\bar{u}^h_{1900}$ for $h = 1$ (time $t = 1900$) in Fig. 3(a) and ρ_{2300} by $\bar{u}^h_{2300}$ (time $t = 2300$) in Fig. 3(b). There are two curves on each figure since both curves $\bar{u}^h_t(1, \cdot)$ (solid line) and $\bar{u}^h_t(0, \cdot)$ (dash dot line) are plotted at times $t = 1900$ and 2300. The approximation of Dirac mass at point t for $\rho_t(1, \cdot)$ (see Eq. (7)) exceeds the vertical scale of the figure at the time $t = 1900$ (the discrete value in the corresponding control volume is equal to $7.7 \cdot 10^{-3}$). At time $t = 2300$, this value decreases until $1.8 \cdot 10^{-4}$, and it is covered by the thickness of the horizontal axis at large times (it is equal to $1.6 \cdot 10^{-5}$ for $t = 2500$).

Example 2.
The probability distribution of the working periods is the same as that of Example 1 but the scale parameter of the failure duration distribution is modified: the probability distribution of the failure periods is defined as a Weibull distribution with a shape parameter β still equal to 3.5 but with a mean value equal to 6. The results are plotted in Figs. 4–6. In this case, the application of Eq. (18) gives: $A(\infty) = 0.99404$.

In this example, Method III of Ref. 2 can give inadmissible results (the implementation of the method should then be discussed) whereas the finite volume method, although it demands more computing time, remains robust and gives admissible results.

Example 3.
The probability distribution of the working periods is a Weibull distribution with a shape parameter $\beta = 2$ and a mean value equal to 886. The probability distribution of the failure periods is a Weibull distribution with a shape parameter $\beta = 1.5$ and a mean value equal

(a)

(b)

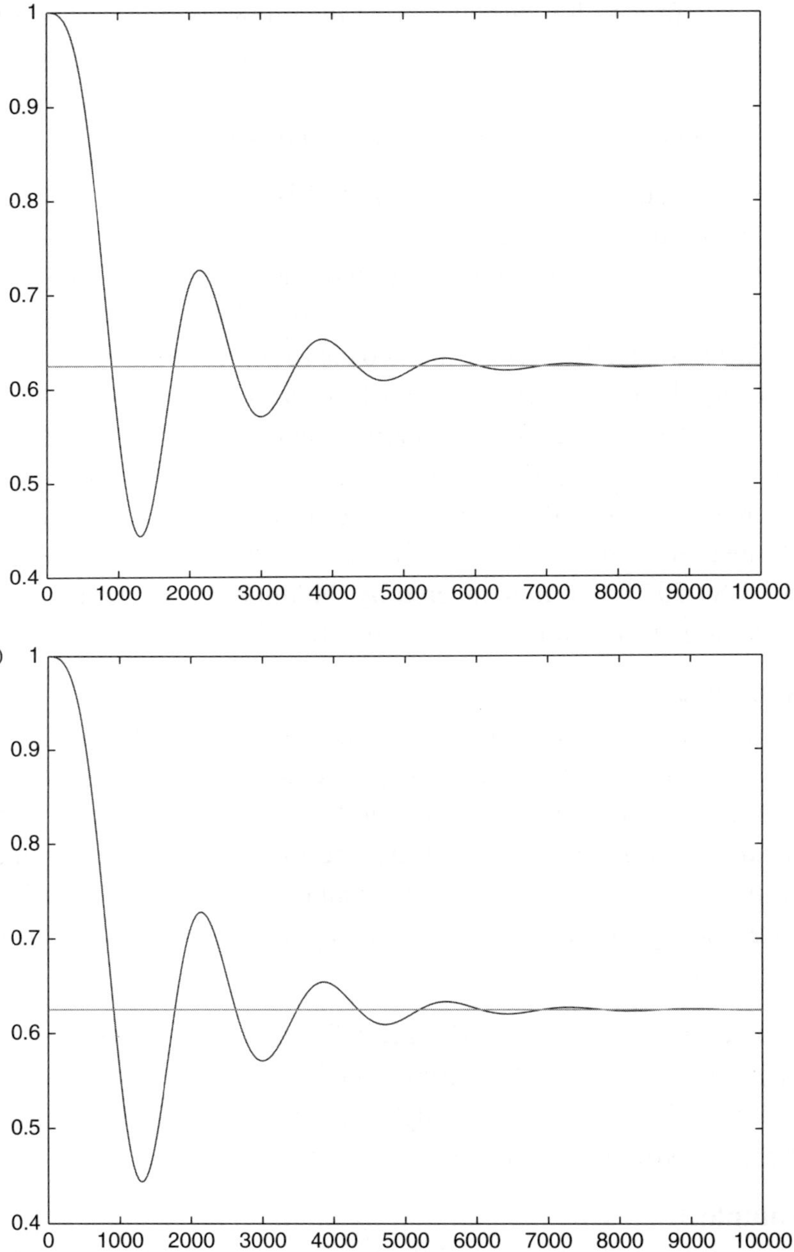

Fig. 1. Availability for Example 1, $N_h = 2000$, $h = 5$.

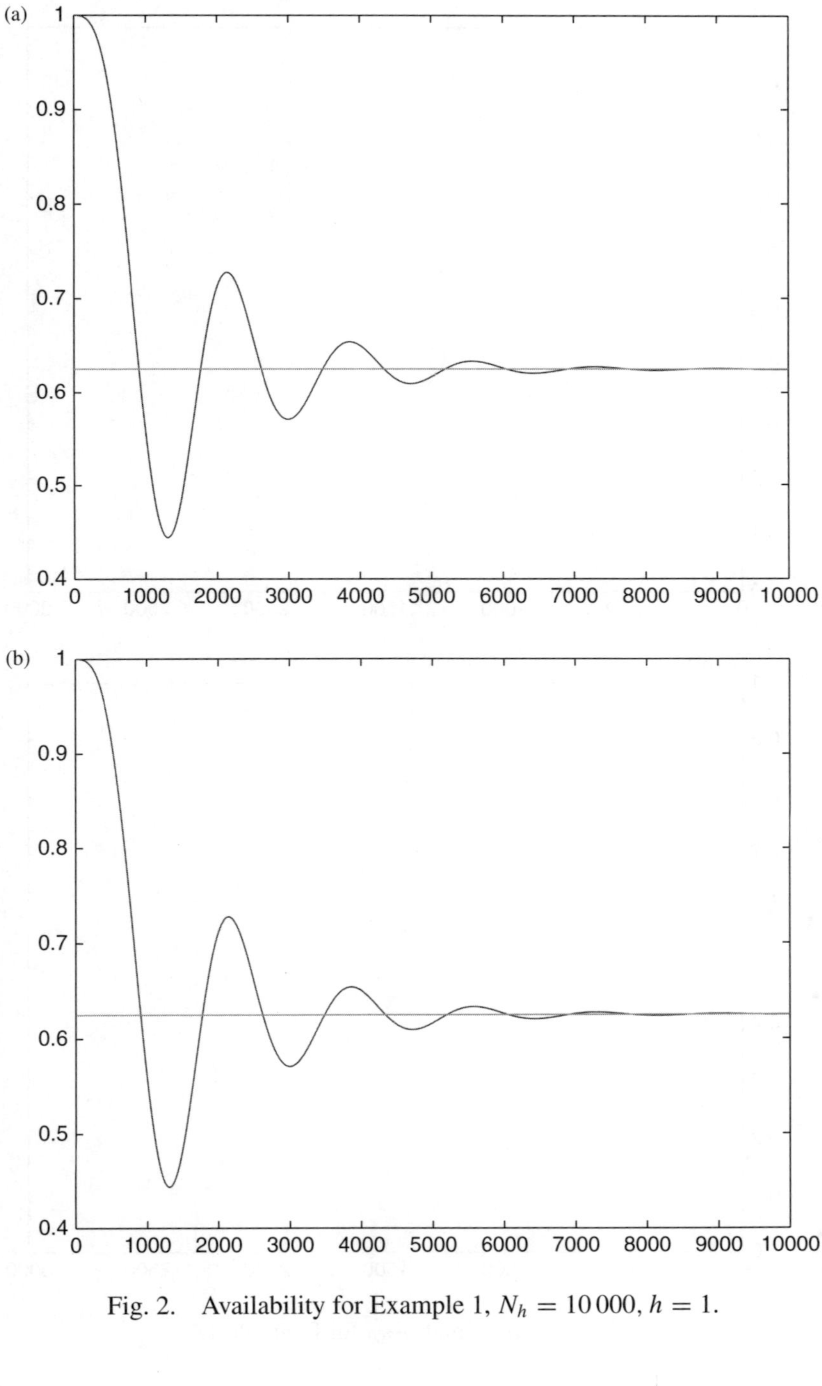

Fig. 2. Availability for Example 1, $N_h = 10\,000$, $h = 1$.

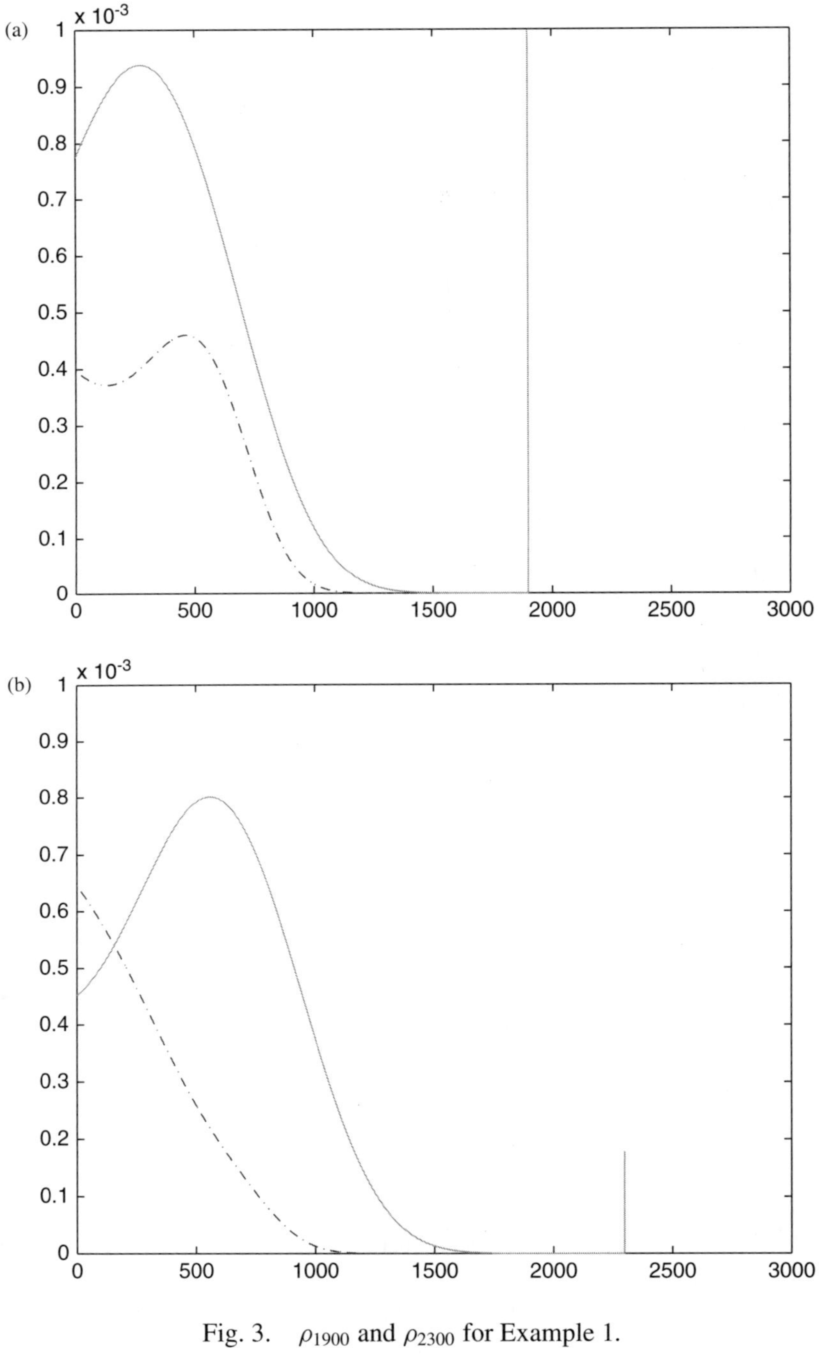

Fig. 3. ρ_{1900} and ρ_{2300} for Example 1.

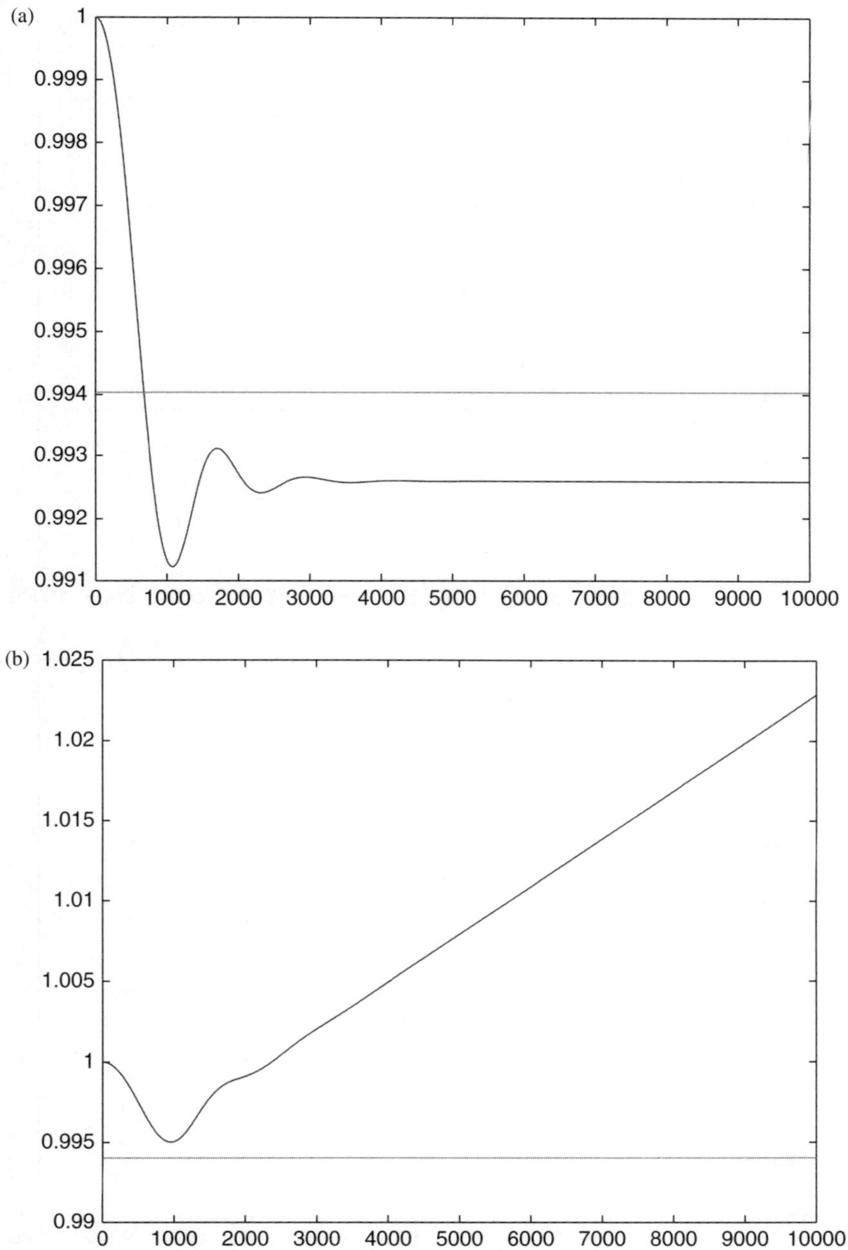

Fig. 4. Availability for Example 2, $N_h = 2000$, $h = 5$.

(a)

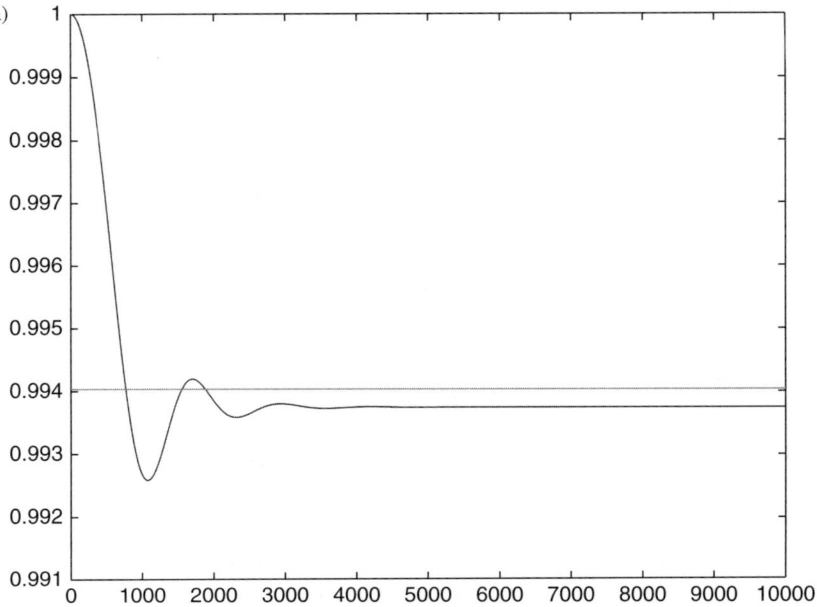

(b)

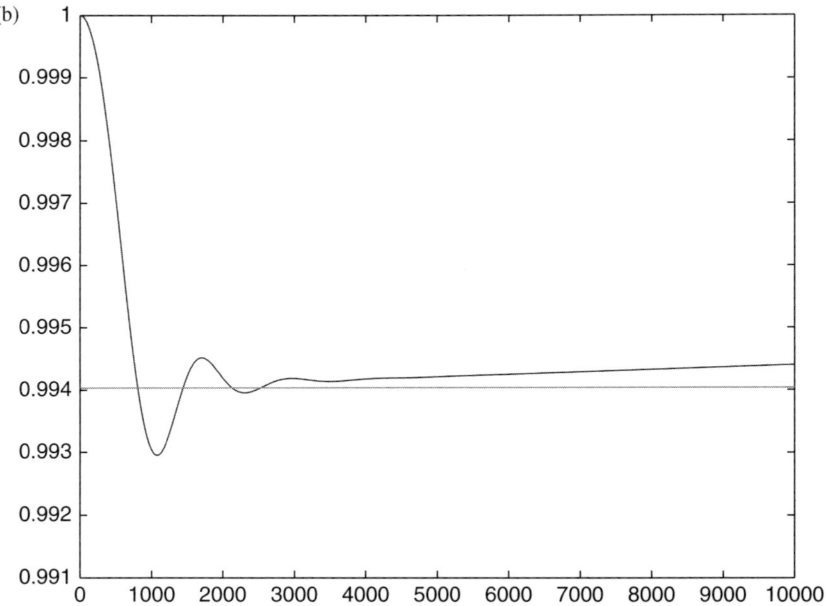

Fig. 5. Availability for Example 2, $N_h = 10\,000$, $h = 1$.

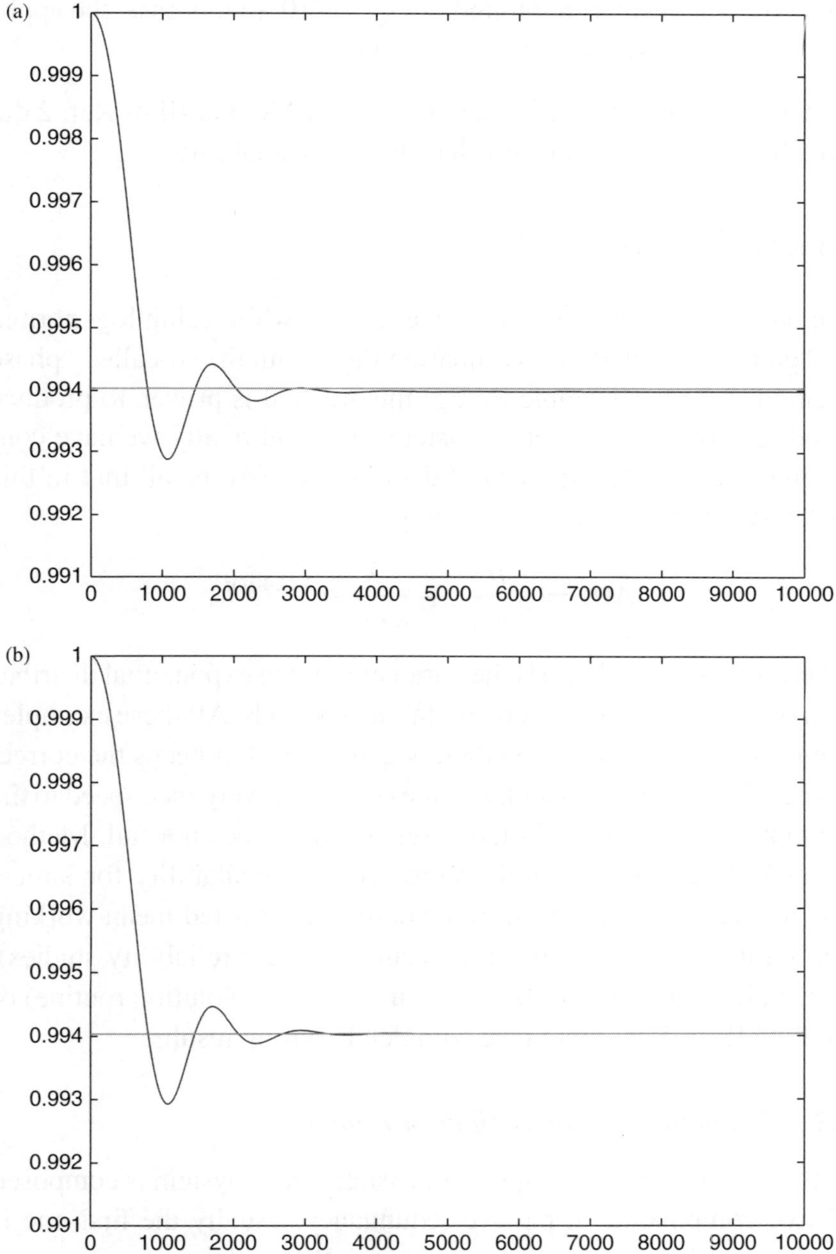

Fig. 6. Availability for Example 2, $N_h = 60\,000$, $h = 0.17$.

to 0.903. The results are plotted in Figs. 7–10. In this case, the application of Eq. (18) gives: $A(\infty) = 0.99898$.

On this example, our implementation of Method III of Ref. 2 did not allow to obtain relevant values for the availability.

Other Experiments

We have also studied the numerical results while using log–normal and gamma distributions, comparing them with the so-called "phase method" (see for example Ref. 5: this method is proven to produce good results but it cannot be systemized), and finally we have considered the case of exponential distributions. We recall that in this last case, the availability is given by:

$$A(t) = \frac{\mu}{\lambda + \mu} + \frac{\lambda}{\lambda + \mu} e^{-(\lambda+\mu)t},$$

where λ (respectively μ) is the parameter of the exponential distribution of the working (respectively failure) periods. All these examples seem to indicate that our method is quite robust, it keeps the correct shape of the graph and it delivers the correct convergence speed to the asymptotic availability. In the cases where it does not fail, Method III of Ref. 2 gives a slightly more accurate availability for large t and it is always faster. But, in the case of contrasted mean working and failure durations (this case occurs in actual reliability studies), our implementation (in MATLAB, using the convolution routine) of Method III of Ref. 2 can give completely wrong results.

4.2. *Examples with more than two states*

In the following two examples, we assume that a system is composed of two components in passive redundancy: usually the first one is working and the second one is at rest. When the first component fails, the second component is started if it is not failed. The second component cannot fail when it is at rest. The system is working if

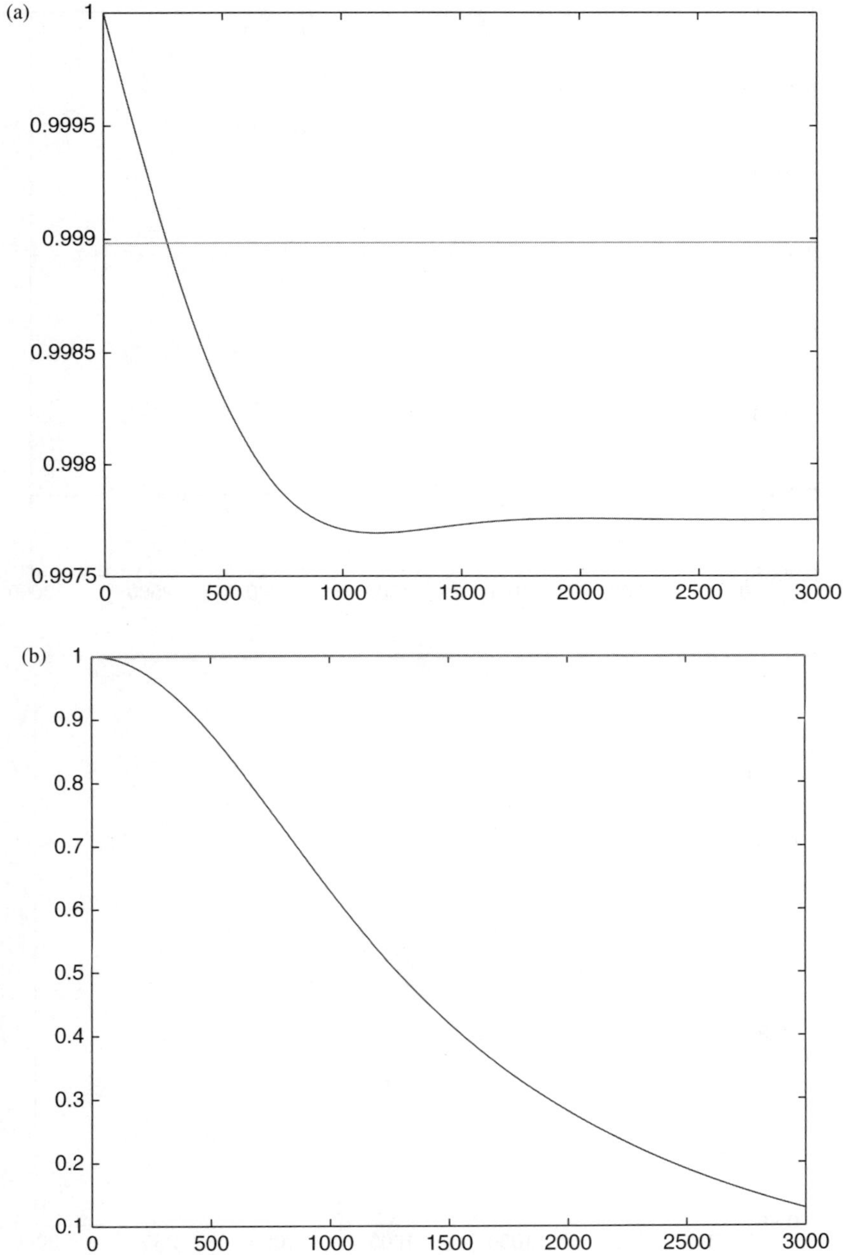

Fig. 7. Availability for Example 3, $N_h = 2000$, $h = 1.5$.

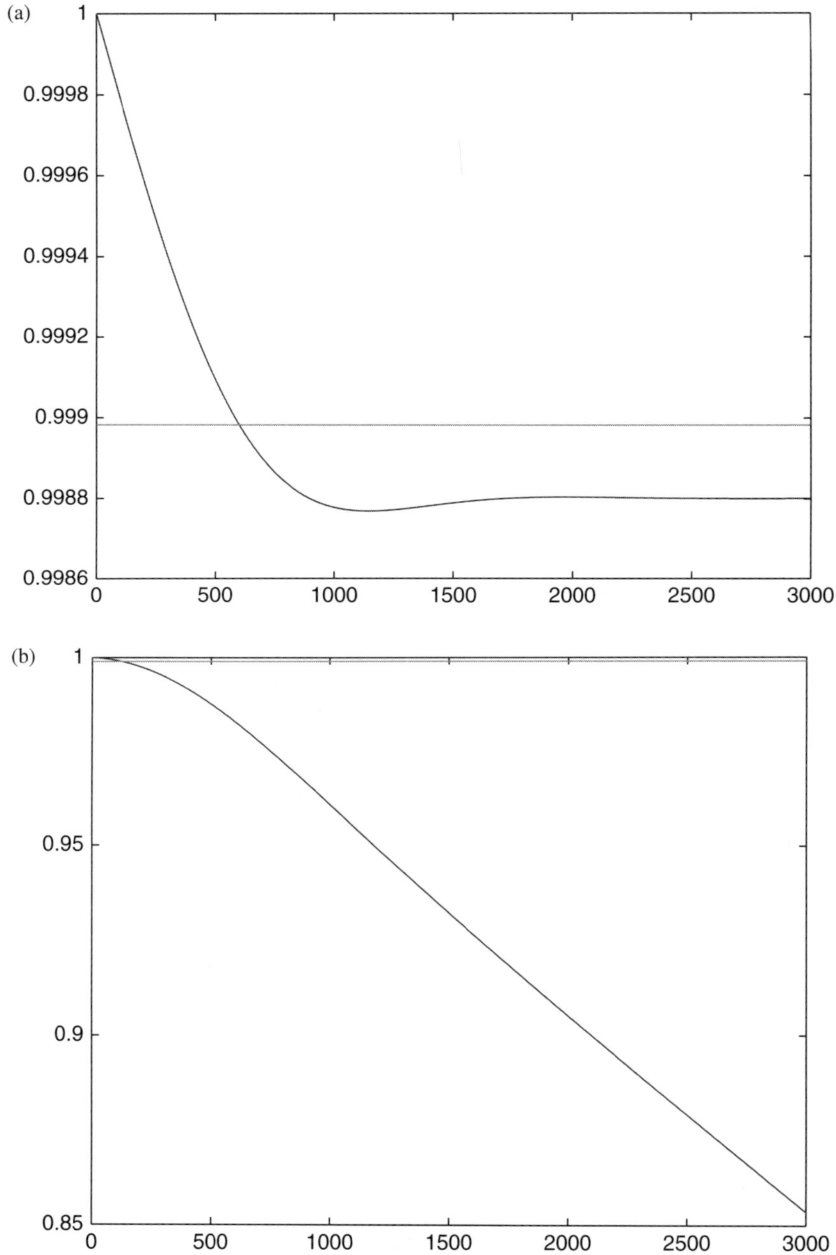

Fig. 8. Availability for Example 3, $N_h = 10\,000$, $h = 0.3$.

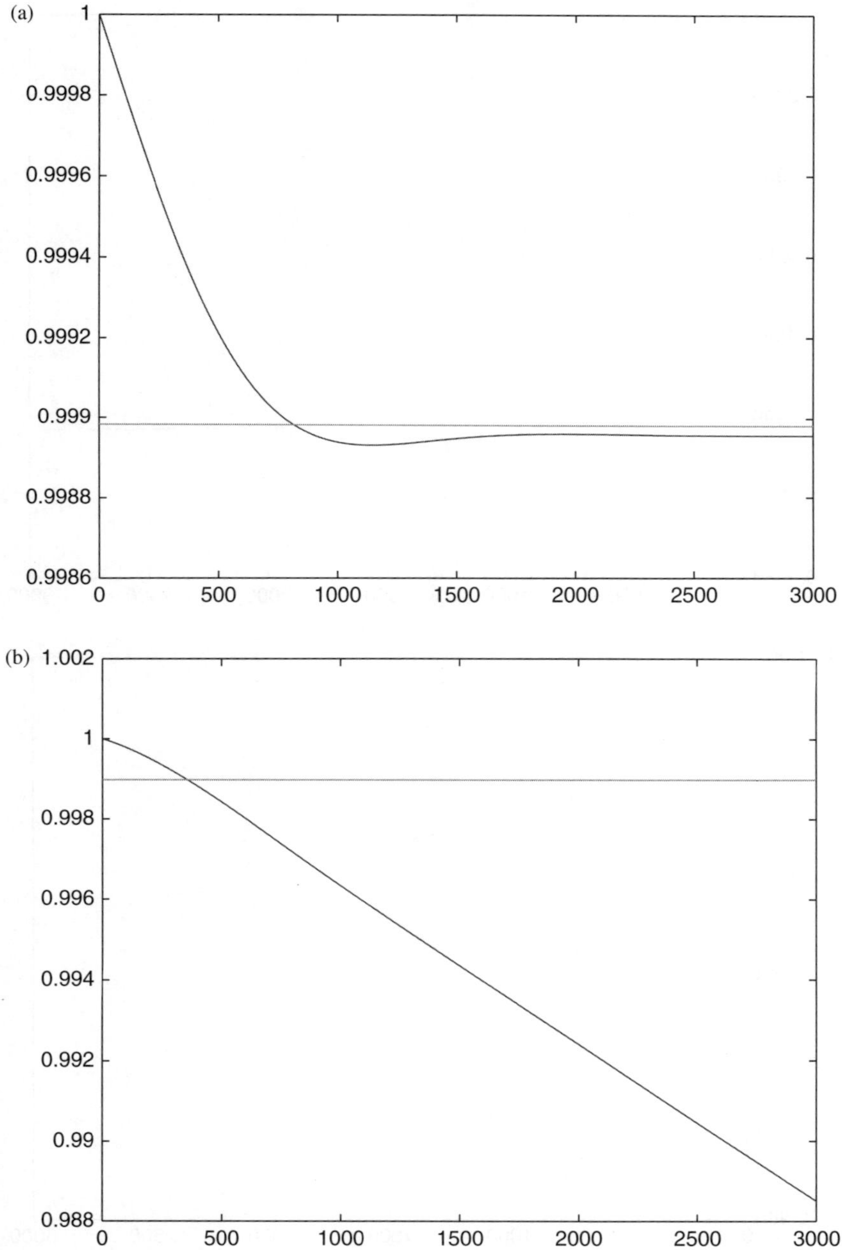

Fig. 9. Availability for Example 3, $N_h = 60\,000$, $h = 0.05$.

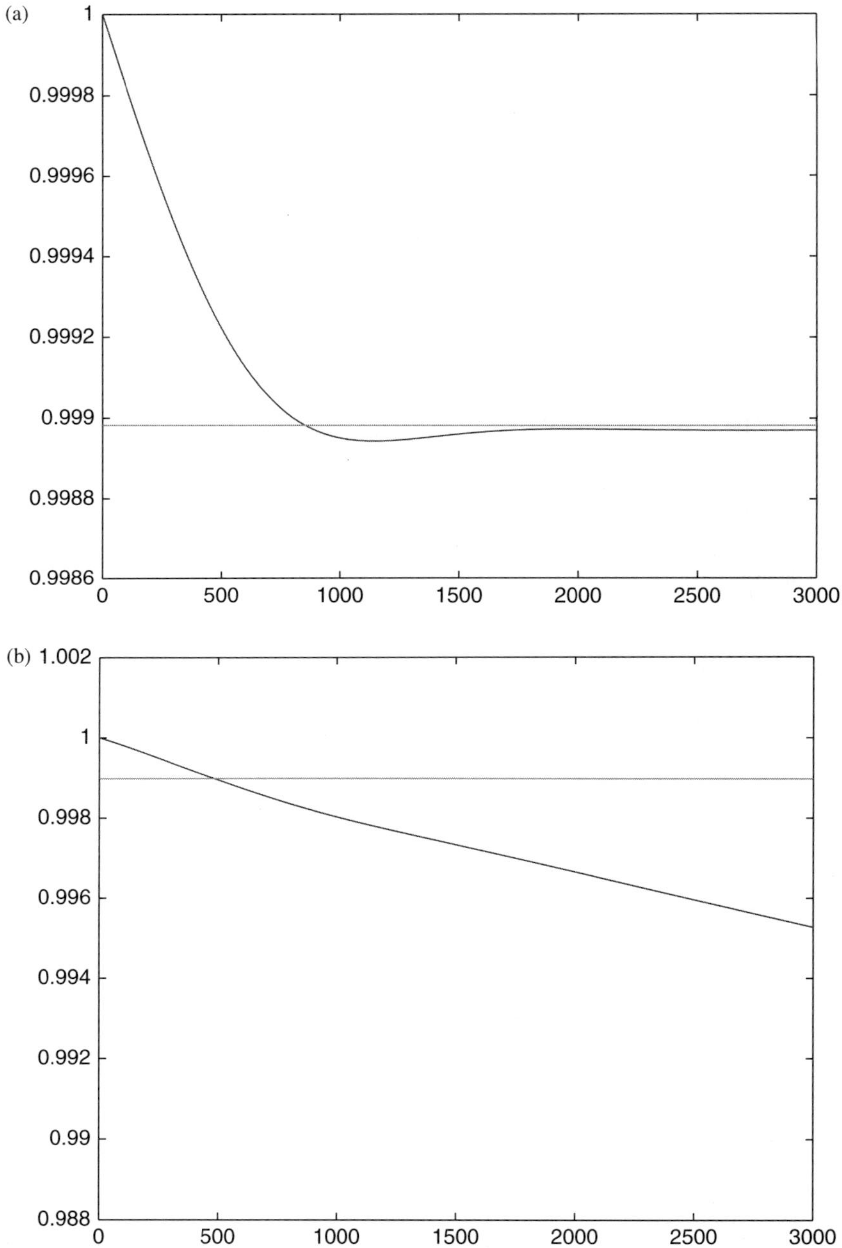

Fig. 10. Availability for Example 3, $N_h = 120\,000$, $h = 0.025$.

and only if one component is working. At the end of its repair, a component is as good as new.

Example 4.
Component 1 has two types of failure. When a failure of the first type occurs, its repair immediately starts. When a failure of the second type occurs, it is not detected, and no repair is planned, thus the second component is not used and the system fails. Let us assume that the first component is being repaired and that the second one is working: at the end of the repair of the first component, this one is immediately used, the second one is stopped and it is instantaneously upgraded so it becomes as good as new. We are interested in the system reliability, i.e., the probability that the system has no failure during the period $[0, t]$ since failure states are supposed to be absorbing states. The system has four states:

- State 1: the first component is working and the second one is at rest,
- State 2: the first component is being repaired and the second one is working,
- State 3: the system is out of order because both components are being repaired,
- State 4: the system is out of order because a second type failure of the first component has occurred; it has therefore not been detected and the second component is at rest.

The positive transition rates are as follows:

$a(1, 2, x) = \lambda_1(x)$: hazard rate of a Weibull distribution with a shape parameter $\beta = 1.5$ and a mean value equal to 2000,
$a(2, 1, x) = \mu(x)$: hazard rate of a log–normal distribution with a mean value equal to 102 and a variation coefficient equal to 0.53,
$a(1, 4, x) = \lambda_1'(x)$: hazard rate of a Weibull distribution with a shape parameter $\beta = 2$ and a mean value equal to 10 000,
$a(2, 3, x) = \lambda_2(x)$: hazard rate of a Weibull distribution with a shape parameter $\beta = 1.5$ and a mean value equal to 1881.

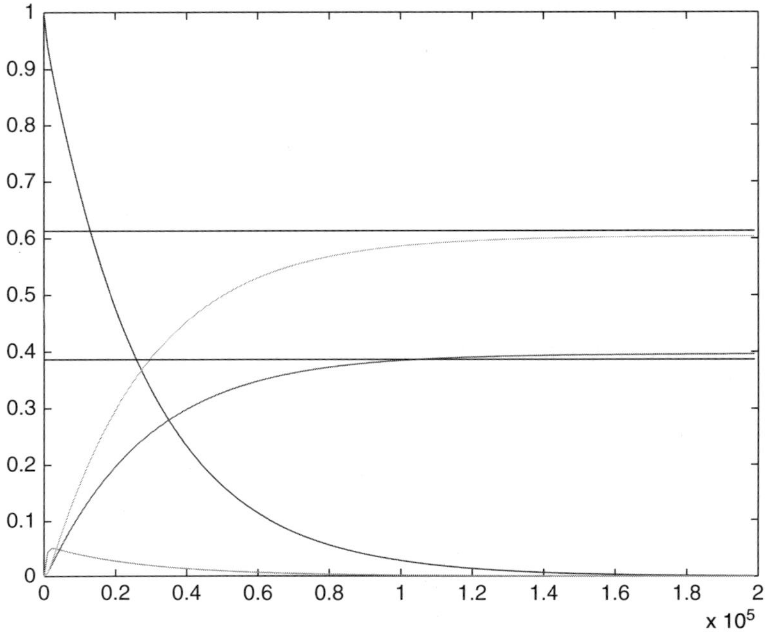

Fig. 11. Example 4.

We assume that at time 0, both components are as good as new, the first one is started and the second one is at rest. In Fig. 11, the probability that the system is in each of the four states is plotted. Horizontal lines again give the asymptotic probability to be in States 3 and 4, respectively (note that the asymptotic probability to be in States 1 and 2 is zero since States 3 and 4 are absorbing states and consequently States 1 and 2 are transient states). As in the previous examples, these asymptotic probabilities indicate the quality of our algorithm since exact formulas are known (see for example Ref. 5, Theorem 10.20 and Remark 10.21).

Example 5.
Each component has only one type of failure. The system must be stopped before any component is being repaired, consequently, components are being repaired only when both components have failed and they are restarted only when both have been repaired.

The system has three states:

- State 1: the first component is working and the second one is at rest,
- State 2: the first component failed and the second one is working,
- State 3: the system is out of order, both components are being repaired.

The positive transition rates are as follows:

$a(1, 2, x) = \lambda_1(x)$: hazard rate of a Weibull distribution with a shape parameter $\beta = 2$ and a mean value equal to 2216,
$a(2, 3, x) = \lambda_2(x)$: hazard rate of a Weibull distribution with a shape parameter $\beta = 1.5$ and a mean value equal to 1881,
$a(3, 1, x) = \mu(x)$: hazard rate of a gamma distribution with a shape parameter $\alpha = 75$ and a mean value equal to 150.

In Fig. 12, the probability that the system is in each of the three states is plotted. Horizontal lines again give the analytical asymptotic probability to be in each state.

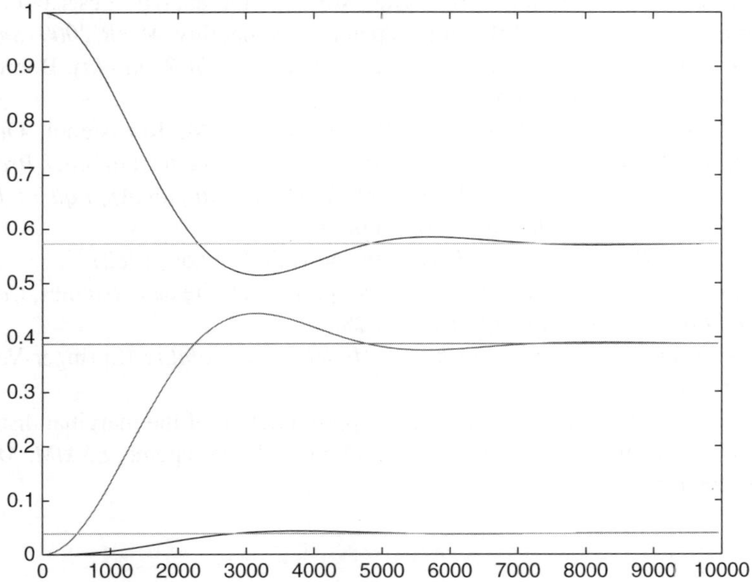

Fig. 12. Example 5.

5. Conclusion

We have proposed a new numerical scheme to compute the availability of a component with general failure and repair rates, and more generally to compute the marginal distributions of a semi-Markov process. It is based on a finite volume scheme which requires a discretization in time and in space; a mathematical proof of the convergence of this algorithm when the discretization step tends to 0 is available.

Numerical examples show that this scheme always gives admissible results, although they are less accurate when the rates are contrasted. However, even in this difficult case, the shape of the graph and the convergence speed remain correct. This new scheme thus appears to be usable in reliability studies.

References

1. C. Cocozza-Thivent and M. Roussignol, A general framework for some asymptotic reliability formulas, *Adv. Appl. Prob.* **32** (2000) 446–467.
2. A. Fritz, P. Pozsgai and B. Bertsche, Notes on the analytic description and numerical calculation of the time dependent availability, *MMR'2000: Second International Conference on Mathematical Methods in Reliability,* Bordeaux, France, 4–7 July 2000, pp. 413–416.
3. C. Cocozza-Thivent, R. Eymard, S. Mercier and M. Roussignol, On the marginal distributions of Markov processes used in dynamic reliability, *Prépublications du Laboratoire d'Analyse et de Mathématiques Appliquées UMR CNRS 8050*, 2/2003, January 2003, submitted.
4. D. R. Cox, *Renewal Theory* (Chapman and Hall, London, 1982).
5. C. Cocozza-Thivent, Processus stochastiques et fiabilité des systèmes, *Collection Mathématiques and Applications* **28** (1997).
6. T. Aven and U. Jensen, *Stochastic Models in Reliability* (Springer-Verlag, New-York, 1999).
7. C. Cocozza-Thivent and R. Eymard, Approximation of the marginal distributions of a semi-Markov process using a finite volume scheme, *ESAIM: M2AN* **38** (2004) 853–875.

CHAPTER 2

Optimal Checkpointing Interval for Task Duplication with Spare Processing

Sayori Nakagawa

Institute of Consumer Sciences and Human Life, Kinjo Gakuin University, 1723 Omori 2-chome, Moriyama-ku, Nagoya 463-8521, Japan

Yoshihiro Okuda

Department of Industrial Engineering, Aichi Institute of Technology, 1247 Yachigusa, Yagusa-cho, Toyota 470-0392, Japan

Shigeru Yamada

Department of Social Systems Engineering, Tottori University, 4-101 Minami, Kozan-cho, Tottori 680-8552, Japan

1. Introduction

In computer systems, some errors often occur due to noises, human errors, hardware faults, etc. To attain the accuracy of computing, it is of great importance to detect such errors by fault tolerant computing techniques.[1] Usually, an error detection of the process can be made by two independent modules where they compare two results at suitable checkpointing times. If their results do not match with each other, we go back to the newest checkpoint and make a retrial of the processes. In such situations, if we compare results frequently, then we could decrease the time of rollback. However, the total overhead of comparisons at checkpoints would be increased. Thus, this is one kind of trade-off problems on how to decide optimal checkpoint frequency.

Several studies of deciding checkpoint frequency have been discussed for the above hardware redundancy. Pradhan and Vaidya[2] evaluated the performance and reliability of the duplex system with a spare processor. Ziv and Bruck[3,4] analytically considered the checkpointing schemes with task duplication and evaluated the performance of schemes. Kim and Shin[5] derived the optimal instruction-retry period minimizing the probability of dynamic failure on the triple modular redundant controller.

In this chapter, we firstly consider a double modular redundancy as redundant techniques of error detection, and analyze an optimal checkpointing interval: when the native execution time of the process is given, we divide it into identical time intervals. Introducing the overhead of comparison by duplication, we obtain the mean time to completion of the process and derive an optimal checkpointing interval which minimizes it. Further, we obtain the mean time to completion for a majority decision redundant system as an error masking system.

However, if permanent faults have occurred, it would be impossible to detect such faults by comparing two results of the processes. When the two results of the processes of some task are not matched, Pradhan and Vaidya[6] prepared another spare module to execute the process. This is helpful to detect permanent faults and useful to reduce the overhead of rollback recovery. We secondly consider a double modular redundancy with one spare module. One main problem of this technique is that the overhead increases in preparing a spare module. Thus, when a finite original execution time of the process is given, we equally divide it into constant intervals. Introducing parameters such as the overheads of comparison and a spare module, we obtain the mean time to completion of the process and derive an optimal checkpointing interval which minimizes it. Further, we numerically compare two schemes in which one is the rollback recovery of two modules and the other is the roll-forward recovery of two and spare modules, and discuss which scheme is better.

2. Double Modular Redundant System

Suppose that S is an original processing time of one task which does not include any overheads of checkpoint generations. Then, to tolerate some faults, we consider the recovery scheme of dual processes with the following assumptions:

(a) The original processing time of one task is S. We divide S equally into N time intervals where $T \equiv S/N$, and create the checkpoint at periodic times kT $(k = 1, 2, \ldots, N)$.

(b) Some errors occur at constant rate λ $(\lambda > 0)$, i.e., the probability that two processes have no error during $(0, T]$ is $e^{-2\lambda T}$.[7]

(c) To detect errors, we provide two independent processes, where they compare two results at checkpointing times kT $(k = 1, 2, \ldots, N)$.

(d) If two results match with each other in case (c), the process is correct and goes forward. However, if two results of making the processing of task I_j do not match, it is judged that some errors have occurred. We rollback to the newest checkpoint and make a retry of task I_j of the processes (Fig. 1).

(e) If two results of making the processing of task I_{j+1} do not match, we roll back to the newest checkpoint and make a retry of task I_{j+1} of the processes in the same ways as case (d).

(f) If two results of making the processing of task I_N match, the process ends.

Let us introduce a constant overhead for the comparison of two results. Further, we neglect any failure of the system caused by common mode faults to make clearer error detection of the processes. The mean time $L_1(N)$ to completion of the process is the sum of the processing times and the overhead C_1 of comparison of two modules.

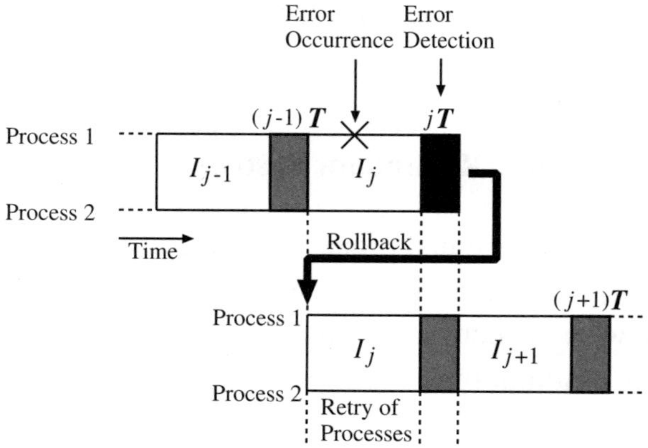

Fig. 1. Recovery scheme of dual processes.

From the assumption that the processes are rolled back to the previous checkpoint when an error has been detected at a checkpoint, the mean execution time of the process for one checkpointing interval $(0, T]$ is given by a renewal equation:

$$L_1(1) = (T + C_1)e^{-2\lambda T} + (T + C_1 + L_1(1))\left(1 - e^{-2\lambda T}\right), \quad (1)$$

and solving it, we have

$$L_1(1) = \frac{(T + C_1)}{e^{-2\lambda T}}. \quad (2)$$

Thus, the mean time to completion of the processes is:

$$L_1(N) \equiv NL_1(1) = N(T + C_1)e^{2\lambda T}, \quad N = 1, 2, \ldots. \quad (3)$$

Since $T = S/N$, we also have

$$L_1(N) = (S + NC_1)e^{2\lambda(S/N)}. \quad (4)$$

We seek an optimal number N_1^* which minimizes $L_1(N)$ for a specified S. Evidently,

$$L_1(1) = (S + C_1)e^{2\lambda S} \quad (5)$$

and

$$L_1(\infty) \equiv \lim_{N\to\infty} L_1(N) = \infty\,. \tag{6}$$

Thus, there exists a finite number $N_1^*(1 \le N_1^* < \infty)$. However, it would be difficult to find N_1^* which minimizes Eq. (4). Putting $T = S/N$ in Eq. (4) and rewriting it by the function T, we have

$$L_1(T) = S\left(1 + \frac{C_1}{T}\right)e^{2\lambda T}\,, \quad 0 \le T \le S\,. \tag{7}$$

It is evident that

$$L_1(0) \equiv \lim_{T\to 0} L_1(T) = \infty \tag{8}$$

and

$$L_1(S) = (S + C_1)e^{2\lambda S}\,. \tag{9}$$

Thus, there exists an optimal $T_1^*(0 < T_1^* \le S)$ which minimizes $L_1(T)$ in Eq. (7). Differentiating $L_1(T)$ with respect to T and setting it to zero, we have

$$T^2 + C_1 T - \frac{C_1}{2\lambda} = 0\,. \tag{10}$$

Solving it with T,

$$T_1^* = \frac{C_1}{2}\left[\sqrt{1 + \frac{2}{\lambda C_1}} - 1\right]. \tag{11}$$

Therefore, we have the following optimal interval number N_1^*:

(i) If $T_1^* < S$, we put $\lfloor S/T_1^* \rfloor = N$, where $\lfloor x \rfloor$ denotes the greatest integer contained in x, and calculate $L_1(N)$ and $L_1(N+1)$ from Eq. (4). If $L_1(N) \le L_1(N+1)$ then $N_1^* = N$, and conversely, if $L_1(N+1) < L_1(N)$ then $N_1^* = N + 1$.

(ii) If $T_1^* \ge S$, i.e., we should make no checkpoint until time S then $N_1^* = 1$, and the mean time is given in Eq. (5).

Note that T_1^* in Eq. (11) does not depend on S. Thus, if S is very large, is changed greatly or is unclear, we may adopt T_1^* as an approximate checkpointing time.

Next, we consider a triple modular redundant system as an error masking system: if more than two results of three modules match with each other, the process is correct, i.e., a triple modular system masks a single error. Then, the probability that the process is correct during $(0, T]$ is:

$$\bar{F}_2(T) = e^{-3\lambda T} + 3e^{-2\lambda T}(1 - e^{-\lambda T}). \tag{12}$$

Let C_2 be the overhead of comparison of three modules. Then, the mean time to completion of the process is

$$L_2(N) = \frac{N(T + C_2)}{\bar{F}_2(T)} = \frac{S + NC_2}{3e^{-2\lambda T} - 2e^{-3\lambda T}} \quad N = 1, 2, \ldots. \tag{13}$$

Further, we consider a redundant system of a majority decision with $(2n + 1)$ modules as an error masking system, i.e., $(n + 1)$-out-of-$(2n + 1)$ system $(n = 1, 2, \ldots)$. If more than $(n + 1)$ results of $(2n + 1)$ modules match, the process is correct. Then, the probability that the process is correct during $(0, T]$ is:

$$\bar{F}_n(T) = \sum_{k=n+1}^{2n+1} \binom{2n+1}{k} \left(e^{-\lambda T}\right)^k \left(1 - e^{-\lambda T}\right)^{2n+1-k}, \quad n = 1, 2, \ldots. \tag{14}$$

Thus, the mean time to completion of the process is:

$$L_n(N) = \frac{N(T + C_n)}{\bar{F}_n(T)}, \quad N = 1, 2, \ldots, \tag{15}$$

where C_n is the overhead of a majority decision of $(2n + 1)$ modules.

3. Roll-Forward and Rollback Recoveries

If permanent faults have occurred, it would be impossible to detect such faults by comparing two results of the processes. When two results of the processes of some task are not matched, we prepare another spare module for executing the process. Then, to tolerate some faults, we consider the roll-forward and rollback recovery schemes with the following assumptions:

(a), (b), (c) The same assumptions as the previous ones.

(d′) If two results match with each other in case (c), the process is correct and goes forward. However, if two results of making the processing of task I_j do not match, it is judged that some errors have occurred. Then, we provide another spare process which make the processing of task I_j, and two processes make the processing of task I_{j+1} (Fig. 2). It is assumed that a spare process has no error.

(e′) If two results of making the processing of task I_{j+1} do not match, a spare module makes its processing in the same ways as case (d′).

(f′) Either if two results of making the processing of task I_N match or a spare module makes its processing, the process ends.

Let C_3 be the overhead of comparison of two modules and C_s be the all overhead of preparing a spare module and of setting a

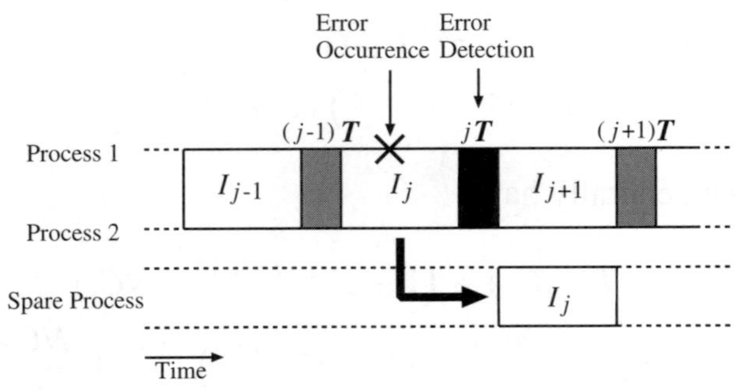

Fig. 2. Recovery scheme with spare process.

correct processing at checkpointing times, where $C_s \geq C_3$. Then, we compute the mean time $L_3(N)$ to complete the process successfully.

In particular, when $N = 1$, we easily have

$$\begin{aligned} L_3(1) &= e^{-2\lambda T}(T + C_3) + \left(1 - e^{-2\lambda T}\right)(T + C_3 + T + C_s) \\ &= T + C_3 + \left(1 - e^{-2\lambda T}\right)(T + C_s)\,. \end{aligned} \tag{16}$$

Further, when $N = 2$ and $N = 3$, we have, respectively,

$$\begin{aligned} L_3(2) &= e^{-2\lambda T}(T + C_3 + L_3(1)) + \left(1 - e^{-2\lambda T}\right) \\ &\quad \times e^{-2\lambda T}(T + C_3 + T + C_s + C_3) \\ &\quad + (1 - e^{-2\lambda T})^2 (T + C_3 + T + C_s + C_3 + T + C_s) \\ &= T + C_3 + \left(1 - e^{-2\lambda T}\right)(T + C_s + C_3) \\ &\quad + (1 - e^{-2\lambda T})^2 (T + C_s) + e^{-2\lambda T} L_3(1) \\ &= 2(T + C_3) + \left(1 - e^{-2\lambda T}\right)(T + 2C_s)\,. \end{aligned} \tag{17}$$

$$\begin{aligned} L_3(3) &= e^{-2\lambda T}(T + C_3 + L_3(2)) + \left(1 - e^{-2\lambda T}\right) \\ &\quad \times e^{-2\lambda T}(T + C_3 + T + C_s + C_3 + L_3(1)) \\ &\quad + (1 - e^{-2\lambda T})^2 e^{-2\lambda T}[T + C_3 + 2\,(T + C_s + C_3)] \\ &\quad + (1 - e^{-2\lambda T})^3\,[T + C_3 + 2\,(T + C_s + C_3) + T + C_s] \\ &= T + C_3 + \left(1 - e^{-2\lambda T}\right)\left(2 - e^{-2\lambda T}\right)(T + C_s + C_3) \\ &\quad + (1 - e^{-2\lambda T})^3\,(T + C_s) + e^{-2\lambda T} L_3(2) \\ &\quad + (1 - e^{-2\lambda T}) e^{-2\lambda T} L_3(1) \\ &= 3(T + C_3) + \left(1 - e^{-2\lambda T}\right)(T + 3C_s)\,. \end{aligned} \tag{18}$$

Thus, we generally have

$$\begin{aligned} L_3(N) &= N(T + C_3) + \left(1 - e^{-2\lambda T}\right)(T + NC_s) \\ &= S\left[1 + \frac{NC_3}{S} + \left(1 - e^{-2\lambda S/N}\right)\left(\frac{1}{N} + \frac{NC_s}{S}\right)\right], \\ &\qquad\qquad N = 1, 2, \ldots\,. \end{aligned} \tag{19}$$

4. Optimal Checkpointing Interval

We seek an optimal number N_3^* which minimizes the mean time $L_3(N)$ in Eq. (19). Since $L_3(\infty) \equiv \lim_{N\to\infty} L_3(N) = \infty$, there exists a finite number N_3^* $(1 \le N_3^* < \infty)$.

Putting that $T = S/N$ in Eq. (19) and rewriting it by the function T, we have

$$L_3(T) = S + \frac{SC_3}{T} + \left(1 - e^{-2\lambda T}\right)\left(T + \frac{SC_s}{T}\right). \tag{20}$$

It is evident that $\lim_{T\to 0} L_3(T) = \infty$ and $\lim_{T\to\infty} L_3(T) = \infty$. Thus, there exists an optimal T_3^* $\left(0 < T_3^* < \infty\right)$ which minimizes $L_3(T)$ in Eq. (20). Further, differentiating $L_3(T)$ with respect to T and setting it equal to zero, we have

$$T^2\left(2\lambda T e^{-2\lambda T} + 1 - e^{-2\lambda T}\right) = SC_3 + SC_s\left[1 - (1 + 2\lambda T)e^{-2\lambda T}\right]. \tag{21}$$

Since $C_s \ge C_3$, we easily have

$$SC_3 \le \frac{T^2\left(2\lambda T e^{-2\lambda T} + 1 - e^{-2\lambda T}\right)}{2 - (1 + 2\lambda T)e^{-2\lambda T}} \le SC_s\,. \tag{22}$$

Further, letting

$$Q(T) \equiv \frac{T^2\left(2\lambda T e^{-2\lambda T} + 1 - e^{-2\lambda T}\right)}{2 - (1 + 2\lambda T)e^{-2\lambda T}},$$

it is easily seen that $Q(T)$ is strictly increasing from 0 to ∞. Thus, denoting T_c and T_s by the solutions of equations $Q(T) = SC_3$ and $Q(T) = SC_s$, respectively, we have that $T_c \le T_3^* \le T_s$.

Therefore, in a similar way of the previous model, we have the following optimal checkpointing number N_3^*:

(i) If $T_3^* < S$, we put $\left[S/T_3^*\right] = N$, and calculate $L_3(N)$ and $L_3(N+1)$ from Eq. (19). If $L_3(N) \le L_3(N+1)$ then $N_3^* = N$, and conversely, if $L_3(N+1) < L_3(N)$ then $N_3^* = N + 1$.

(ii) If $T_3^* \ge S$ then $N_3^* = 1$, and the mean time is given in Eq. (16).

In particular case of $C_3 = C_s$, the optimal time T_3^* is given by a finite and unique solution of equation $Q(T) = SC_3$.

Further, using the approximation of $e^{-at} \approx 1-at$ for small $at > 0$, the mean time in Eq. (20) is simplified as

$$\tilde{L_3}(T) = S + \frac{SC_3}{T} + 2\lambda T^2 + 2\lambda SC_s\,. \qquad (23)$$

Thus, the approximate time, which minimizes $\tilde{L_3}$(T) in Eq. (23), is given by:

$$\tilde{T_3} = \left(\frac{SC_3}{4\lambda}\right)^{1/3}. \qquad (24)$$

5. Numerical Examples

We show the numerical examples of checkpointing intervals when $\lambda S = 10^{-1}$. Table 1 gives λT_1^* in Eq. (11), optimal number N_1^* and $\lambda L_1(N_1^*)$ for λC_1 = 0.5, 1.0, 1.5, 2.0, 3.0, 4.0, 5.0, 10.0, 20.0, 30.0 $(\times 10^{-3})$. For example, when $\lambda = 1.0 \times 10^{-2}$ [1/s],

Table 1. Optimal checkpointing number for a double modular system.

$\lambda C_1 \times 10^3$	$\lambda T_1^* \times 10^2$	N_1^*	$\lambda L_1(N_1^*) \times 10^2$
0.5	1.56	6	10.65
1.0	2.19	5	10.93
1.5	2.66	4	11.14
2.0	3.06	3	11.33
3.0	3.73	3	11.65
4.0	4.28	2	11.94
5.0	4.76	2	12.16
10.0	6.59	2	13.26
20.0	9.05	1	14.66
30.0	10.84	1	15.88

$C_1 = 1.0 \times 10^{-1}$ [s] and $S = 10.0$ [s], the optimal number is $N_1^* = 5$, the optimal interval is $S/N_1^* = 2.0$ [s] and the resulting mean time is $L_1(5) = 10.93$ [s], which is longer about 9% than S. It can be easily seen that the more overheads C_1 increase, the more optimal numbers N_1^* decrease.

Table 2 gives λT_3^*, optimal number N_3^* and $\lambda L_3(N_3^*)$ for $\lambda C_3 =$ 1.0, 2.0, 10.0 $(\times 10^{-3})$ and $\lambda C_s =$ 0.1, 0.5, 1.0, 2.0, 3.0, 5.0, 10.0, 50.0, 100.0, 200.0 $(\times 10^{-2})$ (s). For example, when $\lambda = 1.0 \times 10^{-2}$ (1/s), $C_3 = 1.0 \times 10^{-1}$ (s), $C_s = 1.0$ (s) and $S = 10.0$ (s), the optimal number is $N_3^* = 3$ and the resulting mean time is $L_3(3) = 10.71$ (s), which is shorter about 4% than $L_3(1) = 11.15$ (s). This indicates that the optimal numbers are decreasing slowly with

Table 2. Optimal checkpointing number with a spare module and its approximate time.

	$\lambda C_3 \times 10^3$								
	1.0			2.0			10.0		
λC_s $\times 10^2$	λT_3^* $\times 10^2$	N_3^*	$\lambda L_3(N_3^*)$ $\times 10^2$	λT_3^* $\times 10^2$	N_3^*	$\lambda L_3(N_3^*)$ $\times 10^2$	λT_3^* $\times 10^2$	N_3^*	$\lambda L_3(N_3^*)$ $\times 10^2$
0.1	2.97	3	10.53	—	—	—	—	—	—
0.5	2.98	3	10.61	3.76	3	10.91	—	—	—
1.0	2.98	3	10.71	3.77	3	11.01	6.52	2	12.67
2.0	3.00	3	10.90	3.79	3	11.20	6.54	2	12.86
3.0	3.02	3	11.10	3.81	3	11.40	6.56	2	13.05
5.0	3.06	3	11.48	3.84	3	11.78	6.59	2	13.43
10.0	3.15	3	12.45	3.93	3	12.75	6.68	2	14.38
50.0	4.10	3	20.19	4.83	2	20.39	7.50	1	21.88
100.0	5.85	2	29.71	6.40	2	29.91	8.77	1	30.94
200.0	10.43	1	48.17	10.68	1	48.27	12.20	1	49.07
$\lambda \tilde{T}_3 \times 10^2$	2.92			3.68			6.30		

the increase of C_s, and the approximate time $\tilde{T}_3$ in Eq. (24) shows a good lower bound of optimal time T_3^* for small C_s.

Further, comparing with Tables 1 and 2 when $\lambda C_1 = 1.0 \times 10^{-3}$ and $\lambda C_3 = 1.0 \times 10^{-3}$, if C_s is less than 2.0×10^{-2} then a spare module should be provided, and conversely, if C_s is larger than 3.0×10^{-2} then it should not be done.

6. Conclusions

In this chapter, simple stochastic models are formulated for error detection by redundancy on the finite process execution. We have obtained the mean time to completion of the process for a double modular redundant system. The optimal checkpointing interval which minimizes it is derived analytically. In general, the overhead C_1 and the native execution time S would be estimated easily. Therefore, the establishment of checkpointing scheme would depend on whether we can accurately estimate the error occurrence rate or not.

Further, we have considered a double modular redundancy with spare processing, and obtained the mean time to completion of the process, by dividing it into constant intervals. We have analytically derived the optimal checkpointing number which minimizes the mean time. Comparing with a double modular system with no spare and one spare modules, it has been shown in numerical examples that the system with one spare module is better than with no spare, if the overhead of spare processing is less than a certain value. We need to consider the case where a spare module may have some errors during its processing as further studies.

References

1. T. Anderson and P. Lee, *Fault Tolerance: Principles and Practice* (Prentice-Hall, New Jersey, 1981).
2. D. K. Pradhan and N. H. Vaidya, Roll-forward and rollback recovery: Performance-reliability trade-off, *Proceedings of the 24th International Symposium on Fault-Tolerant Computings*, 1994, pp. 186–195.

3. A. Ziv and J. Bruck, Performance optimization of checkpointing schemes with task duplication, *IEEE Transactions on Computers* **46** (1997) 1381–1386.
4. A. Ziv and J. Bruck, Analysis of checkpointing schemes with task duplication, *IEEE Transactions on Computers* **47** (1998) 222–227.
5. H. Kim and K. G. Shin, Design and analysis of an optimal instruction-retry policy for TMR controller computers, *IEEE Transactions on Computers* **45** (1996) 1217–1226.
6. D. K. Pradhan and N. H. Vaidya, Rollforward checkpointing scheme: Concurrent retry with nondedicated spares, *IEEE Workshop on Fault-Tolerant Parallel and Distributed Systems*, 1992, pp. 166–174.
7. S. Osaki, *Applied Stochastic System Modeling* (Springer-Verlag, Berlin, 1992).

CHAPTER 3

Monitoring Inter-Arrival Times with Statistical Control Charts

P. R. Sharma, M. Xie and T. N. Goh

Department of Industrial and Systems Engineering,
National University of Singapore,
10 Kent Ridge Crescent, Singapore 119260

1. Introduction

All processes suffer from two kinds of variations: chance causes and assignable causes. Chance causes are the causes that are inherently present in the process and thus have to be accepted. On the other hand assignable causes, as the name suggests are induced by the system, i.e., man, machine, material, etc. The main objective of the control chart is to detect the presence of assignable causes and to inform the user by raising an alarm. Usually the control chart has three lines, referred to as the upper control limit, the lower control limit, and the central line. The chart plots the sample statistic of the quality characteristic, which is to be monitored. The presence of some unusual sources of variation results in a point plotting above or below the upper control limit or the lower control limit, respectively. This warrants investigation and removal of such sources to bring the process back to its original state or if possible to improve it.

Failure process monitoring is an important issue for complex or repairable systems. It is also a common problem for a fleet of systems, such as equipment or vehicles of the same type in a company.

Such monitoring can be based on standard Shewhart control charts for attribute data that are used to monitor the number of defects in a Poisson process. However, these charts have several drawbacks which make them less effective in detecting process changes. The traditional Shewhart chart assumes an anticipated false alarm probability of 0.27%, but in reality it could be much higher due to the poor normal approximation of the underlying distribution. Another serious drawback with the Shewhart charts is due to the lower control limit which is usually set at zero. This is not useful because it makes the chart ineffective in picking up process improvements.

One of the alternatives to the *c*- or the *u*-chart is the chart based on cumulative quantity, or *CQC*-chart, proposed by Chan *et al.*[1] This charting procedure is based on the monitoring of cumulative production quantity (or time) to observe a defect in a manufacturing process. This approach has shown to have a number of advantages: it does not involve the choice of a subjective sample size; it raises fewer false alarms; it can be used in any environment irrespective of whether the process is of high quality; and it can detect further process improvement. The discrete version of this chart has also been advocated by Woodall[2]; see also Calvin[3] and Kaminsky *et al.*[4]

As an extension of the *CQC*-chart, Xie *et al.*[5] proposed the CQC_r-chart which advocates monitoring the cumulative quantity (or time) to observe *r* defects. Both charts can be easily employed in reliability monitoring since the quantity produced between the observations of the two defects is related to the time between failures in reliability study. Each of these two approaches has its own advantages and disadvantages. One of the drawbacks of the CQC_r-chart is that the user may have to wait too long to plot a point on the chart, this is especially true for highly reliable systems where the occurrence of a failure is a rare event.

Another alternative to monitor a Poisson process is to use a CUSUM chart. The CUSUM chart has proven to be very useful in detecting small shifts in the process. The time-between-events CUSUM has been studied by many authors.[6–11]

Most of the studies assume that time-between-event is exponentially distributed. An important assumption when exponential distribution is used is that the event occurrence rate is constant. In reliability applications, this implies that the items have no aging property. This assumption is usually violated in reality. Due to wear and tear and other usage condition, items usually have an increasing failure rate. To be able to monitor processes for which the exponential assumption is violated, Weibull distribution is a good alternative and it is a simple generalization of the exponential distribution. This flexibility and its reasonableness have made Weibull distribution probably the most useful distribution model in reliability analysis and it has been widely used by various authors to model the failure times.

There are a couple of papers where the authors have indicated the use of Weibull distribution for process monitoring in reliability,[5,12] but no detailed analysis is carried out. Related to the use of Weibull distribution in statistical process control, Zhang *et al.*[13] studied the economic design of X (bar) chart for monitoring systems with Weibull in-control times with the main objective being the economic performance. Sun *et al.*[14] used the Weibull distribution to model the time to failure of a hard disk drive and came out with a failure percent control chart. Ramalhoto and Morais[15] studied the performance of a control chart for the scale parameter of the three-parameter Weibull Distribution where the location and the shape parameters are assumed to be known. Earlier, Nelson[16] considered Weibull distribution for median and range charge, assuming a fixed subgroup size. The use of Weibull distribution was also investigated in Ref. 17.

Here an overview of the three charts is first given and their properties are discussed. Then the performance of the three charts is compared on the basis of their sensitivity towards the shifts in the process. Next some of the implementation issues are discussed. To illustrate the charting procedure of the three charts, an example is also presented. Finally, the performance of the charts is studied when there is a change in the underlying distribution.

The notations used in this chapter are defined as follows:

λ_0	In-control failure rate
λ	Out-of-control failure rate
T_L	Lower control limit of the *CQC* chart
T_C	Center line of the *CQC* chart
T_U	Upper control limit of the *CQC* chart
α	False alarm probability
β	Type II error probability
ARL	Average run length
$E(T)$	The average (expected) time to plot a point on the *CQC* chart
ATS	Average time to signal of the *CQC* chart
S_i^+	Upper CUSUM statistic
S_i^-	Lower CUSUM statistic
k	Reference value of the CUSUM scheme
λ_d	Out-of-control failure occurrence rate that the CUSUM scheme is designed to detect quickly
h	The decision interval of the CUSUM procedure
t	Number of states in the Markov chain
w	Width of the interval in Markov chain
p_m	Probability of passing from state i to state j, where m is $j - i$
F_m	Probabilityof passing from state i to state 1
P	Transition probability matrix for the Markov chain
I	$h \times h$ identity matrix
R	Matrix obtained from the transition probability matrix P by deleting the last row and column
μ	A vector of factorial moments
T_r	Gamma random variable
r	Shape parameter of the gamma distribution
T_{rU}	Upper control limit of the CQC_r chart
T_{rC}	Center line of the CQC_r chart
T_{rL}	Lower control limit of the CQC_r chart
β_r	Type II error probability of the CQC_r chart

2. Overview: Some Useful Control Charts

2.1. *CQC-charts*

In a Poisson process, if the failures have an in-control rate of occurrence, λ_0, then the time required to observe one defect (event), T, will follow exponential distribution and can be described with the distribution function given by:

$$F(T) = 1 - \exp\{-\lambda_0 T\}\,. \tag{1}$$

Under ideal conditions we would want the control chart to raise less false alarms (to avoid unnecessarily interrupting the process) which in other words means a small Type I error, defined as the probability that a plotted point falls outside the control limit when the process is in-control. While at the same we would like it to detect the process shift as soon as possible, which means that the control chart should also have a small Type II error, defined as the probability that a plotted point falls within the control limits when in fact the reliability of the system has changed. If we widen the control limits, the Type I error decreases but the Type II error increases. Similarly, when we tighten the control limits, the opposite happens, i.e., the Type I error increases while Type II error decreases.

The three-sigma limits concept is based on normal approximation and this approximation does not hold true for skewed distributions like exponential. Thus for control charts based on skewed distributions, it is better to calculate the control limits on exact probabilities.[18–20] So by using the exact probability limits, we actually modify the control chart in such a way that each point has an equal chance of falling above or below the control limits. If α is the acceptable false alarm probability or the Type I error, the lower control limit (T_L), upper control limit (T_U) and the center line (T_C), based on the exponential distribution, can be, respectively, calculated as:

$$T_L = -\frac{\ln(1-\alpha/2)}{\lambda_0}\,, \quad T_U = -\frac{\ln(\alpha/2)}{\lambda_0}\,, \quad T_C = -\frac{\ln 0.5}{\lambda_0} \approx \frac{0.6931}{\lambda_0}\,. \tag{2}$$

The Type II error or the β error of the *CQC*-chart is given by:

$$\beta = F(T_U) - F(T_L)\,. \tag{3}$$

The average run length (ARL) is a commonly used measure of chart performance.[20–22] It is defined as the average number of points that must be plotted on the control chart before a point indicates an out-of-control situation. A good control chart should have a large average run length when the process is in-control and small average run length when the process shifts away from the target. The ARL for the *CQC* chart can be calculated as:

$$ARL = \frac{1}{1-\beta} = \frac{1}{1 + (\alpha/2)^{\lambda/\lambda_0} - (1-\alpha/2)^{\lambda/\lambda_0}}\,, \tag{4}$$

where λ is the out-of-control (shifted) value of the process parameter.

Sometimes the ARL does not give a good idea about the performance of the chart. This is especially true when the emphasis is on the number of items inspected or the total time taken rather than on the number of points plotted. In such case, ATS, average time to signal, is a better measure and is given by:

$$ATS = E(T) \times ARL = \frac{1}{\lambda\left[1 + (\alpha/2)^{\lambda/\lambda_0} - (1-\alpha/2)^{\lambda/\lambda_0}\right]}\,. \tag{5}$$

2.2. *Exponential CUSUM charts*

A time-between-events CUSUM can be defined in the following manner. If X_1, X_2,..., are the inter-arrival times, then the time-between-events CUSUM for detecting an increase or decrease in the inter-arrival times can be, respectively, defined as:

$$\begin{aligned} S_i^+ &= \max\{S_{i-1}^+ + (X_i - k)\}\,, \\ S_i^- &= \min\{S_{i-1}^- + (X_i - k)\}\,, \end{aligned} \tag{6}$$

where k is the reference value. It can be calculated for any given in-control failure occurrence rate of λ_0 and an out-of-control failure occurrence rate of λ_d, that the CUSUM scheme is designed to detect quickly as[9]:

$$k = \frac{\ln \lambda_d - \ln \lambda_0}{\lambda_d - \lambda_0}. \tag{7}$$

The control limits of the CUSUM scheme are denoted by h and the decision on the statistical control of the process is taken depending on whether $S_t^- \leq -h$ or $S_t^+ \geq h$. Once the reference value k has been calculated, a suitable value of h can be found out to give an acceptable in-control average run length. The average run length of the CUSUM scheme can be calculated by the Markov chain approach.[9,23]

When the random variable is continuous, the Markov chain method gives an approximate answer. However, this answer can be brought reasonably close to the exact values by grouping the possible values of the random variable into discrete class intervals. The width of the interval is given by:

$$w = \frac{2h}{2t - 1}, \tag{8}$$

where t is the number of states in the Markov chain process.

The transition probabilities of the Markov chain are defined as:

$$p_m = \Pr\left(mw - \frac{1}{2}w < X - k \leq mw + \frac{1}{2}w\right),$$

and

$$F_m = \Pr\left(X - k \leq mw + \frac{1}{2}w\right), \tag{9}$$

where $m = j - i$, with i as the state before transition and j is the state after transition. The transition probability matrix can then be written as:

$$P = \begin{bmatrix} F_0 & p_1 & \cdots & p_j & \cdots & p_{h-1} & 1 - F_{h-1} \\ F_{-1} & p_0 & \cdots & p_{j-1} & \cdots & p_{h-2} & 1 - F_{h-2} \\ \vdots & \vdots & & \vdots & & \vdots & \vdots \\ F_{-i} & p_{1-i} & \cdots & p_{j-i} & \cdots & p_{h-1-i} & 1 - F_{h-1-i} \\ \vdots & \vdots & & \vdots & & \vdots & \vdots \\ F_{1-h} & p_{2-h} & \cdots & p_{j-(h-1)} & \cdots & p_0 & 1 - H \\ 0 & 0 & \cdots & 0 & \cdots & 0 & 1 \end{bmatrix}.$$

Using the Markov chain result, we have

$$(\mathbf{I} - R)\mu = 1\,, \tag{10}$$

where $\mathbf{I}$ is the $h \times h$ identity matrix and R is the matrix obtained from the transition probability matrix P by deleting the last row and column, and μ is a vector of factorial moments. The first element of the vector μ gives the average run length for the CUSUM chart.

2.3. CQC_r-charts

The sum of r exponentially distributed random variables is the Erlang distribution. The probability density function of T_r is given as:

$$f(T_r, r, \lambda) = \frac{\lambda^r T_r^{r-1}}{(r-1)!} \exp\{-\lambda T_r\}\,. \tag{11}$$

The cumulative Erlang distribution is:

$$F(T_r, r, \lambda) = 1 - \sum_{k=0}^{r-1} \frac{(\lambda T_r)^k}{k!} \exp\{-\lambda T_r\}\,. \tag{12}$$

It should be noted that for $r = 1$, the Erlang distribution reduces to the exponential distribution. Again by using the exact probability limits concept, the upper control limit, T_{rU}, the center line, T_{rC}, and the lower control limit, T_{rL}, can be easily calculated by solving the following set of equations:

$$F(T_{rU}, r, \lambda) = 1 - \sum_{k=0}^{r-1} e^{-\lambda T_{rU}} \frac{(\lambda T_{rU})^k}{k!} = 1 - \frac{\alpha}{2},$$
$$F(T_{rC}, r, \lambda) = 1 - \sum_{k=0}^{r-1} e^{-\lambda T_{eC}} \frac{(\lambda T_{rC})^k}{k!} = 0.5, \quad (13)$$
$$F(T_{rL}, r, \lambda) = 1 - \sum_{k=0}^{r-1} e^{-\lambda T_{rL}} \frac{(\lambda T_{rL})^k}{k!} = \frac{\alpha}{2}.$$

Denote the probability for the time T_r falling within the control limits, even when the process has shifted, of the CQC_r-chart by β_r. Then β_r can be represented as:

$$\beta_r = F(T_{rU}, r, \lambda) - F(T_{rL}, r, \lambda). \quad (14)$$

Using Eq. (12), the probability that the points do not fall between the control limits which is represented as $(1 - \beta_r)$ can be obtained as:

$$1 - \beta_r = 1 - \left[\sum_{k=0}^{r-1} e^{-\lambda T_{rL}} \frac{(\lambda T_{rL})^k}{k!} - \sum_{k=0}^{r-1} e^{-\lambda T_{rU}} \frac{(\lambda T_{rU})^k}{k!} \right]. \quad (15)$$

The ARL of the CQC_r-chart can then be represented as:

$$ARL = \frac{1}{1 - \left[\sum_{k=0}^{r-1} e^{-\lambda T_{rL}} \frac{(\lambda T_{rL})^k}{k!} - \sum_{k=0}^{r-1} e^{-\lambda T_{rU}} \frac{(\lambda T_{rU})^k}{k!} \right]}. \quad (16)$$

Thus, on average, only one out of $1/(1-\beta_r)$ points fall outside the control limits. Now if the process defect rate is λ, then, on average, r defects will occur for r/λ (the mean of the Erlang distribution) items inspected. The average time to signal for the CQC_r-chart can then be represented as:

$$ATS = \frac{r}{\lambda\left[1 + \sum_{k=0}^{r-1} e^{-\lambda T_{rU}} \frac{(\lambda T_{rU})^k}{k!} - \sum_{k=0}^{r-1} e^{-\lambda T_{rL}} \frac{(\lambda T_{rL})^k}{k!}\right]}. \quad (17)$$

3. Comparison Based on ARL and ATS Performance

The case of process deterioration and improvement are considered separately. To detect the process deterioration, the performance of the lower CUSUM is compared to that of the CQC-chart and CQC_r-chart having only a lower control limit. Similarly, the performance of the upper CUSUM chart is compared to that of the CQC-chart and CQC_r-chart with only an upper control limit. An in-control average run length of 370 is used for both cases, which translates to a false alarm probability of approximately 0.0027 for the CQC_r-charts with single limit. The in-control average time to signal (ATS_0) is also fixed as 370 for which the false alarm probability of the CQC_r-charts can be calculated as $0.0027r/\lambda_0$ (where $\lambda_0 = 1$).

3.1. *Process deterioration*

The in-control value of the failure occurrence rate is assumed to be 1. Suppose that the user is interested in quickly identifying a shift to 1.4, 1.9 and 2.5. The reference value k for the three CUSUM charts, now onwards referred to as Lower CUSUM 1 (LC-1), Lower CUSUM 2 (LC-2) and Lower CUSUM 3 (LC-3), respectively, can be calculated using Eq. (7). The appropriate value of h can then be calculated to give an in-control ARL_0 of approximately 370. The k and h values for the three CUSUM charts are found out to be (0.84, 7.16), (0.71, 4.13) and (0.61, 2.783), respectively. The Markov chain approach (with 151 states) is then used to calculate the ARL for different values of defect rate. Table 1 shows the ARL values of the three CUSUM charts along with the ARL values of CQC, CQC_2, CQC_3 and CQC_4 charts.

Table 1. ARL values when the process failure rate deteriorates from $\lambda_0 = 1$.

λ	CQC	CQC_2	CQC_3	CQC_4	LC-1	LC-2	LC-3
1	370.37	370.37	370.37	370.37	370	370.23	370.3
1.1	336.75	307.62	283.89	264.4	164.1	190.74	211.5
1.2	308.73	259.78	223.08	195.1	91.84	111.32	131.2
1.3	285.02	222.46	178.99	148.02	61.07	72.38	87.47
1.4	264.69	192.77	146.19	114.99	45.65	51.47	62.18
1.5	247.08	168.76	121.25	91.17	36.87	39.3	46.7
1.6	231.67	149.07	101.91	73.57	31.33	31.71	36.74
1.7	218.07	132.7	86.66	60.3	27.58	26.69	30.05
1.8	205.98	118.96	74.47	50.11	24.88	23.19	25.38
1.9	195.17	107.3	64.58	42.16	22.85	20.65	21.99
2	185.44	97.32	56.47	35.86	21.27	18.74	19.47
2.1	176.63	88.71	49.75	30.8	20.02	17.26	17.55
2.2	168.62	81.23	44.13	26.7	18.99	16.08	16.03
2.3	161.31	74.69	39.39	23.33	18.13	15.12	14.83
2.4	154.61	68.93	35.36	20.54	17.41	14.33	13.84
2.5	148.45	63.84	31.9	18.21	16.8	13.67	13.03
2.6	142.76	59.32	28.92	16.25	16.26	13.1	12.34
2.7	137.49	55.28	26.33	14.58	15.8	12.61	11.76
2.8	132.6	51.66	24.07	13.15	15.39	12.19	11.26
2.9	128.04	48.39	22.09	11.92	15.02	11.82	10.83
3	123.79	45.44	20.35	10.86	14.7	11.49	10.45

It can be seen that the CQC_r-charts are out-performed by the CUSUM charts. LC-1 chart gives a satisfactorily low ARL for small deteriorations in the failure rate while the LC-2 and LC-3 charts give better performance for moderate and larger shifts. Among the CQC_r-charts, the control chart with large r performs better than those with small r. The performance of CQC_4 chart is quite close to that of the CUSUM charts and in fact is better than LC-1 and LC-2 charts for large shifts.

However it may not be wise to use the ARL as a performance measure as it does not take into account the time needed to plot one point on the control chart. Moreover, the time needed to plot one point on CQC_r-chart is r times the time needed to plot one point on CQC-chart. Thus to give a better picture of the chart performance, average time to signal (ATS) is now used as the yardstick for comparison in place of ARL. Table 2 shows the ATS values for all the seven charts mentioned above. Again, it can be seen that the CQC_r-charts perform worse than the CUSUM chart. For large process deteriorations, however, the performance of CQC_4-chart is somewhat similar to the CUSUM charts.

Table 2. ATS values when the process deteriorates from $\lambda_0 = 1$.

λ	CQC	CQC_2	CQC_3	CQC_4	LC-1	LC-2	LC-3
1	370.37	370.37	370.37	370.37	370	370.23	370.33
1.1	306.13	280.25	260.56	245.48	149.22	173.4	192.27
1.2	257.27	217.4	189.49	169.55	76.54	92.77	109.29
1.3	219.24	172.21	141.68	121.24	46.97	55.68	67.28
1.4	189.07	138.86	108.48	89.28	32.61	36.76	44.42
1.5	164.72	113.7	84.76	67.44	24.58	26.2	31.13
1.6	144.79	94.36	67.42	52.07	19.58	19.82	22.96
1.7	128.28	79.22	54.47	40.99	16.22	15.7	17.68
1.8	114.44	67.21	44.62	32.83	13.82	12.89	14.1
1.9	102.72	57.55	37	26.69	12.03	10.87	11.58
2	92.72	49.69	31.02	22	10.64	9.37	9.74
2.1	84.11	43.23	26.27	18.36	9.53	8.22	8.36
2.2	76.65	37.87	22.45	15.49	8.63	7.31	7.29
2.3	70.14	33.37	19.34	13.2	7.88	6.58	6.45
2.4	64.42	29.58	16.79	11.35	7.26	5.97	5.77
2.5	59.38	26.35	14.67	9.85	6.72	5.47	5.21
2.6	54.91	23.59	12.91	8.61	6.25	5.04	4.75
2.7	50.92	21.22	11.42	7.58	5.85	4.67	4.36
2.8	47.36	19.16	10.16	6.72	5.5	4.35	4.02
2.9	44.15	17.36	9.08	5.99	5.18	4.08	3.73
3	41.26	15.79	8.16	5.37	4.9	3.83	3.48

3.2. *Process improvement*

The in-control value of the defect rate is again assumed to be 1 and say that the user is interested in quickly identifying a shift to 0.9, 0.5 and 0.1. The reference value k and the appropriate value of h were then calculated to give an in-control ARL_0 of approximately 370. The k and h values for the three CUSUM charts, now onwards, referred to as Upper CUSUM-1 (UC-1), Upper CUSUM-2 (UC-2) and Upper CUSUM-3 (UC-3), respectively, are found out to be (1.05, 13.82), (1.39, 6.81) and (2.56, 3.58), respectively.

Table 3 shows the ARL values of the charts when the failure occurrence rate decreases. Clearly the UC-1 chart identifies the shift to

Table 3. ARL values when the failure rate decreases from $\lambda_0 = 1$.

λ	CQC	CQC_2	CQC_3	CQC_4	UC-1	UC-2	UC-3
1	370.37	370.37	370.37	370.37	370.08	370.86	370.21
0.95	275.56	258.21	245.98	236.22	205.1	237.02	268.22
0.9	205.01	180.41	164.08	151.68	126.63	154.19	194.36
0.85	152.52	126.35	109.99	98.11	85.43	102.48	140.91
0.8	113.48	88.72	74.13	63.98	61.62	69.8	102.26
0.75	84.43	62.48	50.25	42.1	46.59	48.83	74.33
0.7	62.81	44.15	34.3	27.99	36.41	35.1	54.14
0.65	46.74	31.31	23.58	18.81	29.12	25.91	39.55
0.6	34.77	22.3	16.35	12.81	23.66	19.59	29
0.55	25.87	15.96	11.45	8.85	19.42	15.13	21.37
0.5	19.25	11.48	8.11	6.21	16.04	11.89	15.83
0.45	14.32	8.32	5.81	4.45	13.28	9.46	11.8
0.4	10.65	6.07	4.23	3.25	10.99	7.6	8.86
0.35	7.92	4.47	3.14	2.44	9.06	6.14	6.7
0.3	5.9	3.33	2.37	1.89	7.41	4.96	5.1
0.25	4.39	2.52	1.85	1.52	5.98	4	3.9
0.2	3.26	1.93	1.48	1.27	4.73	3.2	3
0.15	2.43	1.52	1.24	1.12	3.63	2.52	2.3
0.1	1.81	1.24	1.09	1.03	2.66	1.94	1.77
0.05	1.34	1.07	1.01	1	1.78	1.44	1.34
0.01	1.06	1	1	1	1.15	1.08	1.06

$\lambda = 0.9$ faster than the other charts. In general, UC-1 chart picks up the small changes faster than the rest followed by the CQC_4-chart. For moderate and large shifts, however, the CQC_r-charts perform better than the CUSUM charts.

Table 4 shows the ATS values of the three CUSUM charts listed along with the ATS values of the CQC_r-charts ($r = 1–4$). Clearly the UC-1 chart identifies the shift to $\lambda = 0.9$ faster than the other charts. Once again in general, UC-1 chart picks up the small changes faster than the rest followed by the UC-2 chart. For moderate shifts (the middle portion of the table), UC-2 gives the best performance

Table 4. ATS values for decreasing failure rates ($\lambda_0 = 1$).

λ	CQC	CQC_2	CQC_3	CQC_4	UC-1	UC-2	UC-3
1	370.37	370.37	370.36	370.37	370.08	370.86	370.21
0.95	290.06	282.5	276.7	271.97	215.89	249.49	282.34
0.9	227.79	216.53	208.17	201.55	140.7	171.32	215.96
0.85	179.44	166.87	157.84	150.88	100.51	120.56	165.78
0.8	141.85	129.38	120.71	114.22	77.02	87.25	127.83
0.75	112.57	100.99	93.21	87.55	62.12	65.1	99.1
0.7	89.73	79.43	72.75	68.04	52.02	50.14	77.34
0.65	71.9	63.03	57.49	53.72	44.8	39.86	60.85
0.6	57.95	50.51	46.06	43.16	39.43	32.65	48.34
0.55	47.03	40.95	37.5	35.4	35.31	27.51	38.85
0.5	38.49	33.66	31.11	29.71	32.08	23.78	31.66
0.45	31.82	28.11	26.37	25.63	29.52	21.03	26.23
0.4	26.63	23.95	22.96	22.82	27.48	19.01	22.15
0.35	22.64	20.91	20.64	21.12	25.88	17.54	19.14
0.3	19.65	18.84	19.3	20.47	24.69	16.54	16.99
0.25	17.55	17.69	19.01	21.03	23.91	16	15.6
0.2	16.32	17.59	20.05	23.28	23.65	15.98	14.98
0.15	16.19	19.09	23.34	28.51	24.2	16.79	15.36
0.1	18.07	24.03	31.83	40.76	26.55	19.39	17.66
0.05	26.88	42.24	60.6	80.14	35.66	28.75	26.82
0.01	106.09	200.52	300.03	400	115.02	108.3	106.26

followed by the CQC_4-chart. The table also shows that the ATS of the charts first decreases with decrease in failure rate and then increases. Decrease in failure rate means an improvement in the process and as the process improves, the average time to plot a point on the chart increases. The ATS of the charts is a product of the ARL and the expected time to plot a point. With decrease in failure rate, the ARL decreases its effect less dominant as compared to the other effect. As a result, the average time to signal increases for small values of λ. The effect is more pronounced in the case of CQC_r-charts due to the effect of the term "r" on the ATS, Eq. (17). The CQC-chart and the CUSUM (particularly UC-3) charts, that are free from the effect of r, thus perform better than the CQC_r-charts ($r > 1$).

4. Implementation and Example

This section discusses some of the implementation issues associated with the CQC, CUSUM and the CQC_r-charts (see Table 5). The CQC and the CQC_r-charts plot the time observed till the occurrence of a failure while the CUSUM chart plots the difference of the observed time from the reference value. One of the drawbacks associated with the CUSUM chart is the extensive computing required. In the case of CQC and the CQC_r-charts, the calculation of lower and upper control limits is much easier compared to the calculation of k and h for the CUSUM charts.

Thus if ease of design is an issue, then the CQC_r-charts may turn out to be better alternative compared to the CUSUM charts. Even from the operation point of view, the CQC_r-charts appear more promising due to their resemblance to the Shewhart charts. The optimum CUSUM design discussed in this chapter requires extensive computing. On the other hand, for the case of CQC_r-charts, a simple algorithm can be written to calculate the control limits and the average run length. Most of the calculations in this chapter were done using the statistical software, Mathematica.

Table 5. Implementation issues.

Issues	*CQC*	CQC_r	CUSUM
Information required	Time/quantity between events	Time/quantity between events	Time/quantity between events
Parameters	False alarm probability	• r • False alarm probability	• Reference value (k) • Decision interval (h)
Value plotted	Time/quantity between events	Time/quantity between r events	Deviations from the reference value
Calculation required	Control limits	Control limits	• Reference value (k) • Decision interval (h)
Sensitivity	Comparatively less sensitive	Sensitive to moderate and large shifts	Sensitive to small shifts

The charting procedure of the three charts for times-between-events will be illustrated with an example in the following. Table 6 shows some time-between-events. The first 36 values (across) correspond to a historical in-control failure rate of $\lambda_0 = 1$. The last 24 points were simulated when the average failure rate is shifted to $\lambda = 0.9$, which means that the reliability has improved.

Assuming the user is interested in detecting only the decrease in failure occurrence rate, the upper control limit of the *CQC*-chart (for $\alpha = 0.0027$) can be found as:

$$T_U = -\frac{\ln \alpha}{\lambda_0} = 5.91\,.$$

The reference value of the CUSUM chart designed to detect the shift from $\lambda_0 = 1$ to $\lambda = 0.9$ can be calculated as:

$$k = \frac{\ln 0.9 - \ln 1}{0.9 - 1} = 1.05\,.$$

Once the reference value is known, an appropriate value for the decision interval can be found so that it gives a desired in-control

Table 6. Time between events data (read across for consecutive values).

0.367	1.078	0.732	0.681	0.805	0.373
1.42	0.514	1.649	0.508	2.193	0.368
0.471	0.89	0.095	0.233	0.262	0.727
0.461	0.641	0.318	0.163	1.819	1.304
3.362	0.674	0.384	0.268	0.531	0.197
0.822	1.788	0.927	1.518	1.115	0.744
0.289	0.236	0.967	0.424	7.304	1.249
0.265	2.065	1.439	0.827	0.521	0.137
1.59	0.039	0.063	2.363	0.476	2.15
0.759	0.055	1.515	0.086	1.922	0.823

Table 7. Cumulative time between every three events.

2.177	1.859	3.583	3.069	1.456	1.222	1.42	3.286	4.42	0.996
3.537	3.377	1.492	8.977	3.769	1.485	1.692	4.989	2.329	2.831

ATS performance. The value of h for an in-control ATS of 370 is 13.82.

Since the CQC-chart makes use of a single observation in decision making, a CQC_r-chart could be used if more observations are to be taken into consideration in an easy way. The data shown in Table 6 is converted into the data of Table 7, which shows the cumulative time between every three occurrences, i.e., T_3. The control limits of the CQC_3 chart can be calculated by using Eq. (13) and solving it with the help of some statistical or mathematical package. The upper control limit of the CQC_3 chart can be calculated as 8.67. The CQC-chart, CUSUM chart and the CQC_r-chart are shown in Figs. 1, 2 and 3, respectively.

Interestingly both the CQC and the CQC_3 charts raise an alarm while the CUSUM chart does not. However, the pattern on the CUSUM chart does point out a shift in the process.

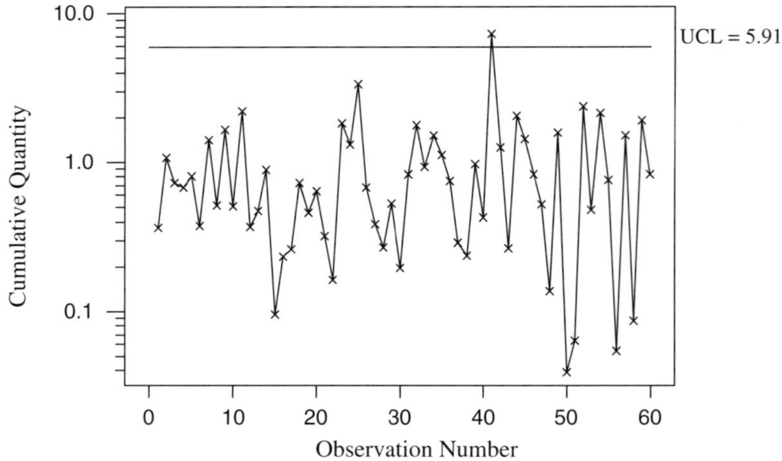

Fig. 1. The *CQC*-chart.

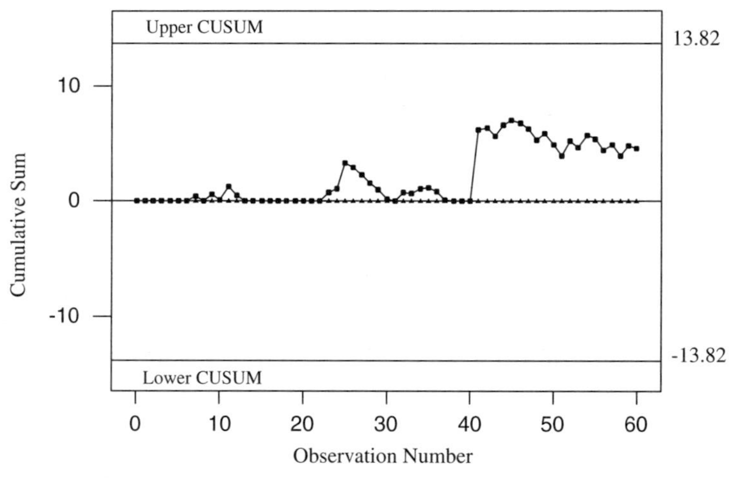

Fig. 2. The CUSUM chart.

5. Detecting Change of Underlying Distribution

In this section, the performance of the CUSUM charts and CQC_r-charts are studied when the underlying distribution can no longer be modeled by the exponential distribution. We assume that the

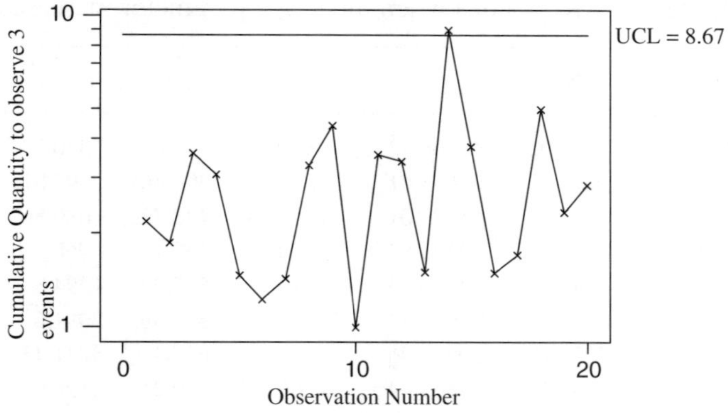

Fig. 3. The CQC_3 chart.

underlying distribution can be modeled by the Weibull distribution. It should be noted that although similar ideas could be used for other distributions, the Weibull distribution is probably the most widely used one and it is very flexible for modeling increasing or decreasing failure rates.

Even though the scale parameter (θ) is more likely to change but sometimes the shape parameter (β), which depends on the material property, can also change. In this study, we have only concentrated on the change in shape parameter, and the scale parameter is fixed as 1. For Weibull distribution, the mean is given by:

$$\mu = E[T] = \theta\Gamma\left(1 + \frac{1}{\beta}\right) \tag{18}$$

and the variance is given by:

$$\sigma^2 = \theta^2\left\{\Gamma\left(1 + \frac{2}{\beta}\right) - \left[\Gamma\left(1 + \frac{1}{\beta}\right)\right]^2\right\}. \tag{19}$$

It can be seen that the mean and the variance are strongly affected by the scale and shape parameter. When the shape parameter increases, both the mean and the variance reduce. However, the decrease in variance is quite significant compared to the decrease in mean.

Table 8. ARL values when the shape parameter increases.

β	CQC	CQC_2	CQC_3	CQC_4	LC-1	LC-2	LC-3
1	370.37	362.02	366.33	368.34	370	370.23	370.33
1.1	668.62	668	683.41	681.19	396.66	495.81	556.8
1.2	1207.36	1233.33	1274.04	1295.18	428.72	680.56	865.39
1.3	2180.54	2343.78	2192.89	2150.86	468.14	960	1391.93
1.4	3938.43	3858.76	4716.23	4076.24	517.32	1394.87	2318.26
1.5	7113.84	8328.59	9077.33	10775.54	579.35	2091.6	3997.87
1.6	12849.8	17053.16	20597.06	15780.21	658.4	3241.43	7134.44
1.7	23210.9	27664.47	23863.44	32366.8	760.21	5196.67	13161.73
1.8	41926.9	45016.36	55224.28	64330.38	893.01	8622.67	25066.73
1.9	75734.8	70173.18	124862	166029.2	1068.74	14807.8	49207.2
2	136804	135287	207234	256074.3	1305.07	26308.73	99394.64

For both the cases of increasing or decreasing shape parameter, the ARL and ATS for CQC_r-charts (except for $r = 1$) were calculated by simulation. For each value of the shape parameter, 100 000, 150 000 and 200 000 points following Weibull distribution were simulated for CQC_2, CQC_3, and CQC_4 charts, respectively. In other words, 50 000 points were plotted on each chart and the mean of the obtained ARLs was used as an estimate of the ARL.

Table 8 shows the ARL values of the charts when the shape parameter increases. As can be seen with the increase in shape parameter, the chances of point falling within the limits increases. The control charts will have larger out-of-control ARLs as compared to the in-control ARLs. This is the same as the case for the average time to signal, shown in Table 9.

When the shape parameter decreases, the variance increases thus resulting in a decrease in the ARL and the control charts will be able to detect the decrease in the shape parameter. Table 10 lists the ARL values when the shape parameter increases and the CQC_4-chart detects the shifts fastest. In fact, in general, all the CQC_r-charts perform better than the CUSUM charts (except for the CQC-chart which outperforms the CUSUM charts only for $\beta \leq 0.3$).

Table 9. ATS when the shape parameter increases.

β	CQC	CQC_2	CQC_3	CQC_4	LC-1	LC-2	LC-3
1	370.37	362.45	367.8	373.98	370	370.23	370.33
1.1	645.16	507.15	503.31	493.4	382.74	478.41	537.27
1.2	1135.71	650.1	635.13	611.01	403.28	640.18	814.03
1.3	2013.9	783.3	755.4	729.48	432.36	886.64	1285.55
1.4	3589.58	917.65	880.04	852.73	471.5	1271.31	2112.92
1.5	6421.98	1067.11	1021.39	976.83	523.01	1888.18	3609.05
1.6	11520.8	1219.61	1150.66	1099.29	590.3	2906.18	6396.56
1.7	20709.81	1332.77	1260.79	1247.81	678.3	4636.7	11743.49
1.8	37285.05	1473.91	1391.62	1375.16	794.15	7668.03	22291.52
1.9	67204.26	1645.28	1534.33	1478.71	948.36	13139.9	43664.65
2	121239.4	1715.03	1665.17	1631.86	1156.59	23315.5	88086.22

Table 10. ARLs when the shape parameter decreases.

β	CQC	CQC_2	CQC_3	CQC_4	UC-1	UC-2	UC-3
1	370.37	366.56	395.84	384.36	370.08	370.28	370.21
0.9	141.37	123.68	120.86	115.35	172.48	145.38	138.05
0.8	63.12	53.6	45.03	42.99	90.61	67.04	61.18
0.7	32.14	23.92	20.77	17.77	51.56	35.27	31.25
0.6	18.27	12.59	10.13	8.73	30.81	20.54	17.92
0.5	11.38	7.41	5.79	4.8	18.98	12.93	11.27
0.4	7.66	4.76	3.64	3.07	11.97	8.65	7.65
0.3	5.5	3.41	2.55	2.15	7.76	6.08	5.52
0.2	4.17	2.51	1.94	1.65	5.21	4.47	4.18
0.1	3.3	2	1.57	1.36	3.66	3.42	3.31

This is again a very good example of the case where the ARL values do not actually represent the correct information. Table 11 shows the ATS values for the control charts when the shape parameter decreases. From the table, it can be noticed that as the shape parameter becomes very small, the average time to signal becomes very large. This can be explained as follows: as the shape parameter decreases, no doubt the variability decreases but at the same time the mean increases, resulting in an increase in the ATS.

Table 11. ATS when the shape parameter decreases.

β	CQC	CQC_2	CQC_3	CQC_4	UC-1	UC-2	UC-3
1	370.37	376.93	356.88	375.82	370.08	370.28	370.21
0.9	148.75	159.35	159.32	169.79	181.48	152.96	145.26
0.8	71.52	82.36	84.55	89.8	102.67	75.96	69.32
0.7	40.68	45.51	50.17	52.08	65.26	44.64	39.56
0.6	27.48	31.4	33.47	36.5	46.35	30.91	26.96
0.5	22.76	26.05	28.62	30.52	37.95	25.86	22.55
0.4	25.46	28.81	32.18	35.62	39.79	28.73	25.42
0.3	50.92	59.56	66.95	73.96	71.84	56.28	51.08
0.2	499.88	555.79	615.58	734.34	625.05	535.83	502.17
0.1	11982152	13283098	11148821	15190426	13291052	12401015	12020158

6. Conclusion

Until now statistical control charts have been mostly used to monitor production processes. Although reliability monitoring, especially that for complex equipment or fleet of systems, is an important subject, little study has been carried out on the applications of traditional control chart for defects such as the c-chart or u-chart. In fact, they might not be suitable unless the number of failures per monitoring interval is large. If the time interval itself is long, such as months or quarters, deteriorating systems will not be detected quickly. The CUSUM charts and the CQC_r-charts are free from the sample size constraint and are thus superior to the c- and the u-chart.

In this chapter, the performance of the time-between-events CUSUM chart has been compared to that of the CQC-chart and the CQC_r-charts. The findings in this chapter suggest that if the focus is on small process deterioration, then the user can select a CUSUM chart while if the concern is on large deteriorations, then a CUSUM or a CQC_r-chart (with large r) can be selected.

In case of process improvements, even though the CQC_r-charts give a superior performance for moderate and large shifts, based on the ARL performance, it is still recommended that a CUSUM chart be used as the ARL performance of the CQC_r-charts can be quite misleading. Again, if the concern is on large process improvements, the CQC-chart can be used. When the underlying distribution changes to Weibull, both the CUSUM as well as the CQC_r-charts turn out to be inadequate in detecting the change in shape parameter.

References

1. L. Y. Chan, M. Xie and T. N. Goh, Cumulative quantity control charts for monitoring production processes, *International Journal of Production Research* **38** (2000) 397–408.
2. W. H. Woodall, Control charts based on attribute data: Bibliography and review, *Journal of Quality Technology* **29** (1997) 172–183.
3. T. W. Calvin, Quality control techniques for 'zero-defects', *IEEE Transactions on Components, Hybrid and Manufacturing Technology* **CHMT-6** (1983) 323–328.
4. F. C. Kaminsky, J. C. Benneyan, R. D. Davis and R. J. Burke, Statistical control charts based on a geometric distribution, *Journal of Quality Technology* **24** (1992) 63–69.
5. M. Xie, T. N. Goh and P. Ranjan, Some effective control chart procedures for reliability monitoring, *Reliability Engineering and Systems Safety* **77** (2002) 143–150.
6. F. F. Gan, Exact run length distributions for one-sided exponential CUSUM schemes, *Statistica Sinica* **2** (1992) 297–312.
7. F. F. Gan, Design of optimal exponential CUSUM control charts, *Journal of Quality Technology* **26** (1994) 109–124.
8. G. Lorden and I. Eisenberger, Detection of failure rate increases, *Technometrics* **15** (1973) 167–175.
9. J. M. Lucas, Counted data CUSUM's, *Technometrics* **27** (1985) 129–144.
10. S. Vardeman and D. Ray, Average run lengths for CUSUM schemes when observations are exponentially distributed, *Technometrics* **27** (1985) 145–150.
11. W. H. Woodall, The distribution of run length of one-sided CUSUM scheme for continuous random variables, *Technometrics* **25** (1983) 295–301.
12. D. Banjevic, A. K. S. Jardine, V. Makis and M. Ennis, A control-limit policy and software for condition-based maintenance optimisation, *INFOR* **39** (2001) 32–50.
13. G. Q. Zhang and V. Berardi, Economic statistical design of X over bar control charts for systems with Weibull in-control times, *Computers and Industrial Engineering* **32** (1997) 575–586.

14. F. B. Sun, J. Yang, R. del Rosario and R. Murphy, A conditional-reliability control-chart for the post-production extended reliability-test, *Proceedings of the Annual Reliability and Maintainability Symposium* (2001), pp. 64–69.
15. M. F. Ramalhoto and M. Morais, Shewhart control charts for the scale parameter of a Weibull control variable with fixed and variable sampling intervals, *Journal of Applied Statistics* **26** (1999) 129–160.
16. P. R. Nelson, Control charts for Weibull processes with standards given, *IEEE Transactions on Reliability* **28** (1979) 283–288.
17. N. L. Johnson, Cumulative sum control charts and the Weibull distribution, *Technometrics* **8** (1966) 481–491.
18. M. Xie and T. N. Goh, Improvement detection by control charts for high yield processes, *International Journal of Quality and Reliability Management* **10** (1993) 23–29.
19. G. B. Wetherill and D. W. Brown, *Statistical Process Control — Theory and Practice* (Chapman & Hall, London, 1991).
20. D. C. Montgomery, *Introduction to Statistical Quality Control* (John Wiley & Sons Inc., New York, 2001).
21. E. L. Grant and R. S. Leavenworth, *Statistical Quality Control* (McGraw-Hill, New York, 1998).
22. C. P. Quesenberry, *SPC Methods for Quality Improvement* (John Wiley & Sons, Toronto, 1997).
23. D. Brook and D. A. Evans, An approach to the probability distribution of CUSUM run length, *Biometrika* **59** (1972) 539–549.

CHAPTER 4

Optimal Interval of CRL Issue in PKI Architecture

Miwako Arafuka and Syouji Nakamura

Department of Human Life and Information, Kinjo Gakuin University, 1723 Omori 2-chome, Moriyama-ku, Nagoya 463-8521, Japan

Toshio Nakagawa

Department of Marketing and Information Systems, Aichi Institute of Technology, 1247 Yachigusa, Yagusa-cho, Toyota 470-0392, Japan

Hitoshi Kondo

Faculty of Economics, Nanzan University, 18 Yamazato-cho, Showa-ku, Nagoya 466-8673, Japan

1. Introduction

In the PKI (Public Key Infrastructure) architecture, the Certificate Management component allows users, administrators and other principals to the request certification of public keys and revocation of previously certified keys. When a certificate is issued, it is expected to be in use for its entire validity period. However, various circumstances may cause a certificate becomes invalid prior to the expiration of its validity period. Such circumstances involve changes of name and association between subject and *Certification Authority* (CA), and compromise or suspected compromise of the corresponding private key. Under such circumstances, the CA needs to revoke a certificate. The X.509 defines one method of certificate revocation. This method involves each CA periodically issuing a signed data structure called *Certificate Revocation List* (CRL).[1]

The issued CRL is stored in a server called *repository*, and is opened to public. A relying party can confirm the effectiveness of a certificate by regularly acquiring CRL of a repository. When a certificate has lapsed, the revoked information is not transmitted to a user because it is issued at the decided cycle. When the cycle time of CRL issue becomes long, it takes a long time to notify the revoked information of a user. Conversely, when the cycle time of CRL issue shortens, the load to acquire CRL increases. It is important to set an appropriate interval corresponding to the business of the security policy and the PKI architecture at the cycle time of CRL issue. As one extention of CRL issue, Delta CRL is actually used in PKI architecture.[2] Delta CRL provides all information about a certificate whose status changed since the previous CRL. So, when Delta CRL is issued, the CA also issues a complete CRL.

We present three stochastic models of Base CRL, Differential CRL and Delta CRL, each of which has different types of CRL issues. Introducing various kinds of costs for CRL issues, we obtain the expected costs of each CRL model and compare them. Further, we analytically discuss optimal intervals of CRL issue which minimize the expected costs per unit of time. Finally, we give numerical examples under suitable conditions and determine which model is the best among these three ones.

2. CRL Models

In order to validate a certificate, a relying party must acquire a recently issued CRL to determine whether a certificate has been revoked. The confirmation method of certificate revocation is assumed to be the usual retrieval from the recent CRL issue. A relying party wishing to make use of the information in a certificate must first validate a certificate. Therefore, the CRL database based on the data downloaded from the CRL distribution point is constructed for a user.

We obtain the expected costs of three models of Base CRL, Differential CRL and Delta CRL, taking into consideration of various costs for different methods of CRL issues, especially, we set an opportunity cost which a user cannot acquire a new CRL infomation. For each model, the CA decides the issue intervals of Base CRL which minimize the expected database construction costs.

The following notations are used:

M_0: Number of all certificates that have been revoked at Base CRL.
T: Interval period between Base CRL ($T = 1, 2, \ldots$).
μ_i: Number of certificates that have been revoked from the previous Base CRL or Delta CRL issue, and μ_i is nondecreasing in i ($i = 1, 2, \ldots$).
c_1: Downloading and communication costs per certificate.
c_2: File handling cost per downloaded Delta CRL.
c_3: Opportunity cost per time in the case where a user cannot acquire a new CRL information.

2.1. *Model 1 — Base CRL operation*

Even if a new revoked certification occurs after Base CRL issue, Base CRL is not issued for T period ($T = 1, 2, \ldots$), that is, Base CRL is issued only at T interval (Fig. 1). A user downloads Base CRL once, and constructs the revoked certificate of CRL database for oneself.

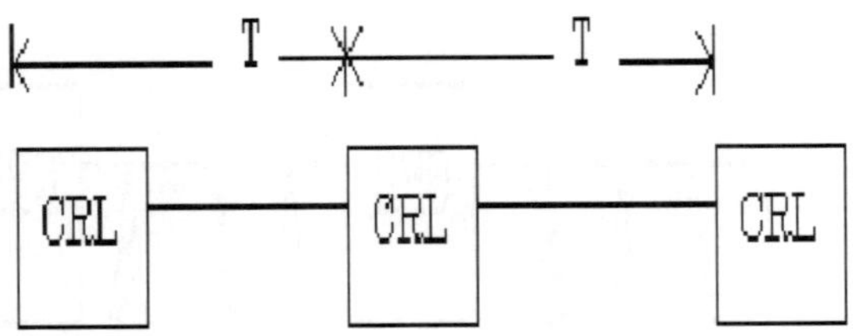

Fig. 1. Base CRL which is downloaded once in the beginning of T period.

There is a possibility that an opportunity cost c_3 may occur if a user cannot acquire a new information for T period. It is assumed that this cost is proportional to the Base CRL issue period.

The expected cost of Model 1 is shown by the total of downloading and opportunity costs of Base CRL for T period as follows:

$$C_1(T) = c_1 M_0 + c_3 \sum_{i=1}^{T} (T - i)\mu_i\,, \quad T = 1, 2, \ldots . \tag{1}$$

2.2. *Model 2 — Differential CRL operation*

In Model 2, Differential CRL is continuously issuing for T period ($T = 1, 2, \ldots$) after Base CRL issue (Fig. 2). The number of revoked certificates in Differential CRL is the total of newly occurred revoked from the previous Differential CRL issue to this Differential CRL issue. To distinguish from Delta CRL of Model 3, we call Model 2 *Differential CRL*. The full CRL database for a user is constructed by Base CRL, which has been downloaded, and by Differential CRL updated every time. Therefore, a handling cost c_2 is needed for the frequencies of downloaded Differential CRL files, i.e., the cost increases with the frequency of downloaded Differential CRL.

In Model 2, it is assumed that a user does not receive the influence of opportunity cost c_3 because a user can acquire the revoked information by Differential CRL issue in a short period. The expected cost of Model 2 is composed of the total of downloading costs of Base CRL, Differential CRL, and handling cost of the number of

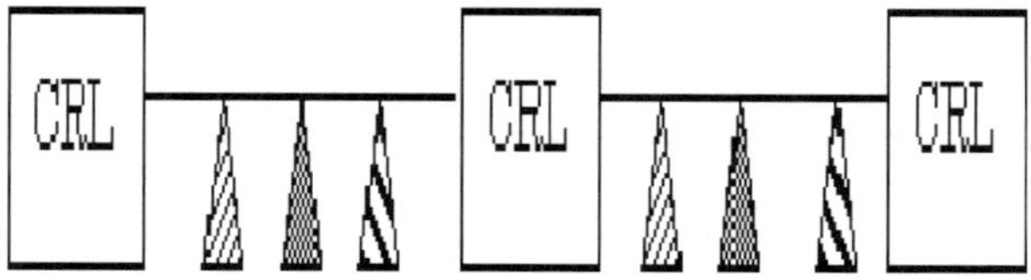

Fig. 2. Differential CRL where Δ means Differential CRL.

Differential CRL files for T period as follows:

$$C_2(T) = c_1 M_0 + c_1 \sum_{i=1}^{T} \mu_i + c_2 \sum_{i=1}^{T} i\,, \quad T = 1, 2, \ldots. \qquad (2)$$

The description about the method of Differential CRL is not given in X.509. However, the Differential CRL exports only files which have changed since the last Differential CRL or Base CRL, imports files of all Differential CRL and the last Base CRL. The reason that the generation of CRL per time increases in proportion to its amount is that if the registration number of Base CRL increases, the operation of Model 2 would be efficient.

2.3. *Model 3 — Delta CRL operation*

In Model 3, Delta CRL is continuously issuing for T period ($T = 1, 2, \ldots$) after Base CRL issue (Fig. 3). To distinguish from Differential CRL of Model 2, we call Model 3 *Delta CRL*. Delta CRL is a small CRL that provides information about certificates whose status changed since the previous Base CRL,[2] that is, the number of revoked certificates in Delta CRL is the total of accumulated revoked certificates from the previous Base CRL issue. The full CRL database for a user is constructed by Base CRL and the previous Delta CRL.

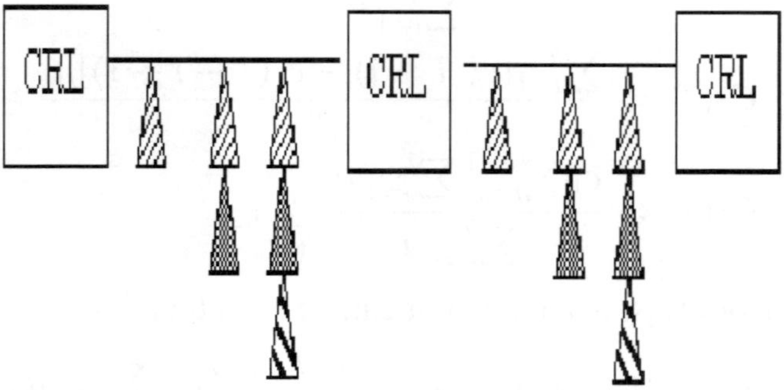

Fig. 3. Delta CRL where Δ means Differential CRL.

It is assumed that an opportunity cost c_3 is not generated because a user can acquire the revoked information by Delta CRL issue in a short period. The expected cost of Model 3 is composed of the total of downloading costs of Base CRL and Delta CRL, and handling cost of files for T period as follows:

$$C_3(T) = c_1 M_0 + c_1 \sum_{i=1}^{T} \sum_{j=1}^{i} \mu_j + c_2 T\,, \quad T = 1, 2, \ldots. \tag{3}$$

The method of operating Delta CRL is introduced in X.509, and has the advantage that full CRL can be always made. A user who needs more up-to-date certificate status obtained by the previous CRL issue, can download the latest Delta CRL. This tends to be significantly smaller than full CRL, and will reduce the load of the repository and improve the response time for a user.[3]

3. Comparisons of Expected Costs

We compare the expected costs $C_1(T)$, $C_2(T)$ and $C_3(T)$ for three models, for a specified T. For $T = 1$, we have

$$C_2(1) = C_3(1) > C_1(1)\,. \tag{4}$$

The following three relations on the expected costs are obtained:

$$C_1(T) \geq C_2(T) \Leftrightarrow \frac{\sum_{i=1}^{T}[c_3(T-i) - c_1]\mu_i}{\sum_{i=1}^{T} i} \geq c_2\,, \tag{5}$$

$$C_1(T) \geq C_3(T) \Leftrightarrow \frac{\sum_{i=1}^{T}[c_3(T-i) - c_1(T-i+1)]\mu_i}{T} \geq c_2, \tag{6}$$

$$C_3(T) \geq C_2(T) \Leftrightarrow \frac{c_1 \sum_{i=1}^{T-1} \sum_{j=1}^{i} \mu_j}{\sum_{i=1}^{T-1} i} \geq c_2\,. \tag{7}$$

Thus, from Eqs. (5) and (6), it can be seen that

$$\frac{c_2}{c_1} \geq \frac{\sum_{i=1}^{T}(M_0 - \mu_i)}{\sum_{i=1}^{T} i}\,, \quad \frac{c_2}{c_1} \geq \frac{\sum_{i=1}^{T}(M_0 - \sum_{j=1}^{i} \mu_j)}{T}\,. \tag{8}$$

If Eq. (8) is satisfied, then the expected cost $C_1(T)$ is minimum. From Eqs. (5) and (7),

$$\frac{\sum_{i=1}^{T}[c_3(T-i)-c_1]\mu_i}{\sum_{i=1}^{T} i} \geq c_2\,, \qquad \frac{c_1\sum_{i=1}^{T-1}\sum_{j=1}^{i}\mu_j}{\sum_{i=1}^{T-1} i} \geq c_2\,. \tag{9}$$

If Eq. (9) is satisfied, then the expected cost $C_2(T)$ is minimum. From Eqs. (6) and (7),

$$\frac{\sum_{i=1}^{T}[c_3(T-i)-c_1]\mu_i}{T} \geq c_2 \geq \frac{c_1\sum_{i=1}^{T-1}\sum_{j=1}^{i}\mu_j}{\sum_{i=1}^{T-1} i}\,. \tag{10}$$

If Eq. (10) is satisfied, then the expected cost $C_3(T)$ is minimum.

3.1. *Special case of* $\mu_i \equiv \mu$

It is appropriate to assume that Differential CRL is constant if there is no special event in PKI operation, i.e., $\mu_i \equiv \mu$.

From Eq. (8), we have

$$\frac{c_2}{\mu} \geq \frac{c_3(T-1)-2c_1}{T+1}\,, \qquad \frac{c_2}{\mu} \geq c_3\frac{T-1}{2} - c_1\frac{T+1}{2}\,. \tag{11}$$

If Eq. (11) is satisfied, then the expected cost $C_1(T)$ is minimum. From Eq. (9),

$$\frac{c_3(T-1)-2c_1}{T+1} \geq \frac{c_2}{\mu}\,, \quad c_1 \geq \frac{c_2}{\mu}\,. \tag{12}$$

If Eq. (12) is satisfied, then the expected cost $C_2(T)$ is minimum. From Eq. (10),

$$c_3\frac{T-1}{2} - c_1\frac{T+1}{2} \geq \frac{c_2}{\mu} \geq c_1\,. \tag{13}$$

If Eq. (13) is satisfied, then the expected cost $C_3(T)$ is minimum.

Equations (11), (12) and (13) indicate that the expected cost $C_1(T)$ is decreasing when c_2 is increasing. Similarly, $C_2(T)$ is decreasing when both c_3 and c_1 are increasing, and $C_3(T)$ is decreasing when c_3 is increasing but c_1 is decreasing.

Moreover, if $c_3(T-1) \geq c_1(T+3)$, we have

$$c_3\frac{T-1}{2} - c_1\frac{T+1}{2} \geq \frac{c_3(T-1)-2c_1}{T+1} \geq c_1\,, \tag{14}$$

and if $c_3(T-1) \leq c_1(T+3)$,

$$c_1 \geq \frac{c_3(T-1)-2c_1}{T+1} \geq c_3\frac{T-1}{2} - c_1\frac{T+1}{2}\,. \tag{15}$$

4. Optimal Intervals of CRL Issue

We discuss the optimal intervals of CRL issue which minimize the expected costs per unit of time $C_i(T)/T$ $(i = 1, 2, 3)$ of three models, using the theory of reliability.[4]

4.1. *Model* 1

We seek an optimal issue interval T_1^* which minimizes $C_1(T)/T$, given by

$$\frac{C_1(T)}{T} = \frac{c_1M_0 + c_3\sum_{i=1}^{T}(T-i)\mu_i}{T}\,, \qquad T = 1, 2, \ldots\,. \tag{16}$$

It is evident that $\lim_{T\to\infty} C_1(T)/T = \infty$ since μ_i is nondecreasing. Thus, there exists a finite $T_1^*(1 \leq T_1^* < \infty)$.

Moreover, from the inequality $C_1(T+1)/(T+1) \geq C_1(T)/T$, we have

$$c_3\sum_{i=1}^{T} i\mu_i \geq c_1M_0\,. \tag{17}$$

Note that the left-hand side of Eq. (17) is a strictly increasing function of T. Thus, we have:

(i) If $c_3\mu_1 \geq c_1M_0$, then $T_1^* = 1$.

(ii) If $c_3\mu_1 < c_1 M_0$, then there exists a finite and unique $T_1^*(1 < T_1^* < \infty)$ which satisfies Eq. (17), and

$$c_3 \sum_{i=1}^{T_1^*} \mu_i \leq \frac{C_1(T_1^*)}{T_1^*} < c_3 \sum_{i=1}^{T_1^*+1} \mu_i \,. \tag{18}$$

4.2. *Model 2*

We seek an optimal issue interval T_2^* which minimizes $C_2(T)/T$, given by:

$$\frac{C_2(T)}{T} = \frac{c_1 M_0 + c_1 \sum_{i=1}^{T} \mu_i + c_2 \sum_{i=1}^{T} i}{T}, \quad T = 1, 2, \ldots . \tag{19}$$

It is evident that $\lim_{T\to\infty} C_2(T)/T = \infty$. Thus, there exists a finite $T_2^*(1 \leq T_2^* < \infty)$.

Moreover, from the inequality $C_2(T+1)/(T+1) \geq C_2(T)/T$, we have

$$c_1 \left(T\mu_{T+1} - \sum_{i=1}^{T} \mu_i \right) + c_2 \sum_{i=1}^{T} i \geq c_1 M_0 \,, \quad T = 1, 2, \ldots . \tag{20}$$

Letting $L_2(T)$ denote the left-hand side of Eq. (20), $L_2(\infty) = \infty$, and

$$L_2(T+1) - L_2(T) = (T+1)[c_1(\mu_{T+2} - \mu_{T+1}) + c_2] > 0 \,. \tag{21}$$

Thus, $L_2(T)$ is a strictly increasing function of T. So, there exists a finite and unique optimal $T_2^*(1 \leq T_2^* < \infty)$ which satisfies Eq. (20), and

$$c_1 \mu_{T_2^*} + c_2 T_2^* \leq \frac{C_2(T_2^*)}{T_2^*} < c_1 \mu_{T_2^*+1} + c_2(T_2^* + 1) \,. \tag{22}$$

4.3. *Model 3*

We seek an optimal issue interval T_3^* which minimizes $C_3(T)/T$, given by:

$$\frac{C_3(T)}{T} = \frac{c_1 M_0 + c_1 \sum_{i=1}^{T} \sum_{j=1}^{i} \mu_j + c_2 T}{T}, \quad T = 1, 2, \ldots . \tag{23}$$

It is evident that $\lim_{T\to\infty} C_3(T)/T = \infty$ since μ_i is nondecreasing. Thus, there exists a finite T_3^* $(1 \le T_3^* < \infty)$. Moreover, from the inequality $C_3(T+1)/(T+1) \ge C_3(T)/T$, we have

$$T \sum_{i=1}^{T+1} \mu_i - \sum_{i=1}^{T} \sum_{j=1}^{i} \mu_j \ge M_0, \quad T = 1, 2, \ldots . \tag{24}$$

Letting $L_3(T)$ denote the left-hand side of Eq. (24), $L_3(\infty) = \infty$, and

$$L_3(T+1) - L_3(T) = (T+1)\mu_{T+2} > 0 . \tag{25}$$

Thus, $L_3(T)$ is a strictly increasing function of T. So, there exists a finite and unique T_3^* $(1 \le T_3^* < \infty)$ which satisfies Eq. (24), and

$$c_1 \sum_{i=1}^{T_3^*} \mu_i + c_2 \le \frac{C_3(T_3^*)}{T_3^*} < c_1 \sum_{i=1}^{T_3^*+1} \mu_i + c_2 . \tag{26}$$

In the particular case of $\mu_i \equiv \mu$, Eq. (17) becomes

$$\frac{T(T+1)}{2} \ge \frac{c_1}{c_3} \frac{M_0}{\mu}, \tag{27}$$

Eq. (20) is

$$\frac{T(T+1)}{2} \ge \frac{c_1}{c_2} M_0 , \tag{28}$$

and Eq. (24) is

$$\frac{T(T+1)}{2} \ge \frac{M_0}{\mu} . \tag{29}$$

When $c_1/c_2 = 1/\mu$ and $c_1 = c_3$, i.e., $\mu c_1 = c_2$ and $c_1 = c_3$, we get $T_1^* = T_2^* = T_3^*$.

5. Numerical Examples

From the statistical information stored in CRL, it would be reasonable to assume that the revocation has occurred daily almost equally and the number of revoked certificates is constant, i.e., $\mu \equiv \mu_i$. When $\mu = 40$ and $M_0 = 10\,000$, we give the optimal issue interval T^* and its expected cost $C_1(T_1^*)/(c_1 T_1^*)$ in Table 1. This indicates that $T_1^* = 1$ for $c_3/c_1 \geq 250.0$, that is, we should issue CRL every day. Evidently, optimal volumes are increasing when the ratio of cost c_3/c_1 is decreasing. For example, when $c_3/c_1 = 16.7$, T_1^* is five days and $C_1(T_1^*)/(c_1 T_1^*) = 3.336$.

Similarly, when $\mu = 40$ and $M_0 = 10\,000$, we give the optimal issue interval T_2^* and its expected cost $C_2(T_2^*)/(c_1 T_2^*)$ in Table 2. This indicates that if the value of c_2/c_1 is very large, we should issue CRL every day. However, since c_2/c_1 is the ratio of the initial construction of the database to its additional handling cost for a PKI user, it would be less than about 21.5. In this case, Base CRL should be done within a month, while Differential CRL would be issued every day.

Table 1. Optimal interval of Base CRL issue in Model 1 when $M_0 = 10\,000$ and $\mu = 40$.

T_1^*	c_3/c_1	$C_1(T_1^*)/(c_1 T_1^*)$
1	250.0	10 000
5	16.7	3336
10	4.5	1810
15	2.1	1255
20	1.2	956
25	0.8	784
30	0.5	623

Table 2. Optimal interval of Differential CRL issue in Model 2 when $M_0 = 10\,000$ and $\mu = 40$.

T_2^*	c_2/c_1	$C_2(T_2^*)/(c_1T_2^*)$
1	10 000	20 040
5	666.7	4040
10	181.8	2040
15	83.3	1373
20	47.6	1040
22	39.5	954
25	30.8	840
30	21.5	707

Table 3. Optimal interval of Delta CRL issue in Model 3 when $M_0 = 10\,000$ and $\mu = 40$.

T_3^*	c_2/c_1	$C_3(T_3^*)/(c_1T_3^*)$
	10 000	10 915
	666.7	1581
	181.8	1096
22	83.3	998
	47.6	962
	39.5	949
	30.8	945
	21.5	936

Finally, when $\mu = 40$ and $M_0 = 10\,000$, the optimal issue interval of Model 3 is $T_3^* = 22$ from Eq. (29), regardless of the costs c_i ($i = 1, 2, 3$). We give $T_3^* = 22$ and its expected cost $C_3(T_3^*)/(c_1T_3^*)$ for the same values as c_2/c_1 in Table 3. Comparing two tables, the expected cost of Model 3 is smaller than that of Model 2 for $c_2/c_1 \geq 47.6$.

If the number μ of certificates becomes larger, the optimal issue interval of Model 3 becomes shorter from Eq. (29). Thus, comparing the optimal intervals of Models 2 and 3, if μ becomes larger, T_3^* becomes shoter and Model 2 is more efective than Model 3. Conversely, if c_2/c_1 becomes larger, T_2^* becomes shorter and Model 3 improves more than Model 2.

6. Conclusions

We have proposed three stochastic models of Base CRL in PKI architecture and have obtained the expected costs of each model, introducing the costs of downloading CRL data and handling of Differential CRL files, and the opportunity cost. We have compared three expected costs and have derived analytically the optimal issue intervals which minimize them. Further, we have discussed in numerical examples which model is the best.

Thus, by estimating the costs of downloading, handling and opportunity, and the amount of revoked certificates from actual data, and by modifying some suppositions, we could practically determine an optimal issue interval of Base CRL.

References

1. R. Housley, W. Ford, W. Polk and D. Solo, Internet X.509 public key infrastructure certificate and CRL profile, The Internet Society (1999).
2. D. A. Cooper, A more efficient use of Delta-CRLs, *Proceedings of the 2000 IEEE Symposium on Security and Privacy*, May 2000, pp. 190–202.
3. D. A. Jordi, Certificate revocation for e-business, e-commerce and m-commerce, http://www.ssgrr.it/en/ssgrr2001/papers/Jordi/20Forne.pdf.
4. R. E. Barlow and F. Proschan, *Mathematical Theory of Reliability* (John Wiley & Sons, New York, 1965).

CHAPTER 5

Discrete-Time Economic Manufacturing Quantity Model with Stochastic Machine Breakdown and Repair

B. C. Giri and T. Dohi*

Department of Information Engineering, Hiroshima University, 1-4-1 Kagamiyama, Higashi-Hiroshima, Japan
**dohi@rel.hiroshima-u.ac.jp*

1. Introduction

Manufacturing infrastructure is rapidly changing with the technological innovations and scientific developments around the world. In this changing environment, production, quality and maintenance performances can be regarded as three important aspects of any manufacturing process. Managers in industry are everyday emphasized to become more productive through shortened product development cycles, increased responsiveness and flexibility. At the same time, they must control the system cost and maintain product quality. In practice, even the highly sophisticated production facilities that are more efficient and reliable than their predecessors, they are not free from deterioration due to aging. An unexpected equipment failure

*Corresponding author.

may result in unnecessary interruption of the production performance and delivery of the products. Such a failure may reduce the equipment performance, causes poor product quality and lowers the product yield. Process inspection and preventive maintenance program can only reduce the probability of random machine failure. However, the loss of revenue due to down time, missed delivery schedules and the cost of repairing upon machine failure can present a significant expense. Thus, the maintenance and production control problem in an unreliable manufacturing environment has been one of the important topics of research during the last decade.

Most of the Economic Manufacturing Quantity (EMQ) models developed in the literature assume that the production facility is perfectly reliable, i.e., the machine never breaks down though, in true sense, failure-free production facilities are rare. The economic lot sizing problem for an unreliable production system has attracted the attention of many researchers because when a machine breakdown takes place in the production phase, the interrupted lot is aborted and as a result, the basic EMQ model loses its usefulness. So, from practical point of view, the development and implementation of the optimal lot sizing policy in an unreliable production environment are significant and meaningful. A number of production/inventory models have been developed in the literature, taking into account the stochastic machine breakdown and repair. In the following, we give a brief review of the relevant literature.

2. Brief Literature Review

McCall,[1] one of the early researchers, raised the issue of interdependence between production and maintenance for a stochastically failing equipment. Bielecki and Kumar[2] showed that there exists a range of parameter values describing an unreliable manufacturing system for which zero inventory policy is exactly optimal even when the production capacity is uncertain. The steady state distribution of the inventory level and some important system characteristics related to both machine utilization and service level to customers in an unreliable production environment were obtained by Posner and Berg.[3] Groenevelt *et al.*[4] analyzed the impacts of machine breakdown and corrective maintenance on an EMQ model, assuming exponentially distributed inter-failure time and instantaneous repair time. They showed that the optimal lot size would be greater than that of the classical EMQ model in order to compensate the production loss due to machine breakdown. In the subsequent article,[5] they investigated the issue of safety stocks required to meet a managerially prescribed service level under a simplified assumption of exponential failure time and randomly distributed repair time.

The stochastic EMQ models of Groenevelt *et al.*[4,5] were extended in the literature by many researchers. Kim and Hong,[6] Kim *et al.*[7] generalized the results of Groenevelt *et al.*[4] assuming arbitrarily distributed inter-failure time. Chung[8] determined the bounds for the optimal production lot size in Groenevelt *et al.*[4] model. He[9] further showed that the long-run average cost function in the steady state, neither convex nor concave, is unimodal. For general failure and general repair time distributions, Dohi *et al.*[10] determined the optimal production policy which can be characterized as an age replenishment like policy. Tse and Makis[11] studied an EMQ problem with preventive replacement and major/minor repair. When a major failure occurs, the failed unit is replaced by a new one and the interrupted lot is aborted. In case of a minor failure, the failed unit is corrected with minimal repair and the production is then resumed. Berg *et al.*[12] analyzed a production system with multiple identical machines devoted to producing a single part type, by employing level crossing techniques. They computed performance measures that characterized the operation of the production-inventory system with respect to its service level to customers, the expected inventory stocked, the machines' utilization, etc. This is the generalization of the work done by Posner and Berg[3] in which a single machine producing a single part type is considered.

Abboud[13] presented an approximate model that describes the production batching problem with Poisson machine breakdowns and general repair times. Makis[14] showed that the optimal preventive replacement is an age replacement if the failed machine is minimally repaired and that the optimal lot size is generally a function of the operating age of the machine. Dohi *et al.*[15] investigated an EMQ model assuming that a finite/infinite number of minimal repairs can be made until the predetermined inventory level is reached, when the machine failure occurs. The joint effect of process deterioration and machine breakdowns on the optimal lot size and the optimal number of inspections in a production cycle were studied by Makis and Fung.[16] Liu and Cao[17] studied an unreliable EMQ model where the demand is a compound Poisson process. Moini and Murthy[18] considered two types (Types I and II) of repair action strategies when a machine failure occurs. After Type I repair, the probability of machine failure remains the same as it was before while after Type II repair, it varies. By analyzing the model, they tried to find a relationship between process uncertainty, repair actions and optimal lot size.

Cheung and Hausman,[19] Dohi *et al.*[20] investigated the joint implementation of preventive maintenance and safety stocks in an unreliable production environment. Recently, Giri *et al.*[21] developed an EMQ model with random machine failure, treating the machine production rate as a decision variable and machine failure rate as a function of the production rate.[22] They further extended their model to the case where certain safety stocks in inventory may be useful to improve customer service level. In another study, Giri and Dohi,[23] implemented the net present value (NPV) approach to compute the EMQ for a failure-prone production facility. They examined the performance of the NPV model and the traditional long-run average cost model in terms of the net present values of the expected total cost based on their respective optimal decisions.

It is noted that all the EMQ models with stochastic machine breakdown and repair cited above are based on continuous failure time distributions. However, the time to failure of a unit might be discrete in many practical situations. For example, consider the failure of switching devices, railroad tracks, ball bearings and tyres of airplane. The time to failure, in each case, would be better measured by the number of cycles to failure rather than the instant of failure since its installation. The idea of discrete time failure distribution was introduced in the classical age replacement models by Nakagawa and Osaki[24] and Nakagawa.[25] Rocha-Martinez and Shaked[26] studied a model of failures and repairs of units with discrete lifetimes. They derived some stochastic comparisons of pairs of such models and obtained results regarding the inheritance of several aging properties by the repaired unit. Abboud[27] modeled an unreliable single machine production-inventory system as a discrete-time Markov chain, assuming geometric failure and geometric repair. Using some results from Markov chain theory, he developed an efficient algorithm to compute the average system cost, which in turn, can be used to find the economic manufacturing quantities.

In this chapter, we develop and analyze an unreliable EMQ model (Dohi *et al.*[10]) in a discrete time framework. Based on discrete probability argument, we derive the optimal production time by minimizing the expected cost per unit time in the steady state. The chapter is organized as follows. The next section deals with the description of the model, the underlying assumptions and notation, and the derivation of the cost function in the steady state under general discrete failure and discrete repair time distributions. In Sec. 4, we formulate the model under geometric failure and geometric repair time distributions and derive the criteria for the existence and uniqueness of the optimal production time. Section 5 treats the model under general discrete failure and constant repair time. As a specific case, geometric failure time and constant repair time are considered in Sec. 6. Section 7 is devoted to numerical computations. Finally, the chapter is concluded with some remarks in Sec. 8.

3. The General Model

3.1. *Notation*

n : discrete time point, $n = 0, 1, 2, \ldots$

$P(n)$: discrete failure time distribution with p.m.f. $p(n)$

$\bar{\psi}(\cdot)$: survivor function of the function $\psi(\cdot)$, i.e., $\bar{\psi}(\cdot) = 1 - \psi(\cdot)$

S : discrete random variable denoting repair time

$G(s)$: discrete repair time distribution with p.m.f. $g(s)$

$p(> 0)$: known production rate

$d(> 0)$: known demand rate

$C_p(> 0)$: fixed production and preventive maintenance cost per unit lot

$C_r(> 0)$: machine repair cost per unit time

$C_i(> 0)$: inventory holding cost per unit product per unit time

$C_s(> 0)$: shortage cost per unit product

ETC : expected total cost

3.2. *Basic assumption*

Without any loss of generality, we may assume that $p > d$. Further, for discrete time setting, we need the following basic assumption.

Assumption 1. p/d is an integer greater than 1.

To check the validity of this assumption, consider the simplest case of a perfect production process (no machine breakdown during production phase) which starts at time $n = 0$ with a uniform production rate p to meet a constant demand rate d for the commodity. Let the optimal production time be one unit. Then, the on-hand inventory carried at time $n = 1$ is $p-d$ (Fig. 1). In order to exhaust the on-hand stock in future k time units, we must have $p - d = kd$, $k > 0$, i.e., $p = (k + 1)d$, $k = 1, 2, 3, \ldots$.

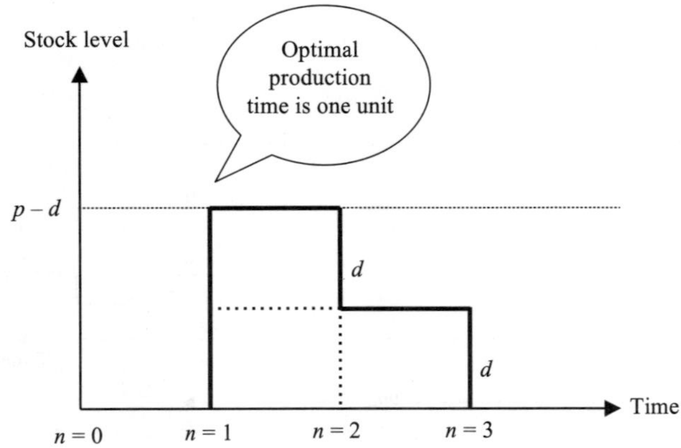

Fig. 1. Production-demand ratio in discrete time setting.

3.3. *Model formulation*

Consider an unreliable one-unit production system in which the production process starts at time $n = 0$. If the machine does not fail up to a prescribed production time $n = n_0 \in (0, \infty)$, then the next production cycle starts after time (pn_0/d). If, however, the failure occurs before time n_0, then the repair is started immediately after machine failure and the demand is met first from the accumulated inventory. If there is sufficient stocks to meet the demand during repair time, then the next production starts only when the on-hand stock is exhausted. On the other hand, if shortage occurs due to longer repair time, then the unsatisfied demands are not delivered after machine repair and are assumed to be lost completely. This may be interpreted as the products are instantly required but the requirement vanishes if it is not met at that instant. The configurations of this EMQ model are depicted in Figs. 2(a)–2(c).

We also assume that when the production is completed at the end of each production phase, the preventive maintenance (with negligible time) is carried out immediately, even if the machine breakdown does not occur. After preventive/corrective maintenance, the machine

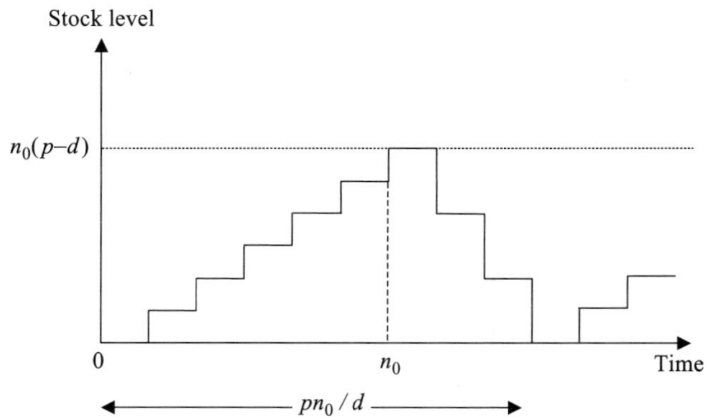

Fig. 2(a). Configuration of the EMQ model — No failure case.

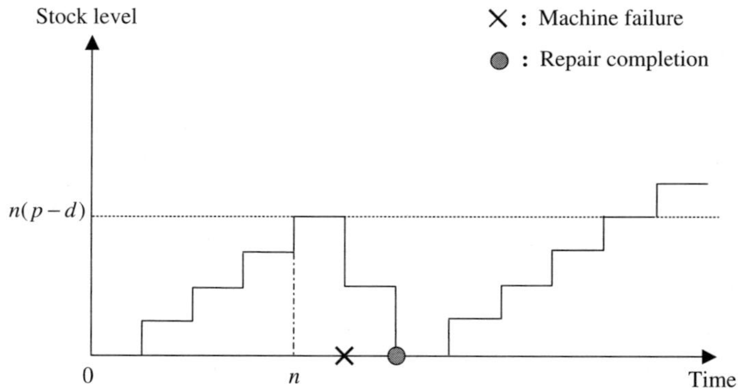

Fig. 2(b). Configuration of the EMQ model — Machine failure and no shortage case.

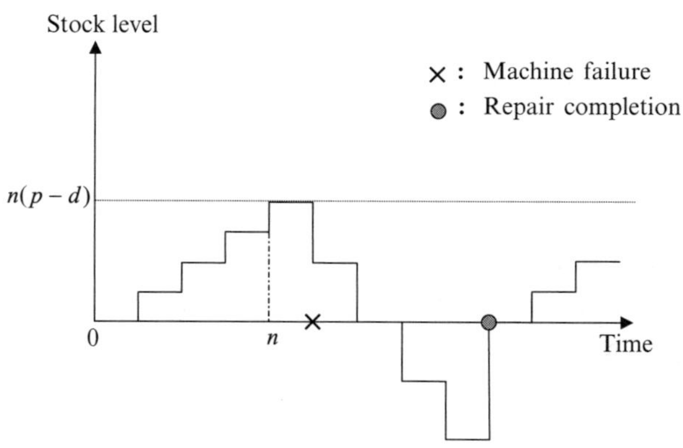

Fig. 2(c). Configuration of the EMQ model — Machine failure and shortage case.

becomes as good as new. We define the time interval between two successive production starting points as a (repeating) cycle. Then, by discrete probability argument, the mean time length of one cycle and the expected cost per cycle are given by:

$$T(n_0) = \sum_{n=0}^{n_0-1} \sum_{s=0}^{n(p/d-1)} n\left(\frac{p}{d}\right) g(s)p(n) + \sum_{n=0}^{n_0-1} \sum_{s=n(p/d-1)+1}^{\infty} (n+s)g(s)p(n) + \sum_{n=n_0}^{\infty} n_0\left(\frac{p}{d}\right) p(n) \qquad (1)$$

and

$$V(n_0) = C_p + C_r \sum_{n=0}^{n_0-1} \sum_{s=0}^{\infty} sg(s)p(n) + C_i \sum_{n=0}^{n_0-1} \frac{p(p-d)}{2d} n^2 p(n) + C_i \sum_{n=n_0}^{\infty} \frac{p(p-d)}{2d} n_0^2 p(n) + C_s d \sum_{n=0}^{n_0-1} \sum_{s=n(p/d-1)+1}^{\infty} \left[s - \frac{(p-d)n}{d}\right] g(s)p(n)\,, \qquad (2)$$

respectively. From the familiar discrete renewal reward theorem, the expected cost per unit time in the steady-state is:

$$C(n_0) = \lim_{n\to\infty} \frac{E[\text{total cost incurred for } (0,n]]}{n} = \frac{V(n_0)}{T(n_0)}\,, \qquad (3)$$

where E denotes the mathematical expectation operator. Our objective is to determine the optimal production time n_0^* which minimizes $C(n_0)$. In order to avoid an unrealistic decision making, we assume that $\underline{n_0} \le n_0^* \le \bar{n}_0$ where $\underline{n_0}$ and $\bar{n}_0$ are the lower and upper limits of the production time, respectively. It is difficult to analyze the model under general discrete time failure and repair distributions. So, in the next section, we treat the model under specific failure and repair time distributions and derive the criteria for the existence and uniqueness of the optimal production time.

4. The Case of Geometric Failure and Geometric Repair

Suppose that the failure and repair time distributions are both geometric. Like exponential distribution, geometric distribution has the memoryless property. We define

$$p(n) = \begin{cases} 0\,, & \text{for } n = 0\,, \\ q_1^{n-1}(1-q_1)\,, & \text{for } n = 1, 2, 3, \ldots;\quad 0 < q_1 < 1\,, \end{cases}$$

and

$$g(s) = \begin{cases} 0\,, & \text{for } s = 0\,, \\ q_2^{s-1}(1-q_2)\,, & \text{for } s = 1, 2, 3, \ldots;\quad 0 < q_2 < 1\,. \end{cases}$$

Then the failure rate function $f(n) = p(n)/[\bar{P}(n-1)] = 1 - q_1$ and the repair rate function $r(s) = g(s)/[\bar{G}(s-1)] = 1 - q_2$ are both constants. For the above failure and repair time distributions, the mean time length of one cycle and the expected cost per cycle can be obtained from Eqs. (1) and (2) as:

$$\begin{aligned} T(n_0) = {} & \frac{p(1-q_1)(1-q_2)}{d} \sum_{n=1}^{n_0-1} n q_1^{n-1} \sum_{s=1}^{n(p/d-1)} q_2^{s-1} \\ & + (1-q_1)(1-q_2) \sum_{n=1}^{n_0-1} n q_1^{n-1} \sum_{s=n(p/d-1)+1}^{\infty} q_2^{s-1} \\ & + (1-q_1)(1-q_2) \sum_{n=1}^{n_0-1} q_1^{n-1} \sum_{s=n(p/d-1)+1}^{\infty} s q_2^{s-1} \\ & + \left(\frac{p}{d}\right) n_0 q_1^{n_0-1} \end{aligned} \tag{4}$$

and

$$\begin{aligned} V(n_0) &= C_p + C_r(1-q_1)(1-q_2)\sum_{n=1}^{n_0-1} q_1^{n-1}\sum_{s=1}^{\infty} s q_2^{s-1} \\ &\quad + \frac{C_i p(p-d)(1-q_1)}{2d}\sum_{n=1}^{n_0-1} n^2 q_1^{n-1} + \frac{C_i p(p-d)}{2d} n_0^2 q_1^{n_0-1} \\ &\quad + C_s d(1-q_1)(1-q_2)\left[\sum_{n=1}^{n_0-1} q_1^{n-1} \sum_{s=n(p/d-1)+1}^{\infty} s q_2^{s-1}\right. \\ &\quad \left. - \frac{(p-d)}{d}\sum_{n=1}^{n_0-1} n q_1^{n-1} \sum_{s=n(p/d-1)+1}^{\infty} q_2^{s-1}\right], \end{aligned} \tag{5}$$

respectively. The difference of $T(n_0)$ with respect to n_0 is:

$$T(n_0+1) - T(n_0) = q_1^{n_0-1}\left[\left(\frac{1-q_1}{1-q_2}\right) q_2^{n_0(p/d-1)} + \frac{pq_1}{d}\right]. \tag{6}$$

Similarly, the difference of $V(n_0)$ with respect to n_0 is:

$$\begin{aligned} V(n_0+1) - V(n_0) = q_1^{n_0-1}&\left[\frac{C_r(1-q_1)}{1-q_2} + \frac{C_i p(p-d)(2n_0+1)q_1}{2d}\right. \\ &\left. + \frac{C_s d(1-q_1)q_2^{n_0(p/d-1)}}{1-q_2}\right]. \end{aligned} \tag{7}$$

Now, define the numerator divided by $\bar{P}(n_0-1) = 1 - q_1^{n_0-1}$ of the difference of $C(n_0) = V(n_0)/T(n_0)$ with respect to n_0 as $w(n_0)$ where

$$\begin{aligned} w(n_0) = &\left[\frac{C_r(1-q_1)}{1-q_2} + \frac{C_i p(p-d)q_1(2n_0+1)}{2d}\right. \\ &\left. + \frac{C_s d(1-q_1)}{1-q_2} q_2^{n_0(p/d-1)}\right] T(n_0) \\ &- \left[\left(\frac{1-q_1}{1-q_2}\right) q_2^{n_0(p/d-1)} + \frac{pq_1}{d}\right] V(n_0)\,. \end{aligned} \tag{8}$$

Proposition 1. *For any arbitrary failure rate and sufficiently high repair rate (as $q_2 \to 0$), the function $w(n_0)$ is increasing in $n_0 \in [\underline{n_0}, \bar{n}_0]$.*

Proposition 2. *Let the opportunity loss per unit demand be sufficiently small so that $C_s d < C(n_0)$. Then $w(n_0)$ is an increasing function of n_0 for any arbitrary failure and repair rates.*

Propositions 1 and 2 follow straightforwardly from the difference of $w(n_0)$ with respect to n_0 which is given by:

$$\begin{aligned}\Delta w(n_0) &= w(n_0+1) - w(n_0)\\ &= \frac{C_i p(p-d)q_1}{d} T(n_0)\\ &\quad + \left(\frac{1-q_1}{1-q_2}\right)\{1 - q_2^{p/d-1}\} q_2^{n_0(p/d-1)}\\ &\quad \times [V(n_0) - C_s d T(n_0)].\end{aligned}$$

We now state the criteria for the existence and uniqueness of the optimal production time n_0^* in the following theorem.

Theorem 1. *Under Proposition* 1 (*or Proposition* 2),

(i) *If $w(\underline{n_0}) < 0$ and $w(\bar{n}_0) > 0$, then there exists* (*at least one, at most two*) *optimal production time $n_0^* (0 < \underline{n_0} \le n_0^* \le \bar{n}_0 < \infty)$ satisfying $w(n_0^* - 1) < 0$ and $w(n_0^*) \ge 0$. The corresponding minimum expected cost satisfies the inequality $\phi(n_0^* - 1) < C(n_0^*) \le \phi(n_0^*)$, where*

$$\phi(n) = \frac{\begin{array}{c}2d(1-q_1)C_r + C_i p(p-d)q_1(1-q_2)(2n+1)\\ + 2C_s d^2(1-q_1)q_2^{n(p/d-1)}\end{array}}{2\{pq_1(1-q_2) + d(1-q_1)q_2^{n(p/d-1)}\}}. \tag{9}$$

(ii) *If $w(\bar{n}_0) \le 0$, then $n_0^* = \bar{n}_0$.*
(iii) *If $w(\underline{n_0}) \ge 0$, then $n_0^* = \underline{n_0}$.*

Proof. The function $w(n_0)$ is increasing in the interval $[\underline{n_0}, \bar{n}_0]$, by Proposition 1 (or Proposition 2). So, if $w(\underline{n_0}) < 0$ and $w(\bar{n}_0) > 0$, then the optimal production time $n_0^*(\underline{n_0} \le n_0^* \le \bar{n}_0)$ must satisfy $w(n_0^*-1) < 0$ and $w(n_0^*) \ge 0$. Now, from Eq. (8) we have, using Eqs. (6) and (7),

$$\begin{aligned}
w(n_0^*-1) &= \left[\frac{C_r(1-q_1)}{1-q_2} + \frac{C_i p(p-d)q_1(2n_0^*-1)}{2d}\right.\\
&\quad \left. + \frac{C_s d(1-q_1)}{1-q_2} q_2^{(n_0^*-1)(p/d-1)}\right] T(n_0^*-1)\\
&\quad - \left[\left(\frac{1-q_1}{1-q_2}\right) q_2^{(n_0^*-1)(p/d-1)} + \frac{pq_1}{d}\right] V(n_0^*-1)\\
&= \left[\frac{C_r(1-q_1)}{1-q_2} + \frac{C_i p(p-d)q_1(2n_0^*-1)}{2d}\right.\\
&\quad \left. + \frac{C_s d(1-q_1)}{1-q_2} q_2^{(n_0^*-1)(p/d-1)}\right]\\
&\quad \times T(n_0^*) - \left[\left(\frac{1-q_1}{1-q_2}\right) q_2^{(n_0^*-1)(p/d-1)} + \frac{pq_1}{d}\right] V(n_0^*)\,.
\end{aligned}$$

Therefore, $w(n_0^*-1) < 0$ implies

$$C(n_0^*) > \frac{\begin{array}{c}2d(1-q_1)C_r + C_i p(p-d)q_1(1-q_2)(2n_0^*-1)\\ +\,2C_s d^2(1-q_1)q_2^{(n^*-1)(p/d-1)}\end{array}}{2\{pq_1(1-q_2) + d(1-q_1)q_2^{(n^*-1)(p/d-1)}\}}\,.$$

Similarly, $w(n_0^*) \ge 0$ gives

$$C(n_0^*) \le \frac{\begin{array}{c}2d(1-q_1)C_r + C_i p(p-d)q_1(1-q_2)(2n_0^*+1)\\ +\,2C_s d^2(1-q_1)q_2^{n^*(p/d-1)}\end{array}}{2\{pq_1(1-q_2) + d(1-q_1)q_2^{n^*(p/d-1)}\}}\,.$$

Hence, the minimum expected cost $C(n_0^*)$ is bounded by the relation $\phi(n_0^*-1) < C(n_0^*) \le \phi(n_0^*)$, where $\phi(n)$ is given in Eq. (9). However, if $w(\bar{n}_0) \le 0$, then $C(n_0)$ is a decreasing function of n_0 in the interval $[\underline{n_0}, \bar{n}_0]$ and therefore, $n_0^* = \bar{n}_0$. If $w(\underline{n_0}) \ge 0$, then

$C(n_0)$ is increasing in the interval $[\underline{n}_0, \bar{n}_0]$ and therefore, $n_0^* = \underline{n}_0$. This completes the proof of the theorem. □

If no machine failure occurs in the production phase, i.e., when $q_1 \to 1$, we have from Eqs. (4) and (5),

$$T(n_0) \to \frac{pn_0}{d} \quad \text{and} \quad V(n_0) \to C_p + \frac{C_i p(p-d)n_0^2}{2d}.$$

In this case, $\Delta w(n_0)$ is always positive. So, for the existence of the optimal production time n_0^*, the restriction on the repair rate or the opportunity loss per unit demand mentioned in the propositions is not required. Then, the corresponding minimum expected cost rate satisfies the inequality

$$\frac{C_i(p-d)(2n_0^*-1)}{2} < C(n_0^*) \leq \frac{C_i(p-d)(2n_0^*+1)}{2}.$$

5. The Model under General Failure and Constant Repair

Suppose that a constant time L (positive integer) is always required to repair the machine upon every failure. In this section, we formulate the model under general discrete failure distribution and a constant repair time L (positive integer). Arguing similarly to that given in Sec. 3.2, we make, for discrete time setting, the following assumption.

Assumption 2. $dL = m(p-d), \quad m = 1, 2, 3, \ldots.$

Assumption 2 means the total demand in the repair time period is a multiple of the incremental stock in the production phase. Under this assumption, if the machine continues to produce up to the time $[dL/(p-d)](< n_0)$, then the accumulated inventory will be sufficient to meet the demand during repair time. If, however, the machine fails at or before the time $dL/(p-d)$, then the expected cycle length will be $\sum_{n=0}^{dL/(p-d)}(n+L)p(n)$, otherwise, $\sum_{n=(dL/(p-d))+1}^{n_0-1}\left[n+\frac{n(p-d)}{d}\right]p(n)$. So, based on discrete probability arguments, the mean time length of one cycle and the expected cost per cycle of the production-inventory system can be obtained as:

$$T_1(n_0) = \sum_{n=0}^{dL/(p-d)} (n+L)p(n) + \sum_{n=(dL/(p-d))+1}^{n_0-1} \frac{np}{d} p(n) + \sum_{n=n_0}^{\infty} \frac{n_0 p}{d} p(n) \tag{10}$$

and

$$V_1(n_0) = C_p + C_r \sum_{n=0}^{n_0-1} Lp(n) + C_i \sum_{n=0}^{n_0-1} \frac{p(p-d)}{2d} n^2 p(n) + C_i \sum_{n=n_0}^{\infty} \frac{p(p-d)}{2d} n_0^2\, p(n) + C_s d \sum_{n=0}^{dL/(p-d)} \left[L - \frac{(p-d)n}{d} \right] p(n)\,, \tag{11}$$

respectively. Our problem is to determine the optimal production time $n_0^* (\underline{n_0} \le n_0^* \le \bar{n}_0)$ which minimizes the long-run average cost $C_1(n_0) = V_1(n_0)/T_1(n_0)$ in the steady state. Taking the difference of $C_1(n_0)$ with respect to n_0, we get

$$\begin{aligned} C_1(n_0+1) - C_1(n_0) &= \frac{1}{T_1(n_0)T_1(n_0+1)} [T_1(n_0)\{V_1(n_0+1) - V_1(n_0)\} \\ &\quad - V_1(n_0)\{T_1(n_0+1) - T_1(n_0)\}] \\ &= \frac{\bar{P}(n_0) w_1(n_0)}{T_1(n_0)T_1(n_0+1)}\,, \end{aligned}$$

where

$$w_1(n_0) = \left[C_r L\, \xi(n_0) + \frac{C_i p(p-d)(2n_0+1)}{2d} \right] T(n_0) - \left(\frac{p}{d}\right) V(n_0)\,, \tag{12}$$

and $\xi(n_0) = p(n_0)/\bar{P}(n_0)$. In fact, $\xi(n)$ is not the failure rate of the failure time distribution. The failure rate should be $p(n)/\bar{P}(n-1)$, see Nakagawa,[25] Nakagawa and Osaki.[24] However, depending on the monotonic characteristics of $\xi(n)$ in the time interval $[\underline{n_0}, \bar{n}_0]$, we can derive the optimal production time n_0^* which minimizes the expected cost per unit time in the steady state $C_1(n_0)$.

Theorem 2. *Suppose that $\xi(n_0)$ is increasing in $n_0 \in [\underline{n_0}, \bar{n}_0]$.*

(i) *If $w_1(\underline{n_0}) < 0$ and $w_1(\bar{n}_0) > 0$, then there exists (at least one, at most two) optimal production time point $n_0^*(0 < \underline{n_0} \le n_0^* \le \bar{n}_0 < \infty)$ satisfying $w_1(n_0^* - 1) < 0$ and $w_1(n_0^*) \ge 0$. Then the corresponding minimum expected cost satisfies the inequality*

$$\psi(n_0^* - 1) < C_1(n_0^*) \le \psi(n_0^*),$$

where

$$\psi(n) = \frac{C_r dL\xi(n)}{p} + \frac{C_i(p-d)(2n+1)}{2}.$$

(ii) *If $w_1(\bar{n}_0) \le 0$, then $n_0^* = \bar{n}_0$ and*
(iii) *If $w_1(\underline{n_0}) \ge 0$, then $n_0^* = \underline{n_0}$.*

Proof. Taking the difference of $w_1(n_0)$ with respect to n_0, we get

$$\Delta w_1(n_0) = \left[C_r L \Delta\xi(n_0) + \frac{C_i p(p-d)}{d}\right] T(n_0),$$

which shows that if $\Delta\xi(n_0) \ge 0$, then $w_1(n_0)$ is strictly increasing in the interval $[\underline{n_0}, \bar{n}_0]$. Therefore, when $w_1(\underline{n_0}) < 0$ and $w_1(\bar{n}_0) > 0$, there exists at least one (at most two) optimal production time $n_0^*(0 < \underline{n_0} \le n_0^* \le \bar{n}_0 < \infty)$ satisfying $w_1(n_0^* - 1) < 0$ and $w_1(n_0^*) \ge 0$, which determine the upper and lower bounds of the optimal cost rate $C_1(n_0^*)$. The second and third parts of the theorem follow directly as $\Delta C_1(n_0) \le 0$ when $w_1(\bar{n}_0) \le 0$ and $\Delta C_1(n_0) \ge 0$ when $w_1(\underline{n_0}) \ge 0$. □

Theorem 3. *Suppose that $\xi(n_0)$ is decreasing in $[\underline{n_0}, \bar{n}_0]$.*

(i) *If $dLC_r\Delta\xi(n_0) + C_i p(p-d) > 0$, then the optimal production policy is the same as given in Theorem 2.*

(ii) *If* $dLC_r\Delta\xi(n_0) + C_i p(p-d) < 0$, *then the optimal production time* n_0^* *is either* $\bar{n}_0$ *or* $\underline{n_0}$.

(iii) *On the other hand, if* $dLC_r\Delta\xi(n_0) + C_i p(p-d) = 0$, *then* $n_0^* = \underline{n_0}$.

The proof of the theorem is simple and therefore, it is left out to the readers. Note that when the machine is repaired instantaneously, i.e., $L \to 0$, the optimal production time n_0^* exists irrespective of the monotonic characteristics of $\xi(n_0)$ in the interval $[\underline{n_0}, \bar{n}_0]$ mentioned in Theorems 2 and 3.

6. The Case of Geometric Failure and Constant Repair

In this section, we assume that the time to failure of the machine is geometrically distributed as described in Sec. 4 and repair time is a constant L (positive integer). Letting $m = dL/(p-d)$, we obtain, after some algebra, the long-run average cost $C_0(n_0) = V_0(n_0)/T_0(n_0)$ in the steady state, where

$$T_0(n_0) = (1-q_1^m)\left(L + \frac{1}{1-q_1}\right) + m\left(\frac{p}{d} - 1\right) q_1^m + \frac{p(q_1^m - q_1^{n_0})}{d(1-q_1)} \tag{13}$$

and

$$\begin{aligned} V_0(n_0) = C_p &+ C_r L\left(1 - q_1^{n_0-1}\right) \\ &+ \frac{C_i p(p-d)}{2d(1-q_1)}\left[\frac{1+q_1}{1-q_1}(1-q_1^{n_0}) - 2n_0 q_1^{n_0}\right] \\ &+ C_s dL\left(1-q_1^m\right) - C_s(p-d)\left[\frac{1-q_1^m}{1-q_1} - m q_1^m\right]. \end{aligned} \tag{14}$$

Similar to the previous cases, taking the difference of $C_0(n_0)$ with respect to n_0 it can be verified that there exists at least one local optimum point n_0^* which minimizes the expected total cost per unit time in the steady state provided $w_0(n_0^* - 1) < 0$ and $w_0(n_0^*) \geq 0$, where

$$w_0(n_0) = \left[C_r L(1-q_1) + \frac{C_i p(p-d)q_1}{2d}(1+2n_0)\right] T_0(n_0) - \left(\frac{pq_1}{d}\right) V_0(n_0)\,.$$

On the other hand, the optimal production time n_0^* would be either $\bar{n}_0$ or $\underline{n_0}$ according to $w_0(\bar{n}_0) \leq 0$ or $w_0(\bar{n}_0) \geq 0$.

If $q_1 \to 1$, i.e., when the machine is perfectly reliable, it is easy to verify from Eqs. (13) and (14), by using L'Hospital's rule, that

$$T_0(n_0) \to \frac{pn_0}{d} \quad \text{and} \quad V_0(n_0) \to C_p + \frac{C_i p(p-d)n_0^2}{2d}\,.$$

7. Numerical Illustration

To derive the numerically optimal production policy, we take the parameter values as $d = 90$, $p = 180$, $C_p = 1500$, $C_s = 1.25$, $C_i = 0.5$, $C_r = 200$, $\underline{n_0} = 3$, $\bar{n}_0 = 12$. For convenience, we use the following abbreviations:

Model A: Model under geometric failure and geometric repair distributions.

Model B: Model under geometric failure distribution and constant repair time.

Tables 1 and 2 present that as the failure rate increases, the minimum ETC per unit time increases. Analogously, for a given failure rate 0.1, the minimum ETC per unit time decreases with the increase in the repair rate, see Table 3. These characteristics are similar to those of the EMQ model having continuous time exponential failure and exponential repair distributions. However, the optimal production time for a perfectly reliable production system is obtained as $n_0^* = 6$ and the corresponding minimum ETC per unit time is 260.

This shows that the expected cost per unit time in the unreliable situation is higher than that of the reliable one though the optimal

Table 1. Dependence of the optimal production policy on the failure rate in Model A when $q_2 = 0.2$.

Failure rate $(1 - q_1)$	n_0^*	$C(n_0^*)$
0.1	6	291.672
0.2	7	330.415
0.3	7	376.714
0.4	8	431.042
0.5	8	491.599
0.6	8	555.788
0.7	7	621.137
0.8	4	685.923
0.9	$3^\dagger$	748.752

$^\dagger n_0^* = \underline{n_0} = 3$, by Theorem 1, as $w(\underline{n_0}) > 0$.

Table 2. Dependence of the optimal production policy on the failure rate in Model B.

Failure rate	$L = 1$		$L = 0$	
$(1 - q_1)$	n_0^*	$C_0(n_0^*)$	n_0^*	$C_0(n_0^*)$
0.1	6	290.070	6	281.330
0.2	7	327.388	7	308.715
0.3	7	373.269	8	344.241
0.4	8	428.519	9	388.931
0.5	9	492.343	10	442.427
0.6	9	562.488	11	502.498
0.7	9	636.785	$12^{\dagger\dagger}$	566.786
0.8	7	713.750	$12^{\dagger\dagger}$	633.750
0.9	$3^\dagger$	792.230	$12^{\dagger\dagger}$	702.500

$^\dagger n_0^* = \underline{n_0} = 3$, as $w_0(\underline{n_0}) > 0$.
$^{\dagger\dagger} n_0^* = \bar{n}_0 = 12$, as $w_0(\underline{n_0}) < 0$.

Table 3. Influence of the repair rate on the optimal production policy for a fixed failure rate 0.1 in Model A.

Repair rate $(1 - q_2)$	n_0^*	$C(n_0^*)$
0.1	6	305.265
0.2	6	301.760
0.3	6	299.149
0.4	6	297.053
0.5	6	295.328
0.6	6	293.893
0.7	6	292.689
0.8	6	291.672
0.9	6	290.809

unreliable production lot size is not necessarily larger than the corresponding optimal lot size of the reliable production system.

We now examine the influence of the parameters involved in Model A on the optimal production policy. We take $q_1 = 0.4$ and $q_2 = 0.2$ in addition to the parameter values given at the beginning of this section. We consider the change in only one parameter value and keep all other parameter values fixed. The computational results are shown in Tables 4–6.

It is easy to observe from Table 4 that the expected cost per unit time decreases drastically with the increase in the production rate. Tables 5 and 6 show that for a 50% change (decrease/increase) in the values of the parameters C_i, C_s, C_r and C_p, the percentage changes (decrease/increase) in the minimum ETC per unit time are approximately 4, 1, 6 and 41, respectively. This implies that the optimal production policy is highly sensitive to changes in the parameters C_p and p whereas moderately sensitive to changes in the parameters

Table 4. Dependence of the optimal production policy on the parameters p and d.

p	n_0^*	$C(n_0^*)$	d	n_0^*	$C(n_0^*)$
100	$12^{\dagger\dagger}$	845.912	30	$3^{\dagger}$	254.802
150	$12^{\dagger\dagger}$	802.572	36	$3^{\dagger}$	288.424
200	7	539.135	45	4	338.329
250	4	492.286	60	5	417.116
300	$3^{\dagger}$	431.618	90	8	555.788

$^{\dagger}n_0^* = \underline{n_0} = 3$, as $w_0(\underline{n_0}) > 0$.
$^{\dagger\dagger}n_0^* = \bar{n}_0 = 12$, as $w_0(\underline{n_0}) < 0$.

Table 5. Dependence of the optimal production policy on the parameters C_i and C_s.

C_i	n_0^*	$C(n_0^*)$	C_s	n_0^*	$C(n_0^*)$
0.1	$12^{\dagger\dagger}$	515.778	1.0	8	554.739
0.3	$12^{\dagger\dagger}$	535.795	1.5	8	556.838
0.5	8	555.788	2.0	8	558.938
0.7	6	575.589	2.5	8	561.038
0.9	5	594.960	3.0	8	563.138

$^{\dagger\dagger}n_0^* = \bar{n}_0 = 12$, as $w_0(\underline{n_0}) < 0$.

C_r and d. In reality, estimation of the shortage cost is quite difficult. But less care can be paid to estimate the parameter C_s because the sensitivity of the minimum expected cost with respect to this parameter is very low.

We have already mentioned in the analysis in Sec. 4 that the failure rate in geometric failure distribution is always constant with respect to time. So, in infant mortality and wear out periods, geometric distribution would not describe a part's life time too well. Instead, a better

Table 6. Dependence of the optimal production policy on the parameters C_p and C_r.

C_p	n_0^*	$C(n_0^*)$	C_r	n_0^*	$C(n_0^*)$
1000	6	412.430	100	9	520.057
2000	11	698.820	200	8	555.788
3000	$12^{\dagger\dagger}$	984.835	300	7	591.478
4000	$12^{\dagger\dagger}$	1270.850	400	6	627.044
5000	$12^{\dagger\dagger}$	1556.860	500	4	662.128

$^{\dagger\dagger}n_0^* = \bar{n}_0 = 12$, as $w_0(\underline{n_0}) < 0$.

choice would be discrete Weibull failure distribution (Nakagawa and Osaki[28]) whose shape parameter α enables it to be applied to any phase (infant mortality, stable, wear out) of a product's life:

$$p(n) = \begin{cases} 0, & \text{for } n = 0 \\ q_1^{(n-1)^\alpha} - q_1^{n^\alpha}, & \text{for } n = 1, 2, 3, \ldots;\quad 0 < q_1 < 1,\quad \alpha > 0\,. \end{cases}$$

The failure rate of this failure time distribution is $\left[1 - q_1^{n^\alpha - (n-1)^\alpha}\right]$, which is a strictly increasing (decreasing) function of n for α greater (less) than 1. When $\alpha = 1$, the Weibull distribution reduces to geometric distribution having constant failure rate $(1 - q_1)$. Since $\xi(n) = p(n)/\bar{P}(n) = \left[q_1^{(n-1)^\alpha - n^\alpha} - 1\right]$ is a strictly increasing (decreasing) function of n for α greater (less) than 1, therefore, the optimal production policy derived in Theorems 2 and 3, in Sec. 5 can be utilized. In the infant mortality period, the ETC per unit time decreases (increases) sharply with a decreasing (increasing) failure rate (see Table 7).

The results given in Tables 8 and 9 indicate that the ETC per unit time increases with the failure rate. More interestingly, a higher value in the shape parameter in Weibull distribution provides lower cost in the wear out period.

Table 7. Optimal results for a discrete Weibull distribution with shape parameter $\alpha = 0.5$ in Model B ($L = 1$).

q_1	n_0^*	$C_0(n_0^*)$	$f(n_0^*)$
0.00909	$12^{\dagger\dagger}$	863.999	0.50
0.01735	$12^{\dagger\dagger}$	855.487	0.45
0.03130	$12^{\dagger\dagger}$	840.107	0.40
0.05387	$12^{\dagger\dagger}$	813.449	0.35
0.08900	$12^{\dagger\dagger}$	769.549	0.30
0.14210	$12^{\dagger\dagger}$	702.052	0.25
0.22010	$12^{\dagger\dagger}$	609.635	0.20
0.36730	10	477.057	0.15
0.56170	8	371.807	0.10

$^{\dagger\dagger}n_0^* = \bar{n}_0 = 12$, as $w_0(\underline{n_0}) < 0$.

Table 8. Optimal results for a discrete Weibull distribution with shape parameter $\alpha = 2$ in Model B ($L = 1$).

q_1	n_0^*	$C_0(n_0^*)$	$f(n_0^*)$
0.9904	6	268.760	0.1
0.9799	6	278.427	0.2
0.9680	6	289.395	0.3
0.9545	6	301.736	0.4
0.9258	5	326.305	0.5
0.9032	5	344.866	0.6
0.8747	5	367.242	0.7
0.7945	4	421.807	0.8
0.6309	3	512.519	0.9

8. Concluding Remarks

The life times are sometimes not measured by the exact instant of failure, but per day, per month, per year and so on. In any case, it

Table 9. Optimal results for a discrete Weibull distribution with shape parameter $\alpha = 3$ in Model B ($L = 1$).

q_1	n_0^*	$C_0(n_0^*)$	$f(n_0^*)$
0.99884	6	264.183	0.1
0.99755	6	268.714	0.2
0.99416	5	278.845	0.3
0.99166	5	285.468	0.4
0.98870	5	292.997	0.5
0.98509	5	301.708	0.6
0.96798	4	332.404	0.7
0.95740	4	348.011	0.8
0.88585	3	413.691	0.9

is appropriate to consider discrete time failure distributions. In this chapter, we have studied an EMQ problem with stochastic machine breakdown and repair in a discrete-time framework. The expected total cost function and the criteria for the existence and uniqueness of the optimal production time are derived under (i) general discrete failure and discrete repair time distributions and (ii) general discrete failure time distribution and constant repair time. Specific formulations of the model under geometric failure and geometric/constant repair time are also derived. The optimal EMQ policy is obtained numerically for both geometric and Weibull failure distributions as geometric failure distribution does not fit well in the infant mortality and wear out periods of a machine's life.

In developing the model, we have assumed that the failure can be detected immediately and perfectly. However, this may be unrealistic in some manufacturing industries. Moreover, if no machine failure occurs during a production phase, the preventive maintenance with negligible time is considered to renew the machine before the start of the next production run. Instead, consideration of a positive random preventive maintenance time may be useful to many real applications. Further, for discrete time setting, we have assumed that the production-demand ratio is an integer greater than 1. Future research may relax these assumptions and treat the problem under a more generalized framework.

Acknowledgments

This work was done when the first author visited the Hiroshima University, Japan, as a JSPS Post-Doctoral Fellow. The authors would like to thank the financial support by a Grant-in-Aid for Scientific Research from the Ministry of Education, Sports, Science and Culture of Japan under Grant No. 02296.

References

1. J. J. McCall, Maintenance policies for stochastically failing equipment: A survey, *Management Science* **11** (1965) 493–524.
2. T. Bielecki and P. R. Kumar, Optimality of zero-inventory policies for unreliable manufacturing systems, *Operations Research* **36** (1988) 532–541.
3. M. J. M. Posner and M. Berg, Analysis of a production-inventory system with unreliable production facility, *Operations Research Letters* **8** (1989) 339–345.
4. H. Groenevelt, L. Pintelon and A. Seidmann, Production lot sizing with machine breakdowns, *Management Science* **38** (1992) 104–123.
5. H. Groenevelt, L. Pintelon and A. Seidmann, Production batching with machine breakdowns and safety stocks, *Operations Research* **40** (1992) 959–971.
6. C. H. Kim and Y. Hong, An extended EMQ model for a failure prone machine with general lifetime distribution, *International Journal of Production Economics* **49** (1997) 215–223.
7. C. H. Kim, Y. Hong and S.-Y. Kim, An extended optimal lot sizing model with an unreliable machine, *Production Planning and Control* **8** (1997) 577–585.
8. K.-J. Chung, Bounds for production lot sizing with machine breakdowns, *Computers and Industrial Engineering* **32** (1997) 139–144.
9. K. J. Chung, Approximations to production lot sizing with machine breakdowns, *Computers and Operations Research* **30** (2003) 1499–1507.
10. T. Dohi, Y. Yamada, N. Kaio and S. Osaki, The optimal lot sizing for unreliable economic manufacturing model, *International Journal of Reliability, Quality and Safety Engineering* **4** (1997) 413–426.
11. E. Tse and V. Makis, Optimization of the lot size and the time to replacement in a production system subject to random failure, *Proc. 3rd International Conference on Automation Technology*, Taipei, Taiwan, 1994, pp. 163–169.
12. M. Berg, M. J. M. Posner and H. Zhao, Production-inventory systems with unreliable machines, *Operations Research* **42** (1994) 111–118.

13. N. E. Abboud, A simple approximation of the EMQ model with Poisson machine failures, *Production Planning and Control* **8** (1997) 385–397.
14. V. Makis, Optimal lot sizing/preventive replacement policy for an EMQ model with minimal repair, *International Journal of Logistics: Research and Applications* **1** (1998) 173–180.
15. T. Dohi, N. Kaio and S. Osaki, Minimal repair policies for an economic manufacturing process, *Journal of Quality in Maintenance Engineering* **4** (1998) 248–262.
16. V. Makis and J. Fung, An EMQ model with inspections and random machine failures, *Journal of the Operational Research Society* **49** (1998) 66–75.
17. B. Liu and J. Cao, Analysis of a production-inventory system with machine breakdowns and shutdowns, *Computers and Operations Research* **26** (1999) 73–91.
18. A. Moini and D. N. P. Murthy, Optimal lot sizing with unreliable production system, *Mathematical and Computer Modelling* **31** (2000) 245–250.
19. K. L. Cheung and W. H. Hausman, Joint determination of preventive maintenance and safety stocks in an unreliable production environment, *Naval Research Logistics* **44** (1997) 257–271.
20. T. Dohi, H. Okamura and S. Osaki, Optimal control of preventive maintenance schedule and safety stocks in an unreliable manufacturing environment, *International Journal of Production Economics* **74** (2001) 147–155.
21. B. C. Giri, W. Y. Yun and T. Dohi, Optimal design of unreliable production/inventory systems with variable production rate, *European Journal of Operational Research* (2004) in press.
22. T. Dohi, W. Y. Yun, N. Kaio and S. Osaki, Optimal design of economic manufacturing process with machine failure rate depending on production speed, *Proc. International Symposium on Manufacturing Strategy*, 1998, pp. 404–409.
23. B. C. Giri and T. Dohi, Optimal lot sizing for an unreliable production system based on net present value approach, *International Journal of Production Economics* (2004) in press.
24. T. Nakagawa and S. Osaki, Discrete time age replacement policies, *Operational Research Quarterly* **28** (1977) 881–885.
25. T. Nakagawa, A summary of discrete replacement policies, *European Journal of Operational Research* **17** (1984) 382–392.
26. J. M. Rocha-Martinez and M. Shaked, A discrete-time model of failures and repairs, *Applied Stochastic Models and Data Analysis* **11** (1995) 167–180.
27. N. E. Abboud, A discrete-time Markov production-inventory model with machine breakdowns, *Computers and Industrial Engineering* **39** (2001) 95–107.
28. T. Nakagawa and S. Osaki, The discrete Weibull distribution, *IEEE Transactions on Reliability* **24** (1975) 300–301.

CHAPTER 6

Applying Accelerated Life Models to HALT Testing

Fabrice Guérin*, Pascal Lantieri and Bernard Dumon

Institut des Sciences et Techniques de l'Ingénieur d'Angers (ISTIA), 62, av. Notre-Dame du Lac, 49000 Angers, France

**fabrice.guerin@istia.univ-angers.fr*

1. Introduction

Current issues of industrial competition about innovation, design, time to market and reliability require more and more efficient qualification strategies. We will particularly focus on so-called maturity design testing with Highly Accelerated Life Testing (HALT) for example. These tests are used during the design step in order to obtain a mature product by showing out weaknesses for which corrective actions are brought in order to be eliminated and, thus, to increase reliability (see Fig. 1).[1–7]

In order to show out weaknesses, the product is submitted to step stress (temperature, vibrations, . . .) by increasing the levels until a failure comes about (see Fig. 2). At each failure, a technological analysis is carried out to show if it is a result of a latent defect. In this case, a corrective action is brought. If the technological limit is reached, the test is ended.

Thus, it can be obtained

- the technological limit concerned,

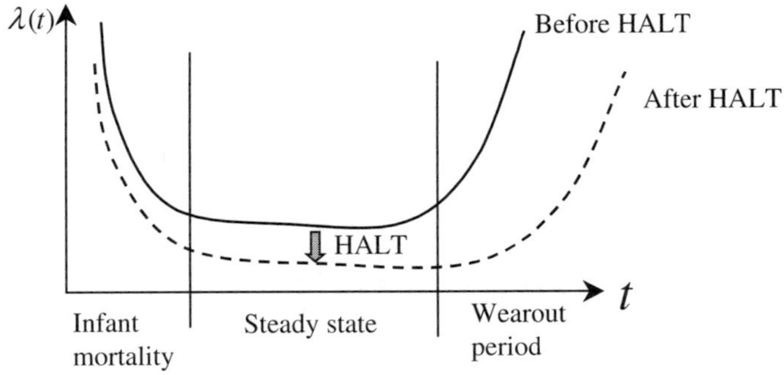

Fig. 1. Maturity of a product through HALT testing.

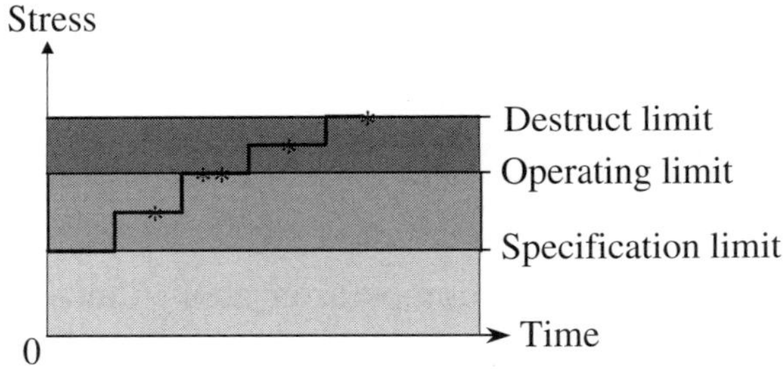

Fig. 2. Test profile.

- a mature product as soon as the beginning of the product cycle,
- an improvement of the operational reliability.

Note: Applied stresses have to be consistent with the technological strength concerned.

Then, reliability assessment tests are carried out knowing that the systems are more and more reliable and thus, the time to failure greater, which leads to testing times inconsistent with industrial requirements.

In order to reduce these times, accelerated tests can be carried out.[1,2,8–11] For this purpose, the product is tested under harder working or environmental conditions in order to accelerate the damaging

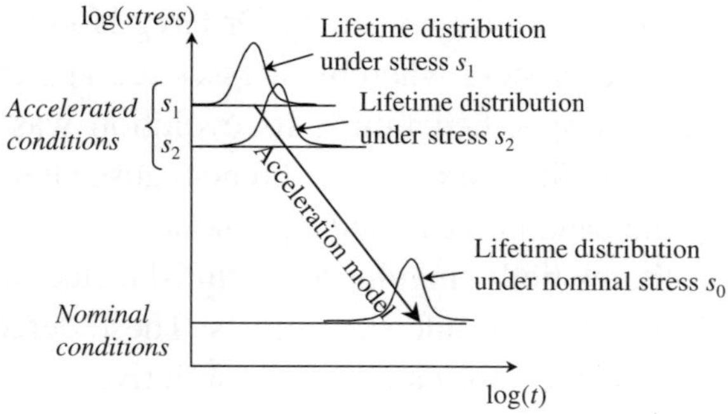

Fig. 3. Reliability assessment with accelerated tests.

mechanism (the failure mechanism has to be specific of nominal conditions) and to cut down the required time for specific estimations under nominal conditions (see Fig. 3).

In this way, it has to be known:

- the analytical model (acceleration model) defining the damaging speed with respect to the range of applied stresses,
- the value of these models' parameters,
- the lifetime distribution.

Thus, product's specific behaviors in nominal conditions can be predicted within delays consistent with calendar requirements of the design period.

We can notice that HALT and accelerated testing methods offer similarities for, in both cases, the products are submitted to amplified stress until failure. Thus, we suggest in this paper, to use step-stress accelerated testing methods to deal with HALT results. We will particularly focus on exponential and Weibull distributions with Arrhenius, Peck and inverse power acceleration models.

2. Maturity Design Testing: HALT

The HALT test is a result of an evolution of product stressing stemming back to the Environmental Stress Screening (ESS) days of the

1960s.[7] The HALT is a term coined by Dr Greg Hobbs in the mid-1980s to describe a process whereby stresses are applied to a new design in excess of specified limits. Its evolution was a result of the discovery that traditional methods did not cause latent (dormant) defects to become patent (active and detectable).

HALT constitutes both singular and multi-faceted stresses that, when applied to a product, uncover defects. These defects are then analyzed, driven to the root cause, and corrective action is implemented. Product robustness is a result of HALT process.

The HALT process utilizes a step stress approach in subjecting products to varied accelerated stresses to discover their design limitations (Fig. 4).

During the HALT process, different stresses are used[5–7]:

- Thermal step stress
- Rapid thermal transitions stress
- Vibration step stress
- Combined environment stress (temperature and vibration or temperature and voltage)
- Voltage step stress
- ...

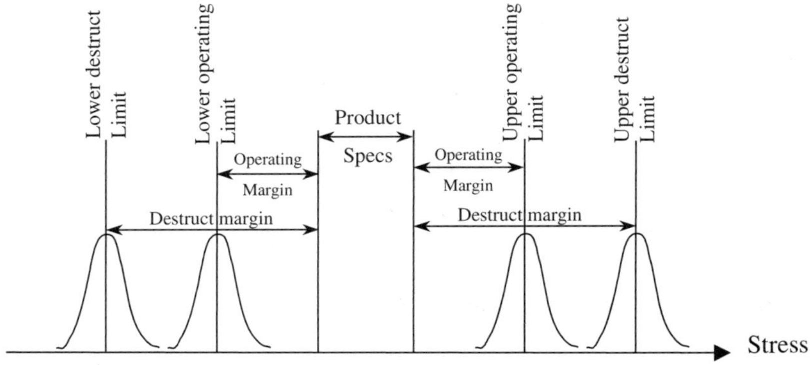

Fig. 4. HALT margin discovery diagram.

In the following sections, the thermal step stress test and rapid thermal transitions stress test are developed. The other parts of the test are built and conducted identically.

2.1. *Thermal step stress test*

Thermal step stress begins at ambient temperature (see Ref. 6). The maximum "high" temperature should be identified prior to the test on the base of the materials phase change limitations. The temperature increments are usually equal to 10°C, but may be increased to 20°C when it is required (see Fig. 5).

The temperature dwell time must be long enough to insure complete stabilization and saturation of the device and its components. This dwell time is usually between five and fifteen minutes following stabilization of the sample at the set point. Complete functional testing immediately follows the dwell period and may also be performed throughout the step.

The thermal stress values increase until the operational limit of the sample is determined. Once the operating limits are determined, temperatures go on increasing beyond the operational limit (with 10°C increments) until the destruction limit is reached.

2.2. *Rapid thermal transitions stress test*

A minimum of three thermal cycles should be performed unless a destructive failure is encountered prior to completion of all three

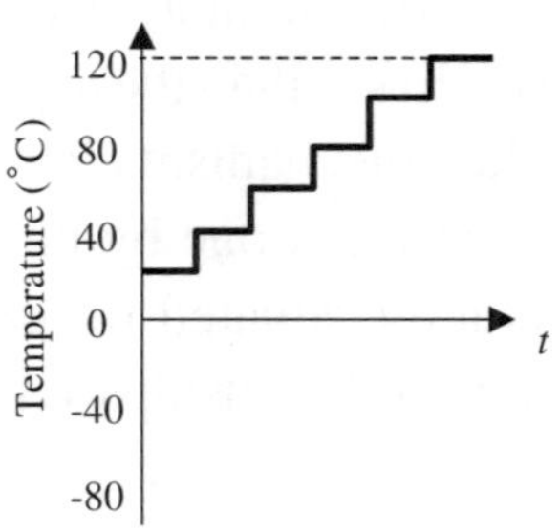

Fig. 5. Thermal step stress test.

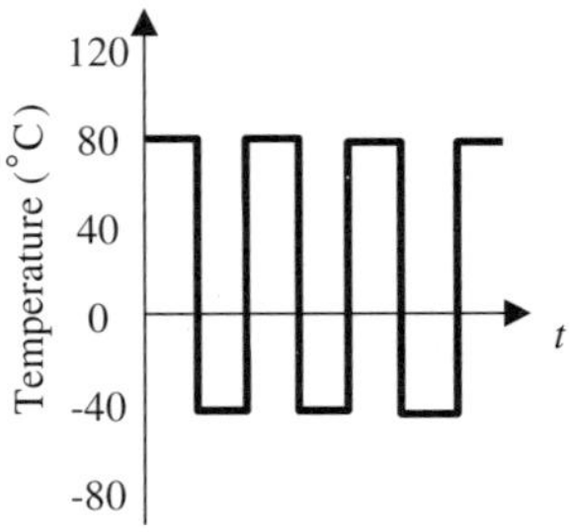

Fig. 6. Rapid thermal transitions stress test.

cycles (see Ref. 6). The thermal transitions are performed at the maximum attainable rate of change (see Fig. 6). Availability of test time and the physics of some products may justify skipping this step as the rapid temperature transitions will occur during the combined thermal-vibration portion of the HALT process.

The range for thermal cycling should be within 5°C of both the Lower Thermal Operating Limit and the Upper Thermal Operating Limit as defined during Thermal Step Stress unless special circumstances occur.

The dwell time is at least five minutes longer than the stabilization time of the sample at the set-point temperature.

3. Accelerated Life Testing

An accelerated life test is a test in which the applied stress is higher than the nominal one in order to shorten the time to failure of the tested product, but it is still lower than the technological limits, to prevent any alteration of failure mechanisms.[1,3,8–11] In accelerated tests, failure mechanisms are activated one by one by increased stresses. The acceleration factors are evaluated for each step of a given test plan thanks to quantitative relationships (Arrehnius, Peck, inverse power, ...).

In the following sections, the common test plans, lifetime distributions and accelerated life test models are presented.

3.1. *Test plan definition*

A detailed test plan is usually designed before conducting an accelerated life test.[1,2,8–15] The test plan requires the determination of:

- The stress type which may be single (temperature, mechanical loading, voltage, vibration, ...) or combined (temperature and humidity, temperature and voltage, ...).
- The stress profile which may be constant (Fig. 7(a)), stepped (Fig. 7(b)) or cyclic (Fig. 7(c)), ...

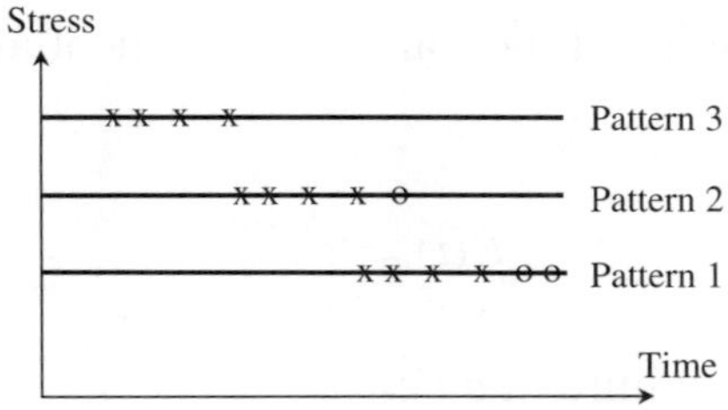

Fig. 7(a). Constant stress test (x failure, o run out).

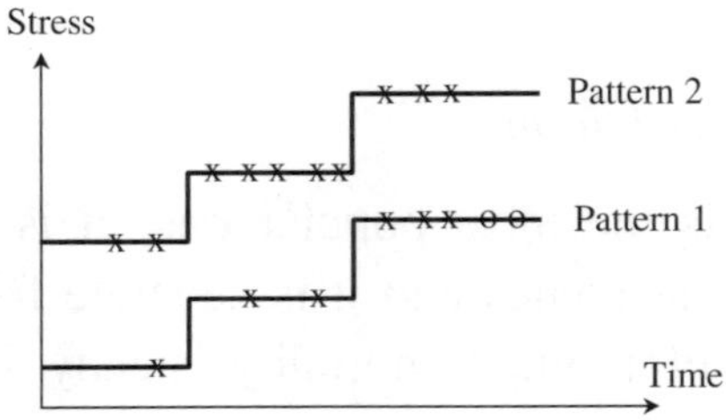

Fig. 7(b). Step-stress test (x failure, o run out).

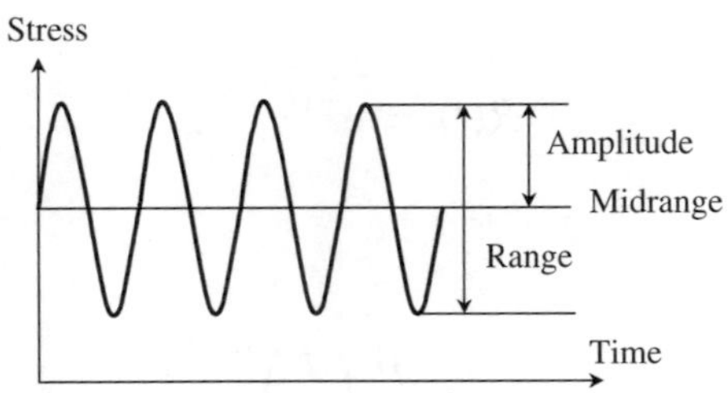

Fig. 7(c). Cyclic-stress loading.

- The sample size.
- The accelerated life test model (to evaluate the lifetime distribution in nominal conditions).

3.2. *Common lifetime distributions*

3.2.1. *Exponential distribution*

This distribution presents a lot of applications in several fields.[1] It is a simple distribution, very common in reliability when the failure rate is constant. It gives the lifetime of equipment submitted to random failures. The reliability function of an exponential distribution with a θ parameter is:

$$R(t) = e^{-t/\theta}. \tag{1}$$

Consequently, the failure rate is:

$$\lambda(t) = \frac{1}{\theta}. \tag{2}$$

3.2.2. *Weibull distribution*

This distribution is the most popular one. It is used in electronics as well as in mechanics and it is accurate for the three stages of the product's life: infant mortality, steady state and wearout period.[1]

The reliability function of a Weibull distribution with η and β parameters is:

$$R(t) = e^{-(t/\eta)^{\beta}}. \tag{3}$$

The failure rate is:

$$\lambda(t) = \frac{\beta}{\eta}\left(\frac{t}{\eta}\right)^{\beta-1}. \tag{4}$$

3.3. *Accelerated life test models*

3.3.1. *Arrhenius model*

It is used when the damaging mechanism is temperature sensitive (especially for dielectrics, semi-conductors, battery cells, lubricant, grease, plastic, incandescent filaments).

The Arrhenius model defines the product lifetime τ by[1]:

$$\tau = Ae^{(+E_a/kT)}, \tag{5}$$

with: A positive constant, E_a activation energy, k Boltzman constant (8.6171×10^{-5} eV/K), and T absolute temperature.

The Arrhenius acceleration factor between the lifetime τ_1 for a temperature T_1 and the lifetime τ_2 for a temperature T_2 is:

$$FA = \frac{\tau_1}{\tau_2} = e^{\frac{Ea}{k}\left(\frac{1}{T_1}-\frac{1}{T_2}\right)}. \tag{6}$$

3.3.2. *Eyring model*

It is used to model accelerated life tests with respect to the temperature and another variable. The model is defined by[1]:

$$\tau = \left(\frac{A}{T}\right) e^{\left(\frac{B}{kT}\right)} e^{\left(V\left(C+\frac{D}{kT}\right)\right)}, \tag{7}$$

with V the stress as voltage, humidity, current density, ..., and A, B, C and D the test and failure specific constants.

3.3.3. *Inverse power model*

It is used when the damaging stress is sensitive to a particular stress (e.g., dielectrics, ball or roller bearing, optoelectronical components, mechanical components submitted to fatigue and incandescent lamp's wires).

The inverse power model defines the damaging rate under a constant stress V. The lifetime is given by[1]:

$$\tau = \frac{A}{V^{\gamma}}, \tag{8}$$

with V the constant stress (V can represent temperature for example), and A and γ the test and failure specific constants.

The acceleration factor between the lifetime τ_1 for a stress level V_1 and the lifetime τ_2 for a stress level V_2 is:

$$AF = \frac{\tau_1}{\tau_2} = \left(\frac{V_2}{V_1}\right)^{\gamma}. \tag{9}$$

A particular inverse power model is the Coffin–Manson model. It defines the number N of cycles to failure by:

$$N = \frac{A}{\Delta T^{\gamma}}, \tag{10}$$

with ΔT the temperature range, and A and γ the model parameters.

It is used to model fatigue failures of metal subjected to thermal cycling. This model is used for mechanical and electronic components. In electronics, it is used for solder joints and other connections.

4. Applying Accelerated Life Models to HALT Testing

In this section, different accelerated life models are applied to HALT results. The particularity is the step duration during step stress tests. Four analyses are dealt with:

- Thermal Step Stress Test,
- Combined Step Stress Test (Temperature and Voltage),
- Rapid Thermal Transitions Stress Test,
- Voltage Step Stress Test.

4.1. *Thermal step stress test*

In this paragraph, an exponential lifetime distribution with an Arrhenius acceleration model is presented.[1,9,11–13,16–20] For that purpose, it is considered that, for each step (indexed by i), the exponential distribution's parameter λ_i is defined by an Arrhenius model, i.e.,

$$\lambda_i = \lambda_0 e^{\left(-\frac{E_a}{k}\left(\frac{1}{T_i}-\frac{1}{T_0}\right)\right)}, \tag{11}$$

with T_i the temperature at step i, T_0 the temperature in nominal conditions, λ_0 the failure rate in nominal conditions, and E_a the activation energy.

Equation (10) can be written:

$$\lambda_i = \lambda_0 e^{(E_a x_i)}, \tag{12}$$

with $x_i = -\frac{1}{k}\left(\frac{1}{T_i} - \frac{1}{T_0}\right)$.

This is the well-known COX's model. To define it, the two unknown parameters E_a and λ_0 have to be evaluated. For this purpose, the approximated failure rate $\hat{\lambda}_i$ will be evaluated for each temperature step i thanks to the following relationship[16]:

$$\hat{\lambda}_i = \frac{k_i}{\sum_{m=1}^{k_i}(t_m - \tau_{i-1}) + (n_i - k_i)\Delta_i}, \tag{13}$$

with k_i the number of failure at i level, n_i the number of tested systems at i level, τ_i the time at the end of step i, t_m the time of the mth failure at step i, and Δ_i the testing time at i level.

Yet, this relationship is not consistent with the hypothesis of a failure rate increase with respect to the temperature. Thus, the following empirical estimator of failure rate has to be used as soon as $\hat{\lambda}_i < \hat{\lambda}_{i-1}$:

$$\hat{\lambda}_i = \hat{\lambda}_{i-1} + \frac{\hat{\lambda}_j - \hat{\lambda}_{i-1}}{j - i + 1}, \tag{14}$$

where j is the first index (greater than i) for which $\hat{\lambda}_j$ is greater than $\hat{\lambda}_i$.

Then, a logarithmic transformation of Eq. (12) leads to:

$$\log(\lambda_i) = E_a x_i + \log(\lambda_0)\,. \tag{15}$$

By plotting the (x_i, $\log(\lambda_i)$) points, a straight line with a slope equal to E_a (activation energy) and a distance to the origin equal to $\log(\lambda_0)$ can be obtained.

Example 1.

For this example, let us consider an electronic board. Data are simulated with the following parameters:

- $E_a = 0.7\,\text{eV}$ (activation energy),
- $N = 50$ (sample size),
- $\lambda_0 = 1 \times 10^{-4}\,\text{h}^{-1}$ (nominal failure rate),
- $\Delta T = 15\,\text{min}$ (testing time for each step),
- $T_0 = 50°\text{C}$ (nominal temperature).

The simulation results are in Table 1.

Table 1. Example's data and results.

T (°C)	x_i	k	n	Cumulated time	λ from Eq. (14)	λ from Eq. (15)	$\log(\lambda)$
120	6.399	0	50	12.500	0.00000	0.0000	
140	7.829	0	50	12.500	0.00000	0.0287	−3.552
160	9.127	2	48	11.715	0.17073	0.0805	−2.519
180	10.311	1	47	11.633	0.08597	0.1324	−2.022
190	10.864	2	45	10.858	0.18420	0.1538	−1.872
200	11.394	1	44	10.838	0.09227	0.1753	−1.741
210	11.902	2	42	10.165	0.19675	0.8695	−0.140
220	12.389	10	32	6.395	1.56367	0.8713	−0.138
230	12.857	6	26	6.872	0.87311	0.8722	−0.137
240	13.307	9	17	3.487	2.58080	2.5808	−0.136

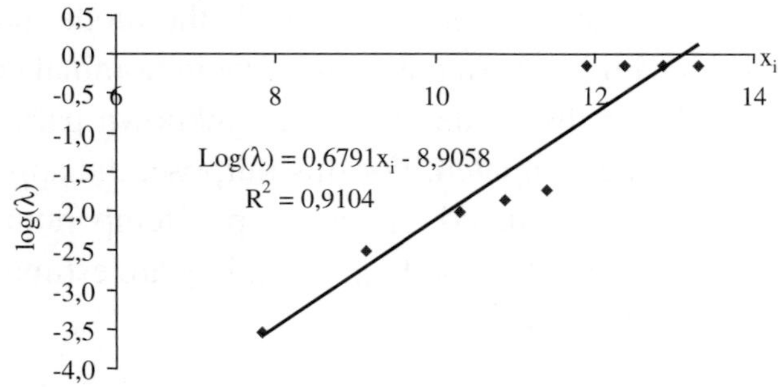

Fig. 8. Evolution plot from Arrhenius model.

From these data, the different values of the failure rate can be evaluated by relationships (14) and (15) (see Table 1). It is possible to plot the straight line (see Fig. 8) defined with Eq. (16) and that enables to estimate the Arrhenius and exponential models parameters (E_a and λ_0).

Thus, the following estimations can be obtained:

- $\hat{E}_a = 0.6791\,\text{eV}$ (instead of 0.7 eV) and
- $\hat{\lambda}_0 = \text{e}^{-8.9058} = 1.36 \times 10^{-4}\,\text{h}^{-1}$ (instead of $1 \times 10^{-4}\,\text{h}^{-1}$).

It can be noted that the estimated values are close to initial data, which proves the efficiency of the method. A determination of confidence intervals for these constants is possible with a Least Square (LS) method.[1]

4.2. *Combined step stress test (temperature and voltage)*

In this paragraph, an exponential lifetime distribution with an Eyring acceleration model is presented.[1,9,11–13,16–21] For that purpose, it is considered that, for each step (indexed by i), the exponential distribution's parameter λ_i is specified by a simplified Eyring model (with $A = 0$, $\beta_1 = B/k$, $\beta_2 = C$ and $D = 0$), i.e.,

$$\lambda_i = \lambda_0 e^{\left(\frac{\beta_1}{T_i} + \beta_2 V_i\right)}, \tag{16}$$

with T_i the temperature at step i, β_1 and β_2 the model parameters, V_i the voltage at step i, and λ_0 the failure rate in nominal conditions.

In order to define the model, the three unknown parameters λ_0, β_1 and β_2 have to be evaluated. For this purpose, the approximated failure rate $\hat{\lambda}_i$ will be evaluated for each step of temperature i thanks to relationships (14) and (15) and $\hat{\lambda}_0$, $\hat{\beta}_1$ and $\hat{\beta}_2$ are estimated by the Least Square (LS) method.[1]

Example 2.

For this example, an electronic board is considered. The data are simulated with the following parameters:

- $N = 50$ (sample size),
- $\lambda_0 = 1.4 \times 10^{-4}\,\mathrm{h}^{-1}$ (nominal failure rate),
- $\Delta T = 15\,\mathrm{min}$ (testing time for each step),
- $\beta_1 = -15$ and $\beta_2 = 0.25$.

The simulation results in Table 2 have been obtained.

The Least Squared method enables to obtain the following estimations:

- $\hat{\lambda}_0 = 1.49 \times 10^{-4}\,\mathrm{h}^{-1}$ (instead of $1.4 \times 10^{-4}\,\mathrm{h}^{-1}$)
- $\hat{\beta}_1 = -15.018$ (instead of -15),
- $\hat{\beta}_2 = 0.265$ (instead of 0.25).

One more time, the efficiency of the method is proven (estimations close to initial data).

4.3. *Rapid thermal transitions stress test*

In this paragraph, an accelerated life test with constant stress is presented to deal with the rapid thermal transition stress test results.[1,10,11,22,23] For that purpose, it is considered that:

- the lifetime is defined by a Weibull distribution (fatigue damage),
- the shape parameter β of the Weibull distribution is constant,

Table 2. Example's data and results.

T (°C)	V	Theorical λ from Eq. (17)	k_i	n	Cumulated time	λ from Eq. (14)	λ from Eq. (15)
60	28	1.47E−01	0	50	12.7066	0.00E+00	0.00E+00
80	28	1.47E−01	0	50	12.5000	0.00E+00	8.32E−02
100	28	1.47E−01	2	48	12.0189	1.66E−01	9.90E−02
120	28	1.48E−01	0	48	12.2483	0.00E+00	9.82E−02
140	28	1.48E−01	1	47	11.7984	8.48E−02	1.29E−01
60	30	2.42E−01	0	47	11.7500	0.00E+00	1.59E−01
80	30	2.43E−01	2	45	11.3366	1.76E−01	1.91E−01
100	30	2.43E−01	2	43	11.2180	1.78E−01	2.23E−01
120	30	2.44E−01	1	42	10.6676	9.37E−02	3.40E−01
140	30	2.44E−01	3	39	10.6238	2.82E−01	3.75E−01
60	32	3.99E−01	3	36	9.6767	3.10E−01	4.12E−01
80	32	4.00E−01	3	33	8.5888	3.49E−01	4.50E−01
100	32	4.01E−01	3	30	7.8023	3.85E−01	4.87E−01
120	32	4.02E−01	6	24	6.4702	9.27E−01	5.24E−01
140	32	4.02E−01	3	21	5.5006	5.45E−01	5.62E−01

- the scale parameter η of the Weibull distribution is defined by a Coffin–Manson model:

$$\eta(\Delta T) = \frac{A}{\Delta T^{\gamma}}, \tag{17}$$

with ΔT the temperature range, and A and γ the unknown parameters.

The unknown parameters A and γ are evaluated by testing two specimens under two levels of temperature ΔT_1 and ΔT_2. The two scale parameters η_1 and η_2 are evaluated from test results and the estimators $\hat{A}$ and $\hat{\gamma}$ are deduced by:

$$\hat{A} = \frac{\eta_1 \Delta T_1^{\gamma} + \eta_2 \Delta T_2^{\gamma}}{2}, \tag{18}$$

and

$$\hat{\gamma} = \frac{\log\left(\frac{\eta_1}{\eta_2}\right)}{\log\left(\frac{\Delta T_2}{\Delta T_1}\right)}. \tag{19}$$

Example 3.

For this example, let us consider an electronic board. The data are simulated with the following parameters:

- $N = 52$ (sample size) decomposed in two subsamples ($N_1 = 26$ and $N_2 = 26$),
- $\eta_0 = 1 \times 10^6$ cycles (nominal scale parameter),
- $\Delta T_0 = 20°\text{C}$ (nominal temperature range),
- $\Delta T_1 = 160°\text{C}$ (first harder temperature; the associated scale parameter $\eta_1 = 30.52$ deducted from Eq. (18)),
- $\Delta T_2 = 120°\text{C}$ (second harder temperature; the associated scale parameter $\eta_2 = 128.6$ deduced from Eq. (18)),
- $\beta = 1.5$,
- $\gamma = 5$,
- $A = 3.2 \times 10^{12}$.

The simulation results in Table 3 have been obtained (in bold censoring time).

From these results, the parameters η_1 and η_2 are estimated by the maximum likelihood method:

- $\hat{\eta}_1 = 30$ cycles (instead of 30.52), $\hat{\beta}_1 = 1.48$ (instead of 1.5),
- $\hat{\eta}_2 = 127.37$ cycles (instead of 128.6), $\hat{\beta}_2 = 1.57$ (instead of 1.5).

Then, the Coffin–Manson parameters $\hat{A}$ and $\hat{\gamma}$ can be evaluated:

- $\hat{A} = 5.026$ (instead of 5),
- $\hat{\gamma} = 3.588 \times 10^{12}$ (instead of 3.2×10^{12}).

Again, it can be noted that these estimations are close to initial data.

Table 3. Example's data and results.

t_1	t_2
0.7	**128**
20.8	60.68
30	128
30	97.91
11.1	10.26
16.6	29.08
29.0	96.66
28.6	113.33
20.5	17.43
18.9	53.03
16.4	49.79
20.5	**128**
30	**128**
28.8	40.34
30	**128**
18.6	**128**
23.9	87.52
30	71.39
18.6	1.90
30	**128**
24.4	86.19
29.9	104.40
24.4	80.63
30	70.05
30	75.67
23.0	121.95

4.4. *Voltage step stress test*

In this section, the Weibull distribution is used for voltage step stress with an inverse power acceleration model.[1,9,11–13,16–21] For that purpose, it is considered that, for each step stress:

- the lifetime is defined by a Weibull distribution,
- the shape parameter β of the Weibull distribution is constant,
- the scale parameter η of the Weibull distribution is defined by an inverse power model:

$$\eta(S) = \eta(S_0)\left(\frac{S_0}{S}\right)^{\gamma}. \tag{20}$$

The test is carried out with several sequential step stresses by recording at each step the time to failure.

Thus, for each step stress, a specific Weibull distribution can be defined (with a constant shape parameter β and a variable scale parameter $\eta(S)$ versus the stress level). When a sample of systems is tested for several successive stress levels, the damaging rate will be conserved, and thus, the cumulative distribution function $F_0(t)$ will be conserved at each transition too.

The cumulative distribution function $F_i(t)$ for a single stress level S_i (see Fig. 9) can be defined by:

$$F_i(t) = 1 - e^{-\left(\frac{t}{\eta(S_0)}\left(\frac{S_i}{S_0}\right)^{\gamma}\right)^{\beta}}, \quad \text{for } i \geq 1\,, \tag{21}$$

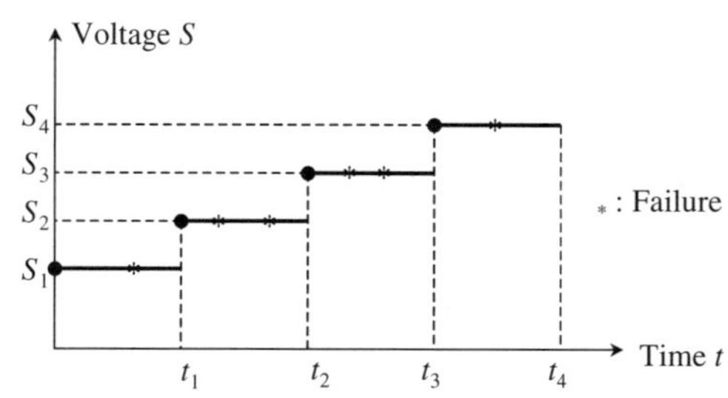

Fig. 9. Test profile.

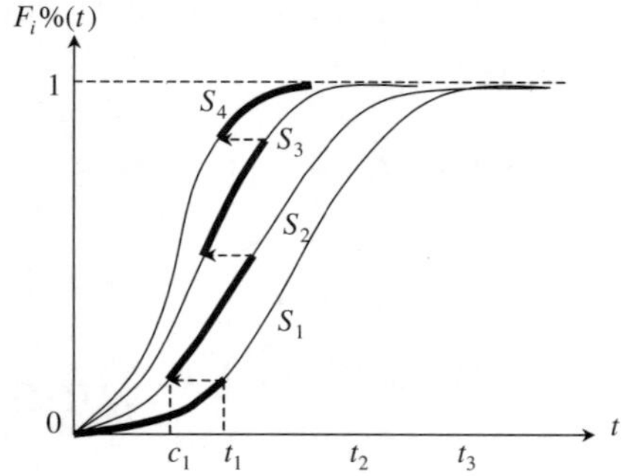

Fig. 10. Cumulative distribution function level change versus stress level.

wherein S_0 is the stress nominal value, $\eta(S_0)$ is the scale parameter in nominal conditions (initially unknown).

At any transition from a step i to a step $i+1$, an equivalent time c_i can be defined as the time to reach the same value of the cumulative distribution function with a single step at a S_{i+1} stress level (see Fig. 10).

During Step 1 under stress S_1.

The cumulative distribution function $F_0(t)$ is given by:

$$F_0(t) = F_1(t)\,, \quad \text{for } 0 \leq t \leq t_1\,. \tag{22}$$

Then

$$F_0(t) = 1 - e^{-\left(\frac{t}{\eta(S_0)}\left(\frac{S_1}{S_0}\right)^{\gamma}\right)^{\beta}}, \quad \text{for } 0 \leq t \leq t_1\,. \tag{23}$$

During Step 2 under stress S_2.

At the beginning of Step 2, the equivalent time c_1 is the duration of single step at a stress level S_2 to reach the cumulative distribution function value $F_0(t_1)$. Thus, c_1 is solution of:

$$F_2(c_1) = F_0(t_1)(= F_1(t_1))\,. \tag{24}$$

Thus,

$$c_1 = t_1\left(\frac{S_1}{S_2}\right)^{\gamma}. \tag{25}$$

The cumulative distribution function $F_0(t)$ defined from failure results at step stress S_2 is

$$F_0(t) = F_2[(t - t_1) + c_1]\,, \quad t_1 \le t \le t_2\,. \tag{26}$$

Thus,

$$F_0(t) = 1 - e^{-\left(\frac{[(t-t_1)+c_1]}{\eta(S_0)}\left(\frac{S_2}{S_0}\right)^{\gamma}\right)^{\beta}}, \quad t_1 \le t \le t_2\,. \tag{27}$$

During Step 3 under stress S_3.

In the same way as for Step 2, the equivalent time c_2 is the duration of single step at a stress level S_3 to reach the cumulative distribution function value $F_0(t_2)$. Thus, c_2 is solution of:

$$F_3(c_2) = F_2(t_2 - t_1 + c_1)\,. \tag{28}$$

Thus,

$$c_2 = (t_2 - t_1 + c_1)\left(\frac{S_2}{S_3}\right)^{\gamma}. \tag{29}$$

Thus,

$$F_0(t) = F_3[(t - t_2) + c_2], \quad t_2 \le t \le t_3\,. \tag{30}$$

$$F_0(t) = 1 - e^{-\left(\frac{[(t-t_2)+c_2]}{\eta(S_0)}\left(\frac{S_3}{S_0}\right)^{\gamma}\right)^{\beta}}, \quad t_2 \le t \le t_3\,. \tag{31}$$

During any Step i under stress S_i.

The equivalent time c_{i-1} is given in the same way by:

$$F_i(s_{i-1}) = F_{i-1}(t_{i-1} - t_{i-2} + c_{i-2})\,, \tag{32}$$

$$c_{i-1} = (t_{i-1} - t_{i-2} + c_{i-2})\left(\frac{S_{i-1}}{S_i}\right)^{\gamma}. \tag{33}$$

Thus, the cumulative distribution function $F_0(t)$ is given from failure results at stress level S_i:

$$F_0(t) = F_i[(t - t_{i-1}) + c_{i-1}], \quad t_{i-1} \le t \le t_i\,. \tag{34}$$

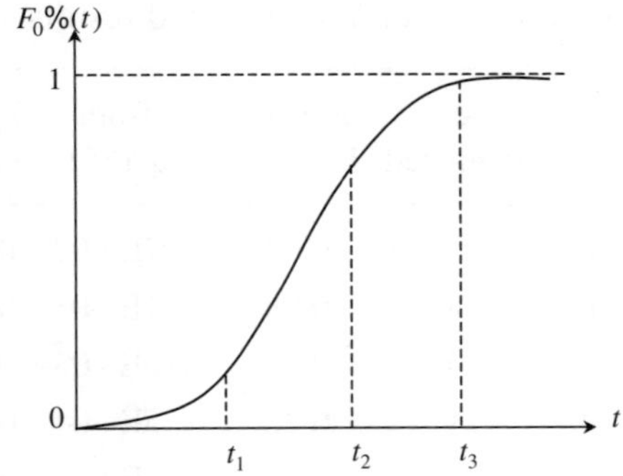

Fig. 11. Cumulative distribution function of Weibull distribution in nominal conditions.

Then

$$F_0(t) = 1 - e^{-\left(\frac{[(t-t_{i-1})+c_{i-1}]}{\eta(S_0)}\left(\frac{S_i}{S_0}\right)^{\gamma}\right)^{\beta}}, \quad t_{i-1} \le t \le t_i\,. \tag{35}$$

It enables to define the Weibull cumulative distribution function in nominal conditions by segments (Fig. 11).

Example 4.

For this example, let us consider an electronic board. The data are simulated with the following parameters:

- $N = 50$ (sample size),
- $\eta_0 = 1 \times 10^5$ hours (nominal scale parameter),
- $S_0 = 12\,\text{V}$ (nominal voltage),
- $\beta = 0.8$ (shape parameter),
- $\gamma = 10$ (inverse power model parameter),
- $\Delta T = 15\,\text{min}$ (testing time for each step).

The simulation results in Table 4 are obtained.

The values in column F_i of Table 4 are defined by the following relationship:

$$F_i = \frac{\left(\sum_{j=1}^{i} k_j\right)}{n}\,. \tag{36}$$

Table 4. Example data and results.

S_i (V)	η_i from Eq. (21)	k_i	n_i	k_i cumulated	F_i from Eq. (37)	t_i from Eq. (35)	c_i from Eq. (35)	$F_0(t_i)$ from Eq. (36)
26	43.8610	1	49	1	0.02	9.07E+02	0.2500	0.0093
28.5	17.5131	3	46	4	0.08	3.31E+03	0.0945	0.0342
31	7.5543	1	45	5	0.1	9.16E+03	0.1413	0.0934
33.5	3.4783	6	39	11	0.22	2.25E+04	0.1719	0.2165
36	1.6935	10	29	21	0.42	5.11E+04	0.1967	0.4295
38.5	0.8654	13	16	34	0.68	1.09E+05	0.2193	0.7034
41	0.4613	14	2	48	0.96	2.23E+05	0.2409	0.9183

k_i: number of failure at i level.
n_i: number of tested systems at i level.
c_i: equivalent time at the end of i level.

The times t_i are evaluated with:

$$t_i = \eta(S_0 = 12\,\text{V}) \times \left(\log\left(\frac{1}{1 - F_i}\right)\right)^{1/\beta}. \tag{37}$$

The values in column $F_0(t_i)$ are given by relationship (36) with parameters γ (from inverse power model), β and η from Weibull distribution in nominal conditions (for $S_0 = 12\,\text{V}$). The total quadratic error between the F_i and the $F_0(t_i)$ can be minimized thanks to the conjugate gradient method.

The minimum has been found for: $\hat{\gamma} = 10.6$ (instead of 10), $\hat{\beta} = 1.016$ (instead of 0.8), $\hat{\eta}(S_0 = 12\,\text{V}) = 9.02 \times 10^4$ (instead of 1×10^5).

The estimations are still close to initial data.

The cumulative distribution function $F_0(t)$ is plotted in nominal conditions ($S_0 = 12\,\text{V}$) (Fig. 12).

5. Conclusion

In this chapter, we have presented a method to define accelerated lifetime models based on HALT results. When stress and failure modes

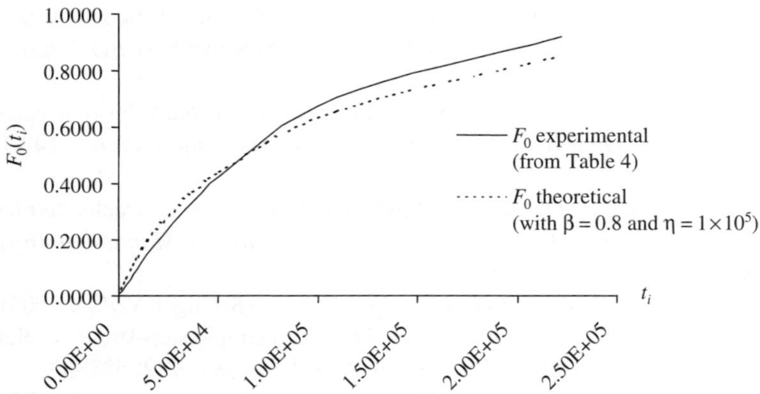

Fig. 12. Cumulative distribution function of Weibull distribution in nominal conditions.

are the same as in nominal conditions and when time intervals are long enough, the calculations carried out have shown a good consistency with input data. Thus, if no important modification of the product comes about during the test, it actually seems possible to define a product's reliability thanks to HALT test results.

References

1. W. Nelson, *Accelerated Testing: Statistical Models, Test Plans and Data Analyses* (Wiley Interscience Publication, 1990).
2. P. O'Connor, Testing for reliability, *Quality and Reliability Engineering International* **19** (2003) 73–84.
3. H. A. Malec, Accelerated stress testing-design, production and field returns, *Quality and Reliability Engineering International* **14** (1998) 449–451.
4. K. Yang and G. Yang, Robust reliability design using environmental stress testing, *Quality and Reliability Engineering International* **14** (1998) 409–416.
5. B. Masotti and M. Morelli, Development of the accelerated testing process at OTIS elevator company, *Quality and Reliability Engineering International* **14** (1998) 381–384.
6. *Highly Accelerated Life Testing*, Test Procedure Analysis-General Motor, GMW8287.
7. H. W. McLean, *HALT, HASS & HASA Explained: Accelerated Reliability Techniques* (ASQ Quality Press, 2000).

8. H.-J. Shyur, E. A. Elsayed and J. T. Luxhoj, A general model for accelerated life testing with time-dependent covariates, *Naval Research Logistics* **46** (1999) 303–321.
9. J. A. McLinn, Ways to improve the analysis of multi-level accelerated life testing, *Quality and Reliability Engineering International* **14** (1998) 125–137.
10. H. Caruso and A. Dasgupta, A fundamental overview of accelerated testing analytical models, *Proceedings Annual Reliability and Maintainability Symposium*, 1998, pp. 389–393.
11. H. Pham, *Handbook of Reliability Engineering* (Springer-Verlag, 2003).
12. D. S. Bai, M. S. Kim and S. H. Lee, Optimum simple step-stress accelerated life tests with censoring, *IEEE Trans. Reliability* **38** (1989) 528–532.
13. I. H. Khamis and J. J. Higgins, Optimum 3-step step-stress tests, *IEEE Transaction on Reliability* **45** (1996) 341–345.
14. R. R. Barton, Optimal accelerated life-time plans that minimize the maximum test-stress, *IEEE Transaction on Reliability* **40** (1991) 166–172.
15. G.-B. Yang, Optimum constant-stress accelerated life-test plans, *IEEE Transaction on Reliability* **43** (1994) 575–581.
16. E. Gouno, An inference method for temperature step-stress accelerated life testing, *Quality and Reliability Engineering International* **17** (2001) 57–64.
17. C. Xiong and G. A. Milliken, Step-stress life testing with random stress change times for exponential data, *IEEE Transaction on Reliability* **48** (1999) 141–148.
18. C. Xiong, Inference on a simple step-stress model with type-II censored exponential data, *IEEE Transaction on Reliability* **47** (1998) 142–146.
19. L. C. Tang, Y. S. Sun and H. L. Ong, Analysis of step-stress accelerated life test data: A new approach, *IEEE Transaction on Reliability* **45** (1996) 69–74.
20. V. B. Bagdonavicius, L. Gerville-Réache and M. S. Nikulin, Parametric inference for step stress models, *IEEE Transaction on Reliability* **51** (2002) 27–31.
21. K.-P. Yeo and L. C. Tang, Planing step-stress life-test with a target acceleration-factor, *IEEE Transaction on Reliability* **48** (1999) 61–67.
22. J. R. Van Dorp, T. A. Mazzuchi, G. E. Fornell and L. R. Pollock, A Bayes approach to step-stress accelerated life testing, *IEEE Transaction on Reliability* **45** (1996) 491–498.
23. A. J. Watkins, On the analysis of accelerated life-testing experiments, *IEEE Transaction on Reliability* **40** (1991) 98–101.

CHAPTER 7

A Poisson Regression Model of Software Quality: A Comparative Study

Taghi M. Khoshgoftaar
Department of Computer Science and Engineering, Florida Atlantic University, Boca Raton, FL 33431, USA
taghi@cse.fau.edu

Robert M. Szabo
IBM Corporation, 8051 Congress Avenue, Boca Raton, FL 33487, USA
rszabo@us.ibm.com

1. Introduction

The study and measurement of software quality has contributed to the advancement of software engineering by providing ways of quantifying software systems which can lead to objective management decision making processes. Prior work has shown that a usable relationship exists between software measures and software quality.[1–4] Models exhibiting high levels of predictive quality can exert some measurable influence on the overall quality of a software system.

It is important to remember that the results obtained from such research are often difficult to apply in environments other than the one in which they were originally developed. So we caution the practitioner to be acutely aware when attempting to apply a model developed in one specific environment to another environment without validating it first. In cases where a model cannot be directly applied, we can still use the modeling methodology. The idea is to develop a model specific to the new environment provided that the model assumptions are not violated. For example, the data collected from software systems used to model software faults often violate the normality assumption of multiple linear regression (MLR) modeling. Applying such a model to this particular data set may not be a good choice. To analyze such data, researchers investigate other modeling methods whose assumptions, or lack of assumptions, better fit the data being collected. This may lead to improvements in predictive quality which in turn should improve the software development process.

In this chapter, we investigate the application of Poisson regression analysis to software quality data known to have a Poisson distribution. We first give an overview of MLR and Poisson regression modeling. Then, using software measures collected from a large military telecommunications software system, we develop MLR and Poisson regression fault models. The independent variables of both models are principal components derived from the observed software measures. Next, we compare the predictive quality of the two competing models and explore the ability of the Poisson regression model to classify the software system into low and high-risk groups with respect to the number of expected faults. We show that for this system, the predictive quality of the Poisson regression model is no better than the predictive quality of the MLR model. Furthermore, we show that the ability of the Poisson model to classify data into groups, rivals a discriminant model.

2. Statistical Modeling Methodologies

The study and measurement of software is essential to software engineers. Typically, one would like to predict the quality of a software system based on some quantifiable measures. For example, it is quite common to try to predict the number of faults remaining in a software system based on program size and other measures of the software. Many of the common software measures used today tend to be highly correlated. For example, program size could be expressed as lines of code, the number of executable statements, and many of the software science metrics.[5] This correlation of the measures is called *multicollinearity*. In an MLR model, multicollinearity can lead to parameter estimates that are not stable, and violates an assumption of MLR modeling.[6]

To avoid this problem, many researchers limit their study to a few carefully selected software complexity metrics. On the other hand, taking as many metrics as possible into consideration should lead to a more complete model, since each measurement assesses a particular and sometimes overlapping aspect of the software. It is difficult to analyze such a data set. To address this issue, principal components analysis may be applied prior to developing an MLR model.[6] Published case studies provide instances where MLR and neural network models using principal components perform better than models that use only the observed data.[7,8] We follow this approach by defining the model independent variables to be principal components of the observed measures. In Sec. 2.1, we discuss Poisson regression modeling, and in Sec. 2.2, we discuss discriminant modeling.

2.1. *Poisson regression modeling*

Software engineering quality data often violates the assumptions of the MLR model.

- The distribution of y is usually not normal.
- The response, y, can be discrete, not continuous.
- The variance of the MLR error terms is *heterogeneous*. This means the variance is not constant for y.

Therefore, modeling methodologies immune to these assumptions should be explored. Poisson regression is such a methodology.

Poisson regression is founded on the Poisson distribution given as:

$$P(y;\,\mu) = \frac{e^{-\mu}(\mu)^y}{y!} \quad (y = 0, 1, 2, \ldots)\,.$$

This methodology assumes that the response variable is discrete and has a Poisson distribution with mean μ. μ depends on a specified time unit or period of interest. For example, the probability of y events in time period t is given by:

$$P(y;\,\mu) = \frac{e^{-\mu t}(\mu t)^y}{y!}\,,$$

where the mean number of incidents is μt. Of course, these assumptions should be validated prior to applying this methodology; otherwise, the results will likely suffer from the same problems inherent with MLR.

The regression model may be written as:

$$y_i = \mu_i + e_i \quad (i = 1, 2, \ldots, n)\,,$$

where n is the number of observations, y_i is the response, e_i is the error, and μ_i is the mean number of incidents in time period t_i. By using the Poisson distribution and modeling the mean as a linear combination of the independent variables, we have

$$P(y_i;\,\boldsymbol{\beta}) = \frac{e^{-t_i[\mu(\mathbf{x}_i,\boldsymbol{\beta})]}[t_i\mu(\mathbf{x}_i,\,\boldsymbol{\beta})]^{y_i}}{y_i!} \quad (i = 1, 2, \ldots, n)\,, \qquad (1)$$

where

- $\mu(\mathbf{x}_i,\,\boldsymbol{\beta})$ is the Poisson mean,
- x_i is a vector of independent variables, and $\mathbf{x}_i'$ is its transpose,
- $\boldsymbol{\beta}$ is a vector of parameters to be estimated using the maximum likelihood estimation technique.

To ensure the Poisson mean is nonnegative, a *link* function, $\mu(\mathbf{x}_i,\,\boldsymbol{\beta})$, is chosen by the analyst. This function represents a relationship between the mean and the independent variables. A common choice is the log link function, $\ln(\mu_i) = \mathbf{x}_i'\boldsymbol{\beta}$. After the parameters are estimated, the mean may be modeled as:

$$\hat{\mu}_i = t_i\mu(\mathbf{x}_i, \hat{\boldsymbol{\beta}}),$$

where $\hat{\boldsymbol{\beta}}$ is the vector of estimated values. For the log link function at period t,

$$\hat{\mu} = te^{\mathbf{x}'\hat{\boldsymbol{\beta}}}.$$

In this chapter, t is set to one, indicating that the basis is a single project. The probability of y incidents may be estimated by substituting y for y_i, $\hat{\mu}$ for $\mu(\mathbf{x}_i, \hat{\boldsymbol{\beta}})$, and 1 for t_i in Eq. (1). Thus, the Poisson regression model may be used to classify data into membership classes in addition to predicting the response variable y. Refer to Myers for further details regarding Poisson regression and maximum likelihood estimation.[9]

The model quality of fit is measured by the *deviance*, defined to be

$$D(\mathbf{y}, \hat{\mathbf{y}}) = 2(l(\mathbf{y}, \mathbf{y}) - l(\mathbf{y}, \hat{\mathbf{y}})),$$

where $\mathbf{y}$ is a vector of response values, $\hat{\mathbf{y}}$ is a vector of predicted values, $l(\mathbf{y}, \mathbf{y})$ is the log likelihood of a perfect fit, and $l(\mathbf{y}, \hat{\mathbf{y}})$ is the log likelihood of the model being fit. The distribution of D is χ^2 with $n - p$ degrees of freedom, where n is the number of observations, and p is the number of model parameters. The fit of a model is poor when $D > \chi^2_{\alpha,n-p}$, where α is the level of significance.

Outliers in the data may be identified by computing *deviance residuals*. A deviance residual is defined as:

$$r_{D_i} = \text{sign}(y_i - \hat{y}_i)\sqrt{d_i},$$

where $D = \sum d_i$. So, d_i represents the contribution of the ith observation to the deviance, i.e.,

$$d_i = 2(l(y_i, y_i) - l(y_i, \hat{y}_i)).$$

The sign is determined by the raw residual. Deviance residuals are approximately normally distributed. At a significance level of 5%, deviance residuals outside this range, $|d_i| > 1.96$, are suspected outliers.[10] Unfortunately, it is common for the estimated variance of a Poisson regression model to be larger than the expected variance, $\mathrm{Var}(y) = \mu$. Thus, some of the outliers identified by the deviance residuals may in fact not be outliers. To model this situation, the variance may be modeled as $\mathrm{Var}(y) = \phi\mu$ where ϕ is a constant *dispersion* factor. ϕ may be estimated by the model deviance divided by the degrees of freedom. To compensate for over-dispersion, Mayer and Sykes suggest dividing the deviance residuals by $\sqrt{\phi}$ before identifying suspected outliers.[10] This should lead to a more realistic set of outliers and hopefully, a better fitting model.

Deviance is also useful in identifying which combination of independent variables best estimates the dependent variable. We chose to fit all possible combinations of the independent variables and selected the model that had the smallest deviance. This model was then refined by identifying outliers as described earlier. Note that this method is only practical when the number of independent variables is small.[9]

2.2. *Discriminant modeling*

Discriminant analysis is a statistical technique concerned with the optimum assignment of observations to two or more distinct groups based upon one or more quantitative measurements. These measurements are assumed to differ from group to group. Given a set of observations with known group memberships, i.e., a *fit data set*, the methodology develops an assignment rule such that the chance of misclassification is minimized. The resulting model may be used to classify future observations based on the observed quantitative measures.

In this chapter, we apply discriminant analysis to build a model that classifies program modules as either fault-prone or not fault-prone. We fit a two group discriminant model in which the observations are program files. The quantitative measurements upon which classification is based (independent variables) are principal components derived from a set of software product measurements extracted directly from the source code.

The classification (dependent) variable, *Fault*, is a measure of the number of errors that will be detected at the end of a specific development phase. The modules are divided into two groups based on a cutoff value. Modules exceeding the cutoff point are assigned to the fault-prone group. Those less than or equal to the cutoff are assigned to the not fault-prone group. The cutoff value clearly determines the size of each group and varies from environment to environment. Typically, the cutoff value determination is based on the past history of projects developed in a similar environment.

The results of this classification are then compared with those derived from the Poisson classification model. For more details on stepwise discriminant analysis and model selection, see Seber.[11]

3. Experimental Methods

3.1. *Data collection*

We applied the methodologies described here to several different projects and achieved similar results. Our aim is to focus the discussion on the merits of the Poisson regression methodology. Therefore, we chose to illustrate the method by reporting the results for one project only.

MLR and Poisson regression models were developed using data collected from the Command and Control Communications System, *CCCS*, a large military telecommunications system written in Ada. A set of software measures was collected from the program source files, or *modules*. In addition, the number of faults, *Fault*, was collected from problem tracking reports generated during the system integration phase, test phase, and first year of operation.

Fault served as our dependent variable. In addition, *Fault* was used to classify the modules into two groups, or *classes*. For this environment, modules with more than four faults are defined to be high-risk modules. Conversely, modules with less than five faults were defined to be low risk. To ensure a fair comparison of the two modeling methodologies, outliers were identified and removed. This allowed each model to perform as well as possible, given the available data.

The study was comprised of 282 program modules. For each module, 14 software measures were collected. Since we do not have access to the source code, we are limited to the measures collected for us. Table 1 lists the eight software measures that served as our independent variables.

Note that the software complexity metrics we selected are not special. Other measures, if available, could be used as well. The goal of this chapter is to investigate Poisson regression modeling and not to justify a specific subset of software measures. For details regarding the selection and validation of software metrics, see Refs. 13 and 14. Furthermore, we wished to compare our results with another classification methodology.[4] In that paper, a statistical classification technique called discriminant modeling was used on the same data set.

3.2. *Evaluating predictive and classification quality*

We applied the technique of data splitting to evaluate the predictive and classification quality of our models since a data set from a similar

Table 1. Software product measures for *CCCS*.

Measure	Description
η_1	Number of unique operators[5]
η_2	Number of unique operands[5]
N_1	Total number of operators[5]
N_2	Total number of operands[5]
LOC	Lines of code
XQT	Number of executable statements
$V_1(G)$	McCabe's cyclomatic number[12]
$V_2(G)$	Extended cyclomatic number, $V_1(G) +$ the number of logical operators

project was not available. The data set was split randomly into a fitting data set and a testing data set. Two-thirds of the observations (188) were assigned to the fitting data set, $CCCS_{\text{Fit}}$. The remaining one-third (94) of the observations were assigned to the testing data set, $CCCS_{\text{Test}}$. The fitting data set was used to develop the models while the testing data set was used to evaluate the predictive and classification quality. Therefore, the testing data set simulated the application of the models to a similar project with unknown results.

By definition, all the modules must be classified into one of the two groups. In a classification study using neural networks, Khosgoftaar *et al.* defined low risk modules as those having zero faults while high-risk modules were defined as having five or more faults.[4] By removing those modules with one to four faults, the fitting and testing data sets were biased. This led to an understatement of the misclassification rates. In this chapter, we used all the data and did not bias the fitting and testing data sets. Table 2 provides a set of descriptive statistics for *Fault*.

There are many ways to quantify the predictive quality of a model. In this chapter, we chose to compute the model's average relative error based on the distribution of *Fault*.[15] Let n be the number of observations and y_i be the desired output where $1 \leq i \leq n$. The corresponding estimated value is $\hat{y}_i$. The average relative error, ARE, is defined to be

$$\text{ARE} = \frac{1}{n}\sum_{i=1}^{n}\left|\frac{y_i - \hat{y}_i}{y_i + 1}\right|.$$

Table 2. Descriptive statistics for *Fault*.

System name	Number of Obs.	Average	Std dev	Quantiles (%) 0	25	50	75	100
$CCCS_{\text{Fit}}$	188	2.27	4.65	0	0	0	2	29
$CCCS_{\text{Test}}$	94	2.56	5.88	0	0	0	2	42

Since y_i may be zero, we add one unit to y_i when computing ARE.[16] Lower values of ARE indicate better predictive quality. We used the data from $CCCS_{\text{Test}}$ to compute ARE.

To evaluate the classification quality of the Poisson regression model, we measured the misclassification rate and the uncertainty of those successfully classified. Classification errors are divided into two classes: Type 1 and Type 2. A Type 1 misclassification error occurs when a low-risk module is classified as high-risk. Such errors lead to wasted time by unnecessarily focusing development resources on low-risk modules. A Type 2 misclassification happens when a high-risk module is classified as low risk. Type 2 errors can lead to quality problems and slipped schedules by ignoring modules that are truly high-risk. This suggests that the cost of a Type 2 error is somewhat higher than a Type 1 error.

When classifying modules, the Poisson regression model will place a given module into one of two classes based upon its probability of membership exceeding a cutoff value. Remember that the probability of membership is a function of the independent variables. In our case, they are the principal components derived from the observed software measures. A module with a probability of five or more faults will be assigned to the high-risk group when that probability is greater than the cutoff value. Otherwise, the module will be assigned to the low-risk group. For some modules, the membership probability will be much greater that the cutoff indicating a high probability of correct classification. Conversely, some modules will have probabilities close to the cutoff indicating a lower probability of correct assignment. For those modules correctly classified, the model probability of membership to the opposite class is a measure of the *uncertainty* of the classification. In this chapter, we used 0.5 for the cutoff value.

3.3. *Deriving principal components*

We extracted two significant principal components from the observed measures. They accounted for about 94% of the explained variance.

Table 3. Rotated component pattern for system $CCCS_{\text{Fit}}$.

	Principal component	
Metric	PC_1	PC_2
η_1	**0.8413**	0.2615
N_2	**0.8315**	0.5072
N_1	**0.8267**	0.5281
LOC	**0.8118**	0.5108
η_2	**0.7855**	0.5569
XQT	**0.7522**	0.6408
$V_1(G)$	0.3956	**0.9109**
$V_2(G)$	0.4512	**0.8849**
Eigenvalues	4.2844	3.1989
% Variance	53.5550	39.9863
Cumulative % Variance	53.56	93.54

This is a reduction from the original eight software measures. Table 3 gives the loading pattern of the two principal components found. PC_1 loads strongly on η_1, N_2, N_1, LOC, η_2, and XQT. These measures are related to program size. PC_2 loads strongly on $V_1(G)$ and $V_2(G)$. These measures are derived from the program control flow graph.

Component PC_1 accounted for the largest portion of the variability in the software measures at about 54%. Component PC_2 accounted for the remaining 40%. Table 4 shows the standardized transformation matrix, **T**. This matrix is used to convert the standardized complexity metrics from the testing data set, **z**, into principal components, $\mathbf{PC} = \mathbf{zT}$. **PC** is used to develop the regression models.

4. The MLR Model *Fault*$_{\text{mreg}}$

We developed an MLR model to predict *Fault*. Before fitting the model, we used the standardized transformation matrix and the

Table 4. Standardized transformation matrix for system $CCCS_{\text{Fit}}$.

Metric	Principal component	
	PC_1	PC_2
η_1	0.5109	−0.4248
N_2	0.2885	−0.1275
N_1	0.2661	−0.0988
LOC	0.2680	−0.1061
η_2	0.2047	−0.0289
XQT	0.1025	0.0987
$V_1(G)$	−0.4456	0.7266
$V_2(G)$	−0.3742	0.6477

vectors of standardized software measures to derive principal components for the 188 modules in $CCCS_{\text{Fit}}$. We identified and removed 14 outliers from the data set. This left us with 174 observations to fit the model. Predictive quality was evaluated using the 94 program modules from $CCCS_{\text{Test}}$.

After selecting the model, both principal components were significant at 5%. The model was found to be significant at less than 0.01% and had a coefficient of determination, R^2, of 0.53. The regression model based on the principal components is given as:

$$Fault_{\text{mreg}} = 1.4558 + 1.6947PC_1 + 0.4945PC_2\,.$$

Table 5 summarizes the predictive quality of the model.

Table 5. Predicitive quality for system $CCCS_{\text{Test}}$.

Model	$\left\lvert \frac{y_i - \hat{y}_i}{y_i + 1} \right\rvert$			
	Average	Std dev	Min	Max
$Fault_{\text{mreg}}$	0.57	0.59	0.02	3.21
$Fault_{\text{preg}}$	0.57	0.46	0.03	3.17

5. The Poisson Regression Model $Fault_{\text{preg}}$

As described in the MLR case, we used the standardized transformation matrix and the vectors of standardized software measures to derive principal components for the 188 modules in $CCCS_{\text{Fit}}$. We identified and removed 16 outliers from the data set as described in Sec. 2.1 leaving 172 observations to fit the model. Eight outliers were common to the MLR model.

Next, we ensured that Poisson modeling was appropriate for the system under study. In this case, the dependent variable (*Fault*) in the fit data set should have a Poisson distribution. Furthermore, the distribution of the dependent variable should not be heavily skewed to zero (zero-inflated).[17] When these assumptions are violated, alternative modeling methods are indicated. For example, when the dependent variable is zero-inflated, a zero-inflated Poisson regression model should be considered.[18] We have seen too many software quality case studies with zero-inflated data sets. To our knowledge, this study is the first to test software quality data for zero-inflation.

Consider the distribution of the dependent variable in the fit data set as shown in Fig. 1. A hypothesis test that the fit data set has a Poisson distribution could not be rejected at 5%.[10] Furthermore, a score test for a zero-inflated Poisson regression model[18] was rejected at 1%.[17] Therefore, Poisson modeling was an appropriate methodology for this study. Predictive quality was evaluated using the 94 program modules from $CCCS_{\text{Test}}$.

After selecting the model, both principal components were significant at 5%. The model quality of fit was good at the 5% significance level. The scale parameter, ϕ, was 1.13 which indicates a modest amount of over dispersion. A scale value of 1 would indicate no dispersion. The Poisson regression model based on the principal components is given as:

$$Fault_{\text{preg}} = e^{-0.1939+1.1248PC_1-0.2277PC_2}.$$

It is interesting to note that $Fault_{\text{preg}}$ increases as PC_1 increases. On the other hand, $Fault_{\text{preg}}$ decreases as PC_2 increases. Table 5

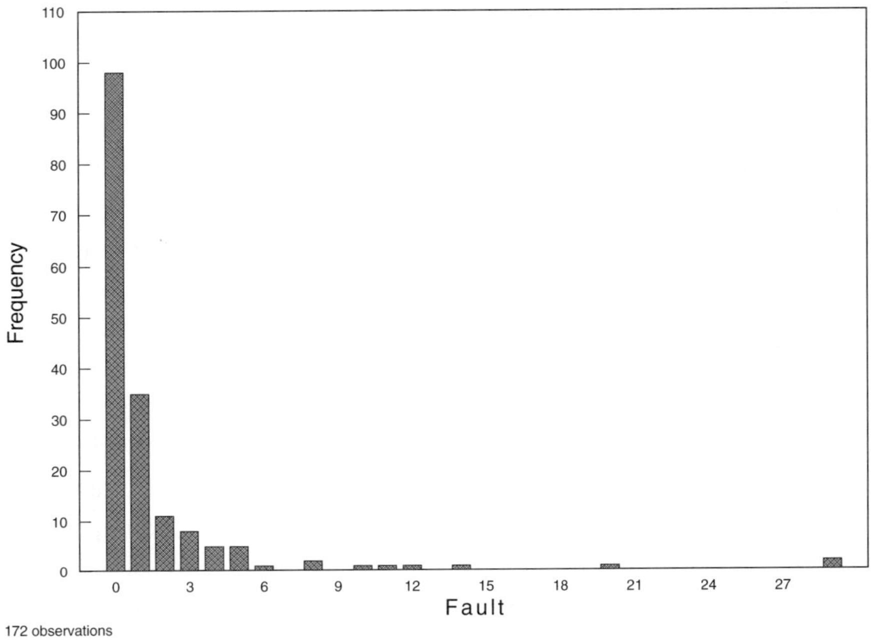

Fig. 1. Distribution of *Fault*.

summarizes the predictive quality of the model. This table shows that as measured by ARE, there is no practical difference between the MLR and Poisson regression model in this environment. However, the variance of the Poisson regression model was smaller than the MLR model indicating its predictions were slightly more stable. In fact, similar results have been found with other data sets we analyzed.

Table 6 gives the classification results of the Poisson regression model and Table 7 shows the classification results of the discriminant model. For each module in $CCCS_{\text{Test}}$, the tables give the principal component values, the actual number of faults, the predicted classification, the probabilities of class membership, and the uncertainty of correct classifications. Together, these tables show that with a cutoff value of 0.5, the Poisson regression model misclassified none of the low-risk modules while there were seven misclassifications of high-risk modules. This gives a Type 1 error rate of 0% and a Type 2 error

Table 6. Poisson regression classification data.

Module number	PC_1	PC_2	Faults	Poisson regression model: Pred. Class	Prob. Class 1	Prob. Class 2	Uncert.
1	−0.76	0.14	0	1	1.00	0.00	0.00
2	−0.80	−0.03	0	1	1.00	0.00	0.00
3	−0.24	−0.42	0	1	1.00	0.00	0.00
4	−0.84	−0.01	0	1	1.00	0.00	0.00
5	−0.79	−0.04	0	1	1.00	0.00	0.00
6	−0.90	0.04	0	1	1.00	0.00	0.00
7	−0.58	−0.17	0	1	1.00	0.00	0.00
8	−0.54	−0.10	0	1	1.00	0.00	0.00
9	1.13	−0.32	0	1	0.79	0.21	0.21
10	−0.38	−0.27	0	1	1.00	0.00	0.00
11	−0.16	−0.48	0	1	1.00	0.00	0.00
12	0.33	−0.33	0	1	0.99	0.01	0.01
13	−0.79	−0.04	0	1	1.00	0.00	0.00
14	−0.22	−0.30	0	1	1.00	0.00	0.00
15	−0.65	−0.13	0	1	1.00	0.00	0.00
16	0.16	−0.48	0	1	0.99	0.01	0.01
17	−0.72	−0.09	0	1	1.00	0.00	0.00
18	−0.55	−0.23	0	1	1.00	0.00	0.00
19	−0.43	−0.26	0	1	1.00	0.00	0.00
20	−0.79	−0.04	0	1	1.00	0.00	0.00
21	0.01	−0.07	0	1	1.00	0.00	0.00
22	−0.46	−0.21	0	1	1.00	0.00	0.00
23	−0.31	−0.32	0	1	1.00	0.00	0.00
24	0.53	−0.87	0	1	0.96	0.04	0.04
25	−0.38	−0.27	0	1	1.00	0.00	0.00
26	−0.63	−0.10	0	1	1.00	0.00	0.00
27	−0.10	−0.41	0	1	1.00	0.00	0.00
28	−0.37	−0.29	0	1	1.00	0.00	0.00
29	−0.55	−0.19	0	1	1.00	0.00	0.00
30	−0.18	−0.39	0	1	1.00	0.00	0.00
31	−0.91	0.03	0	1	1.00	0.00	0.00
32	−0.90	0.23	0	1	1.00	0.00	0.00

Table 6. (*Continued*)

Module number	PC_1	PC_2	Faults	Poisson regression model: Pred. Class	Prob. Class 1	Prob. Class 2	Uncert.
33	−0.89	0.02	0	1	1.00	0.00	0.00
34	0.69	−0.41	0	1	0.95	0.05	0.05
35	−0.02	−0.26	0	1	1.00	0.00	0.00
36	−0.89	0.00	0	1	1.00	0.00	0.00
37	−0.92	0.03	0	1	1.00	0.00	0.00
38	−0.21	−0.37	0	1	1.00	0.00	0.00
39	0.70	−0.30	0	1	0.95	0.05	0.05
40	−0.13	−0.41	0	1	1.00	0.00	0.00
41	−0.21	0.13	0	1	1.00	0.00	0.00
42	−0.36	−0.04	0	1	1.00	0.00	0.00
43	−0.35	0.25	0	1	1.00	0.00	0.00
44	0.51	−0.65	0	1	0.97	0.03	0.03
45	−0.80	−0.00	0	1	1.00	0.00	0.00
46	−0.65	0.02	0	1	1.00	0.00	0.00
47	−0.82	−0.04	0	1	1.00	0.00	0.00
48	−0.37	−0.27	0	1	1.00	0.00	0.00
49	−0.28	−0.13	1	1	1.00	0.00	0.00
50	−0.05	−0.34	1	1	1.00	0.00	0.00
51	−0.75	−0.09	1	1	1.00	0.00	0.00
52	−0.24	−0.02	1	1	1.00	0.00	0.00
53	−0.69	−0.12	1	1	1.00	0.00	0.00
54	−0.38	−0.26	1	1	1.00	0.00	0.00
55	0.14	−0.49	1	1	0.99	0.01	0.01
56	−0.44	−0.07	1	1	1.00	0.00	0.00
57	−0.42	−0.25	1	1	1.00	0.00	0.00
58	−0.38	−0.25	1	1	1.00	0.00	0.00
59	−1.06	0.12	1	1	1.00	0.00	0.00
60	−0.14	−0.32	1	1	1.00	0.00	0.00
61	−0.94	0.03	1	1	1.00	0.00	0.00
62	−0.94	0.03	1	1	1.00	0.00	0.00
63	−0.89	0.13	1	1	1.00	0.00	0.00

Table 6. (*Continued*)

Module number	PC_1	PC_2	Faults	Poisson regression model: Pred. Class	Prob. Class 1	Prob. Class 2	Uncert.
64	−0.12	−0.40	1	1	1.00	0.00	0.00
65	−0.81	1.34	1	1	1.00	0.00	0.00
66	0.68	−0.46	1	1	0.95	0.05	0.05
67	−0.45	1.14	2	1	1.00	0.00	0.00
68	0.70	0.39	2	1	0.97	0.03	0.03
69	0.77	0.09	2	1	0.95	0.05	0.05
70	1.23	−0.09	2	1	0.75	0.25	0.25
71	−0.03	0.66	2	1	1.00	0.00	0.00
72	0.35	−0.27	2	1	0.99	0.01	0.01
73	0.57	−0.01	3	1	0.98	0.02	0.02
74	−0.04	−0.30	3	1	1.00	0.00	0.00
75	−0.87	0.04	3	1	1.00	0.00	0.00
76	0.41	−0.26	4	1	0.99	0.01	0.01
77	0.47	−0.23	4	1	0.98	0.02	0.02
78	0.11	0.61	4	1	1.00	0.00	0.00
79	1.22	−0.21	4	1	0.74	0.26	0.26
80	−0.14	−0.37	5	1	1.00	0.00	—
81	2.10	0.48	5	2	0.11	0.89	0.11
82	0.46	0.35	5	1	0.99	0.01	—
83	0.07	−0.17	6	1	1.00	0.00	—
84	2.38	−0.75	6	2	0.00	1.00	0.00
85	1.49	−0.53	7	2	0.44	0.56	0.44
86	2.74	0.28	8	2	0.00	1.00	0.00
87	0.23	−0.35	9	1	0.99	0.01	—
88	0.93	0.19	10	1	0.92	0.08	—
89	1.14	−0.51	12	1	0.75	0.25	—
90	1.59	−0.14	12	2	0.43	0.57	0.43
91	1.20	−0.25	15	1	0.75	0.25	—
92	1.86	0.72	19	2	0.33	0.67	0.33
93	2.38	0.56	25	2	0.02	0.98	0.02
94	4.35	0.87	42	2	0.00	1.00	0.00

Table 7. Discriminant model classification data.

Module number	PC_1	PC_2	Faults	Discriminant model: Pred. Class	Prob. Class 1	Prob. Class 2	Uncert.
1	−0.76	0.14	0	1	0.99	0.01	0.01
2	−0.80	−0.03	0	1	0.99	0.01	0.01
3	−0.24	−0.42	0	1	0.96	0.04	0.04
4	−0.84	−0.01	0	1	0.99	0.01	0.01
5	−0.79	−0.04	0	1	0.99	0.01	0.01
6	−0.90	0.04	0	1	0.99	0.01	0.01
7	−0.58	−0.17	0	1	0.98	0.02	0.02
8	−0.54	−0.10	0	1	0.98	0.02	0.02
9	1.13	−0.32	0	2	0.22	0.78	—
10	−0.38	−0.27	0	1	0.97	0.03	0.03
11	−0.16	−0.48	0	1	0.95	0.05	0.05
12	0.33	−0.33	0	1	0.78	0.22	0.22
13	−0.79	−0.04	0	1	0.99	0.01	0.01
14	−0.22	−0.30	0	1	0.95	0.05	0.05
15	−0.65	−0.13	0	1	0.98	0.02	0.02
16	0.16	−0.48	0	1	0.86	0.14	0.14
17	−0.72	−0.09	0	1	0.99	0.01	0.01
18	−0.55	−0.23	0	1	0.98	0.02	0.02
19	−0.43	−0.26	0	1	0.97	0.03	0.03
20	−0.79	−0.04	0	1	0.99	0.01	0.01
21	0.01	−0.07	0	1	0.89	0.11	0.11
22	−0.46	−0.21	0	1	0.97	0.03	0.03
23	−0.31	−0.32	0	1	0.96	0.04	0.04
24	0.53	−0.87	0	1	0.73	0.27	0.27
25	−0.38	−0.27	0	1	0.97	0.03	0.03
26	−0.63	−0.10	0	1	0.98	0.02	0.02
27	−0.10	−0.41	0	1	0.93	0.07	0.07
28	−0.37	−0.29	0	1	0.97	0.03	0.03
29	−0.55	−0.19	0	1	0.98	0.02	0.02
30	−0.18	−0.39	0	1	0.95	0.05	0.05
31	−0.91	0.03	0	1	0.99	0.01	0.01

Table 7. (*Continued*)

				Discriminant model			
Module number	PC_1	PC_2	Faults	Pred. Class	Prob. Class 1	Prob. Class 2	Uncert.
32	−0.90	0.23	0	1	0.99	0.01	0.01
33	−0.89	0.02	0	1	0.99	0.01	0.01
34	0.69	−0.41	0	1	0.54	0.46	0.46
35	−0.02	−0.26	0	1	0.91	0.09	0.09
36	−0.89	0.00	0	1	0.99	0.01	0.01
37	−0.92	0.03	0	1	0.99	0.01	0.01
38	−0.21	−0.37	0	1	0.95	0.05	0.05
39	0.70	−0.30	0	1	0.51	0.49	0.49
40	−0.13	−0.41	0	1	0.94	0.06	0.06
41	−0.21	0.13	0	1	0.93	0.07	0.07
42	−0.36	−0.04	0	1	0.96	0.04	0.04
43	−0.35	0.25	0	1	0.95	0.05	0.05
44	0.51	−0.65	0	1	0.71	0.29	0.29
45	−0.80	−0.00	0	1	0.99	0.01	0.01
46	−0.65	0.02	0	1	0.98	0.02	0.02
47	−0.82	−0.04	0	1	0.99	0.01	0.01
48	−0.37	−0.27	0	1	0.97	0.03	0.03
49	−0.28	−0.13	1	1	0.95	0.05	0.05
50	−0.05	−0.34	1	1	0.92	0.08	0.08
51	−0.75	−0.09	1	1	0.99	0.01	0.01
52	−0.24	−0.02	1	1	0.94	0.06	0.06
53	−0.69	−0.12	1	1	0.99	0.01	0.01
54	−0.38	−0.26	1	1	0.97	0.03	0.03
55	0.14	−0.49	1	1	0.87	0.13	0.13
56	−0.44	−0.07	1	1	0.97	0.03	0.03
57	−0.42	−0.25	1	1	0.97	0.03	0.03
58	−0.38	−0.25	1	1	0.97	0.03	0.03
59	−1.06	0.12	1	1	0.99	0.01	0.01
60	−0.14	−0.32	1	1	0.94	0.06	0.06
61	−0.94	0.03	1	1	0.99	0.01	0.01
62	−0.94	0.03	1	1	0.99	0.01	0.01

Table 7. (*Continued*)

				Discriminant model			
Module number	PC_1	PC_2	Faults	Pred. Class	Prob. Class 1	Prob. Class 2	Uncert.
63	−0.89	0.13	1	1	0.99	0.01	0.01
64	−0.12	−0.40	1	1	0.94	0.06	0.06
65	−0.81	1.34	1	1	0.97	0.03	0.03
66	0.68	−0.46	1	1	0.56	0.44	0.44
67	−0.45	1.14	2	1	0.94	0.07	0.07
68	0.70	0.39	2	2	0.40	0.60	—
69	0.77	0.09	2	2	0.39	0.61	—
70	1.23	−0.09	2	2	0.15	0.85	—
71	−0.03	0.66	2	1	0.84	0.16	0.16
72	0.35	−0.27	2	1	0.75	0.25	0.25
73	0.57	−0.01	3	1	0.57	0.43	0.43
74	−0.04	−0.30	3	1	0.91	0.09	0.09
75	−0.87	0.04	3	1	0.99	0.01	0.01
76	0.41	−0.26	4	1	0.72	0.28	0.28
77	0.47	−0.23	4	1	0.67	0.33	0.33
78	0.11	0.61	4	1	0.78	0.22	0.22
79	1.22	−0.21	4	2	0.17	0.83	—
80	−0.14	−0.37	5	1	0.94	0.06	—
81	2.10	0.48	5	2	0.01	0.99	0.01
82	0.46	0.35	5	1	0.59	0.41	—
83	0.07	−0.17	6	1	0.87	0.13	—
84	2.38	−0.75	6	2	0.01	0.99	0.01
85	1.49	−0.53	7	2	0.09	0.91	0.09
86	2.74	0.28	8	2	0.00	1.00	0.00
87	0.23	−0.35	9	1	0.83	0.17	—
88	0.93	0.19	10	2	0.27	0.73	0.27
89	1.14	−0.51	12	2	0.24	0.76	0.24
90	1.59	−0.14	12	2	0.06	0.94	0.06
91	1.20	−0.25	15	2	0.18	0.82	0.18
92	1.86	0.72	19	2	0.01	0.99	0.01
93	2.38	0.56	25	2	0.00	1.00	0.00
94	4.35	0.87	42	2	0.00	1.00	0.00

rate of 46.67%. The overall error rate was 7.45%. The average uncertainty of the low-risk modules is 1.41% and 16.63% for the high-risk modules. Overall, the average uncertainty is 2.81%.

By comparison, the discriminant model committed five Type 1 and four Type 2 errors. This yields a Type 1 error rate of 6.33% and a Type 2 error rate of 26.67%. The overall misclassification rate is 9.57%. The average uncertainty of the low-risk and high-risk modules is 8.16% and 7.95%, respectively. Overall average uncertainty was 8.13%. Table 8 summarizes the classification performance of the Poisson regression and discriminant model.

It is useful to consider the cost of misclassification when comparing two models.[4] Let C_1 and C_2 be the costs of Type 1 and Type 2 errors, respectively. By disregarding the uncertainty effects, the cost of the Poisson regression model is $M_p = 7C_2$ and the cost of the discriminant model is $M_d = 5C_1 + 4C_2$. If we equate M_p to M_d and solve for C_2 we see that the models have equal cost when $C_2 = (5/3)C_1$. So, when the ratio C_2/C_1 is less than $5/3$, the misclassification cost of the Poisson regression model is lower. When the ratio exceeds $5/3$, the misclassification cost of the discriminant model is lower. This could serve as a management guide to help determine which classification method should be used on a project given some understanding of the actual costs involved.

Table 8. Classification performance for system $CCCS_{\text{Test}}$.

	Poisson regression			Discriminant model		
Error type	Count	Rate (%)	Average uncertainty (%)	Count	Rate (%)	Average uncertainty (%)
Type 1	0	0.00	1.41	5	6.33	8.16
Type 2	7	46.67	16.63	4	26.67	7.95
Total	7	7.45	2.81	9	9.57	8.13

6. Conclusions

From the practitioner's point of view, this study shows that the predictive quality of a Poisson regression fault model versus an MLR fault model are similar, even though the observed data match the Poisson model assumptions better than the MLR model assumptions. Note that these results are specific to this system only. We cannot generalize beyond this environment without additional empirical studies. Undoubtedly, there will be systems where the Poisson model clearly outperforms an MLR model.

Statistical models always have an error term. As such, the modeling errors may result from a variety of reasons. Some of the potential error sources are as follows:

- sampling error
- measurement error
- unmeasured factors
- violation of model assumptions

Thus, the violation of the normality assumption by an MLR model is but one component of the total error. This error may be acceptable as shown by this and other studies we have performed.

Estimating the parameters for an MLR model is computationally straightforward in comparison to the Poisson regression model's maximum likelihood estimators which require solving a system of nonlinear equations. Furthermore, consider that personal computer spreadsheet programs have MLR analysis capabilities built in while sophisticated, and expensive, mathematical packages are typically required to solve the nonlinear equations needed for the Poisson regression. Given the similarity of the predicted results, the simpler solution would be preferred.

However, the Poisson regression model does have one clear advantage over the MLR model: its ability to classify data by estimating the probability $P(y = i)$ $(i = 0, 1, 2, \ldots)$. As this study shows, the classification quality of the Poisson regression model is comparable to that achieved by discriminant modeling. Unlike the discriminant model, the Poisson model is able to predict values in addition to classifying them by groups. In cases where it is desired to have both predictions and classifications available for use, the Poisson regression model is ideal since it has the ability to do both. For example, given two modules classified as high-risk, how do you decide to assign limited development resources to these two apparently equivalent modules? If we can predict that one of the high-risk modules will have 7 faults and the other 12, it is easier to determine how to maximize the return of the development resources. Therefore, the increased level of functionality offered by the Poisson model should help offset the added difficulty in implementing Poisson regression.

In the future, we plan to extend this work by applying a generalized classification rule to the Poisson regression model.[19] By varying the parameter of the generalized classification rule, we can balance the misclassification rates of the model and provide an even more useful and practical tool for software management.

Acknowledgments

We acknowledge Kehan Gao for reviewing this chapter and appreciate the numerous discussions with Edward Allen, and David Lanning. We also acknowledge the useful comments made by anonymous referees. These discussions and comments contributed significantly to the quality of this chapter.

References

1. S. Henry and S. Wake, Predicting maintainability with software quality metrics, *Journal of Software Maintenance: Research and Practice* **3** (1991) 129–143.
2. T. M. Khoshgoftaar and R. M. Szabo, Improving code churn predictions during the system test and maintenance phases, *IEEE International Conference on Software Maintenance '94*, Victoria, British Columbia, Canada, September 1994, pp. 58–67.

3. T. M. Khoshgoftaar, A. S. Pandya and H. B. More, A neural network approach for predicting software development faults, *Third IEEE International Symposium on Software Reliability Engineering*, Research Triangle Park, NC, October 1992, pp. 83–89.
4. T. M. Khoshgoftaar, D. L. Lanning and A. S. Pandya, A comparative study of pattern recognition techniques for quality evaluation of telecommunications software, *IEEE Journal of Selected Areas in Communications* **12** (1994) 279–291.
5. M. H. Halstead, *Elements of Software Science* (Elsevier North-Holland, New York, 1977).
6. W. R. Dillon and M. Goldstein, *Multivariate Analysis* (John Wiley and Sons, New York, 1984).
7. T. M. Khoshgoftaar and J. C. Munson, Predicting software development errors using software complexity metrics, *IEEE Journal of Selected Areas in Communications* **8** (1990) 253–261.
8. T. M. Khoshgoftaar and R. M. Szabo, Predicting software quality, during testing, using neural network models: A comparative study, *International Journal of Reliability, Quality, and Safety Engineering* **1** (1994) 303–319.
9. R. H. Myers, *Classical and Modern Regression with Applications* (Duxbury Press, Boston, MA, 1990).
10. A. Mayer and A. Sykes, A probability model for analyzing complexity metrics data, *Software Engineering Journal* **26** (1989) 254–258.
11. G. A. F. Seber, *Multivariate Observations* (John Wiley and Sons, New York, 1984).
12. T. J. McCabe, A complexity metric, *IEEE Transactions on Software Engineering* **SE-2** (1976) 308–320.
13. N. E. Fenton, *Software Metrics: A Rigorous Approach* (Chapman & Hall, London, 1992).
14. N. F. Schneidewind, Methodology for validating software metrics, *IEEE Transactions on Software Engineering* **18** (1992) 410–421.
15. V. Y. Shen, T. Yu, S. M. Thebaut and L. R. Paulsen, Identifying error-prone software — An empirical study, *IEEE Transactions on Software Engineering* **SE-11** (1985) 317–324.
16. T. M. Khoshgoftaar, J. C. Munson, B. B. Bhattacharya and G. D. Richardson, Predictive modeling techniques of software quality from software measures, *IEEE Transactions on Software Engineering* **18** (1992) 979–987.
17. J. van den Broek, A score test for zero inflation in a poisson distribution, *Biometrics* **51** (1995) 738–743.
18. D. Lambert, Zero-inflated poisson regression, with an application to defects in manufacturing, *Technometrics* **34** (1992) 1–14.
19. T. M. Khoshgoftaar and E. B. Allen, A practical classification-rule for software-quality models, *IEEE Transactions on Reliability* **49** (2000) 209–216.

CHAPTER 8

Measurement of Object-Oriented Software Understandability Using Spatial Complexity*

Jitender Kumar Chhabra

Department of Computer Engineering,
National Institute of Technology (formerly R.E.C.),
Kurukshetra 136119, India
jitenderchhabra@rediffmail.com

K. K. Aggarwal

GGS Indraprastha University, Delhi 110006, India
aggarwal_krishan@hotmail.com

Yogesh Singh

School of Information Technology, GGS Indraprastha University,
Delhi 110006, India
ys66@rediffmail.com

*The concept of object-oriented spatial complexity was accepted as a paper in 9th ISSAT International Conference, Honolulu, USA, 2003 and its revised form has been communicated to Information and Software Technology Journal.

1. Need of Measurement

A critical distinction between software engineering and other, more well-established branches of engineering is the shortage of well-accepted measures, or metrics, of software development. Without metrics, the tasks of planning and controlling software development and maintenance will remain stagnant in a craft-type mode, wherein greater skill is acquired only through greater experience, and such experience cannot be easily communicated to the next system for study, adoption, and further improvement. With metrics, software projects can be quantitatively described, and the methods and tools used on the projects to improve productivity and quality can be evaluated.[1] In order to control, manage, and maintain software, the software complexity needs to be measured. If you cannot measure it, you cannot control it.[2]

2. Concept of Complexity and Understandability

There are many aspects of the software complexity. Some of them contribute towards the design and algorithmic complexity, some contribute towards readability and understandability of the software, and some other aspects have an influence on the debugging and testability of the software. No single metric of complexity is adequate to indicate all of these aspects of the software.[3] For example, McCabe's cyclomatic complexity is a measure of control flow complexity,[4] Halstead's science metrics concentrate on size of the software, average number of live variables per statement and program weakness indicate the design complexity.[5–7] But these types of metrics do not indicate the complexity related to understandability, and readability of the software. Understandability of the software is very important from maintenance point of view. The more understandable the source code is, the more quickly and accurately a programmer can obtain critical information about a program by reading the code. Increased understanding also leads to better management of software projects. Some measures of understandability of software-documents have been proposed,[8,9] but these are not directly applicable to source code, as the understandability of the source code demands the knowledge of the corresponding programming language also. This type of complexity is related to psychological complexity. A program that is psychologically complex is difficult to understand.

3. Spatial Complexity of Object-Oriented Software

The theory of working memory is very useful to measure psychological complexity and directly affects the understandability of source code.[10] Spatial measures for object-oriented software, proposed in this chapter are based on this theory of working memory. Spatial ability is a term that is used to refer to an individual's cognitive abilities relating to orientation, the location of objects in space, and the processing of location related visual information. Spatial ability has been correlated with the selection of problem solving strategy, and has played an important role in the formulation of an influential model of working memory.[11] In order to debug and maintain the software, programmer must understand the code, have an understanding of the application domain, and establish an appreciation of the relationships that can exist between the two.[12] Program comprehension and software maintenance are considered to substantially use programmers' spatial abilities.[11] The amount of these spatial abilities needed to understand the source code is measured with help of a complexity measure named as spatial complexity. Henceforth in this chapter, wherever the word complexity is referred, it basically denotes the spatial complexity.

The object-oriented software can be better understood, if one is able to correlate objects with their classes, attributes with their usage, and methods calls with their definitions respectively. Douce *et al.* have tried to define spatial complexity of object-oriented software by proposing two categories of measures — function related and inheritance related measure. The function related measure has been proposed as method location rating, which is a count of how close the definition of a member function is to its class declaration. The second category of measure concentrates on inheritance with help of two metrics — class relation measure and object relation measure. Class relation measure computes the distance (in LOC) of derived class from the inherited class, while object relation measure examines the usage of objects of other classes (if any) with in a class.[11] But these proposed measures are inadequate as they are not able to capture all spatial abilities needed to understand the working of object-oriented software. For example, if an object is being defined immediately after its class definition, the understanding will be easier as no searching for that class is to be done, and the details of that class are present in the working memory of the human being. On the other hand, if an object is defined and used after 1000 lines of its class definition, lot of searching/thinking has to be done, as many classes/objects details appearing in those 1000 lines will get their place in the working memory of the human mind, and recalling the details of a class read 1000 lines earlier may not be easy. Similarly if an attribute is used by class's own method very close to its class declaration, the comprehension of purpose of that attribute will be much easier than the possible use of that attribute after few hundred/thousand lines of code. Many such aspects of spatial complexity of object-oriented software need to be measured, which have not been considered at all in Ref. 11. The concept of object-oriented programming revolves around classes, objects and their interactions. Thus, the understandability of any object-oriented software requires comprehending of the definition and usages of various classes (as an encapsulation of attributes and methods) and objects. This aspect of encapsulation has also not been considered at all in Ref. 11. Douce *et al.* have tried to extend the definition of spatial complexity of procedure-oriented software to object-oriented software without considering the conceptual difference between the two. The design of object-oriented software differs a lot from procedure-oriented software because of encapsulation, polymorphism, and inheritance. But the authors have not paid any attention to these features of object-oriented software, except slightly touching the inheritance, and that also for methods only. No consideration to data members has been provided at all in any of the proposed spatial complexity metrics of object-oriented software. The use of these metrics for any types of conclusions/results has also not been pointed out.[11] The concept of code spatial and data spatial complexity proposed in Ref. 13 is also proposed for procedure-oriented software and is not directly applicable to object-oriented software as those measures do not take care of concepts like encapsulation, polymorphism etc. as mentioned by the authors themselves.[13]

4. Proposed Spatial Complexity Measures

In this chapter, we have proposed two categories of measures of spatial complexity of object-oriented software — class spatial complexity, and object spatial complexity. To the best knowledge of the authors, these measures are being proposed for the first time in literature. These proposed metrics are not just the extension of the spatial complexity metrics of procedure-oriented software, but these measure do take care of salient features of object-oriented software. The above pointed out shortcomings of the existing metrics have been removed. The understandability of the object-oriented software starts with comprehending the concept of classes as an encapsulation of data and methods. This conceptual difference of object-oriented software from procedure-oriented software has been the principle of our definition of proposed metrics. We have given equal attention to data members also, which was totally missing earlier. The proposed measures have been defined such that they automatically take care of inheritance and polymorphism also. The significance of these metrics has also been very clearly listed in this chapter. The class spatial complexity measures the spatial complexity of both parts of the classes — methods and attributes. To understand the behavior of any class, one needs to comprehend both of these entities. The method's code helps in understanding the processing logic and the attributes help in recognizing the properties of the class. The second category of proposed spatial complexity is based on the definition and usages of objects. The classes do not directly execute normally, but their instances are used in form of objects in the object-oriented software. The proposed object spatial complexity estimates the spatial abilities needed to correlate various definitions of the objects with their respective classes and various methods calls to their respective definitions. The spatial complexity of object-oriented software is integration of class-spatial and object-spatial complexity.

5. Class Spatial Complexity

The basic entity of any object-oriented software is class. While computing the class-spatial complexity, the aim is to measure the effort needed by the programmer in understanding the behavior of the class. For that the programmer needs to establish relation between attributes definition and their usage, and between methods specification and their definitions. So the class spatial complexity consists of two parts — class attribute spatial complexity and class method spatial complexity.

5.1. *Class attribute spatial complexity*

Almost all of the classes consist of some attributes, which are used by various methods of that class. The functionality of the class can be easily understood, if the programmer is able to comprehend the role of attributes. The basic aim of paradigm shift from procedure-oriented programming to object-oriented programming was to give importance to data (i.e., attributes) also. The attributes (along with the methods) are encapsulated into the class, on which the methods of the class operate. Thus, the cognitive effort needed to understand the purpose of every attribute must be measured, which is being considered in this chapter for the first time in the literature. These efforts are measured in terms of class attribute spatial complexity.

The concept behind the measurement of class attribute spatial complexity is to measure the distance between use and definition of the attributes. If an attribute is being used close to its definition, then the details about that attribute will be available in working memory of the programmer and thus he/she will be able to comprehend the purpose of the attribute. On the other hand, if an attribute has been defined in a class, but it is being used after, say, 500 lines of source-code, then most likely, the programmer must have forgotten the details about that attribute and the corresponding class because those details would have been overwritten in the working memory of the programmer by more recently defined/used attributes and classes. In that case, the programmer has to probably search for the definition of that attribute and class and then he/she has to comprehend the purpose of that attribute. This process will require more cognitive effort than the previous case, where the use of the attribute was very close to its definition. But definition of the attribute is not the sole important factor. The attribute definition does not tell anything else than its data type (and may be initial value). More details about the attribute are understood through its use in a particular sequence with in any method. Within a method, when an attribute is used for the first time, its definition/initial value may be of use, but after that if the attribute is used again with in the method, the previous use of the attribute is more important than its initial value. If an attribute is being used at some place in a method, understanding the processing being done at that place is dependent on the previous use of that attribute, instead of its original definition. So the class attribute spatial complexity of any attribute is measured using the distance between first use within the method and definition, and then between two successive uses within the same method. The greater the distance in lines of code between the successive uses of the attributes, more is the cognitive effort required to understand the purpose and data flow of that attribute. If an attribute is successively used very close to its definition or at very small intervals, the details about that attribute remain in the working memory of the programmer, and thus, he/she will be able to comprehend that use of the attribute easily. This concept of the class attribute spatial complexity very closely resembles with average span of a variable in procedure-oriented software, which has been already accepted as a good complexity measure.[6,7,14] Based on the above discussion, we define the Class Attribute Spatial Complexity of an attribute (CASC) as the average of distances of various use of that attribute from its definition/previous use.

$$\text{CASC} = \frac{\sum_{i-1}^{p} \text{Distance}_i}{p},$$

where p represents count of use of that attribute and Distance_i is equal to the absolute difference in number of lines of the current use of the attribute from its just previous use with in the same method. If an attribute is being used for the first time in that method, then the distance is defined as absolute difference (in lines of code) of the current use from the definition of the attribute. If the attribute is defined and used in the same source-code file, the distance can be calculated as above. Many a times the software is written using multiple source-code files, then an attribute may be defined in one file and used in some other file. In that case, the above definition of distance will be incomplete. When an attribute is used for the first time in a file, where it is not defined, the programmer first tries to find that class and attribute in the starting of the current file, because classes are usually declared at the start of any file. If the definition is not present in the current file, the programmer tries to find the details of that class and attribute in the other file. If he is unable to find the definition in that file also, he searches for that class and attribute in another file, and so on. In that case, understanding of such use takes more cognitive effort. The effort is dependent on the file in which the attribute is being used and all other files, in which the programmer searches for its definition. If the definition is not present in the file, where it is being used, then the programmer usually has got some idea about the possible file, in which that class and attribute may have been defined. So he/she will immediately search within that file. In that case, the effort needed get dependent on those two files. Based on our experience, we have found that in more than 90% of the cases, the definition of the attribute is found in either the file, where it is being used, or in the next file, which the programmers looks into. But in remaining cases, the programmer has to keep on searching in the other source files, till he/she does not get the definition. One alternate to measure the distance is then to consider all such cases, but then the formula becomes quite complex unnecessarily. It may be noted that the probability of not getting the definition in 3rd file, 4th file, and so on keeps on decreasing. Thus, to simplify the definition, distance for these files is considered on average basis as half of the distance of all such files. So if the attribute definition is not present in the same file of its use, then we propose the distance as:

$$\begin{aligned}\text{Distance} = {} & (\text{Distance of first use of the attribute from top of} \\ & \quad \text{current file}) \\ & + (\text{Distance of definition of the attribute from top of file} \\ & \quad \text{containing definition}) \\ & + (0.1 * (\text{total lines of code of remaining files})/2).\end{aligned}$$

Here, if the attribute is present either in the same file or in the next searched file, then the third factor does not come into picture. As already pointed out, in more than 90% of the case, this happens. But in remaining less than 10% of the cases, one may have to search in the remaining files also one by one. For those cases, we have taken the average distance of remaining files and that has been multiplied by the worst-case probability, i.e., 0.1 corresponding to remaining 10% cases.

Total Class Attribute Spatial Complexity of a class (TCASC) is defined as average of class attribute spatial complexity of all attributes (variables as well as constants) of that class.

$$\text{TCASC} = \frac{\sum_{i-1}^{q} \text{CASC}_i}{q},$$

where q is count of attributes in the class.

5.2. *Class method spatial complexity*

Every class consists of many methods. A method basically means a function/subroutine in any language containing some processing steps. The purpose and functionality of the class can be better understood, if all methods of the class are defined close to the class declaration. The greater the distance in lines of code between the definition of the methods from the declaration of the method within the corresponding class and use of the modules, more is the cognitive effort required to comprehend the connections of those methods in the class. If a method is being defined within the class declaration, the understanding will be easier as no searching for that class is to be done, and the details of that class are present in the working memory of the human being. On the other hand, if a method is defined after, say, 1000 lines of its declaration with in the class, lot of searching/thinking has to be done by the programmer, as many other class details appearing in those 1000 lines will get their place in the working memory of the human mind, and recalling the details of a class read 1000 lines earlier may not be easy. Thus, the Class Method Spatial Complexity of a method m (CMSC) is defined as distance (in terms of lines of code) between the declaration and definition of that method. The distance can be easily computed as long as the method declaration and definition belong to same file, but if source code of the software is written in multiple files and a method is declared in one file and defined in some other file, then the programmer first tries to find that class (which definitely contains the corresponding method-declaration also) in the current file, and then looks for that class's declaration in the other files, as discussed above in case of class attribute spatial complexity. Thus, understanding of such definitions takes more cognitive effort. The effort is dependent on the file in which the method is being defined and the files searched for its declaration. In that case, we define the distance for that particular definition of the method in the similar way as done above, i.e.,

$$\begin{aligned}\text{Distance} = {} & (\text{Distance of definition from top of file containing definition}) \\ & + (\text{Distance of declaration of the method from top of the file containing declaration}) \\ & + (0.1 * (\text{total lines of code of remaining files})/2).\end{aligned}$$

Total Class Method Spatial Complexity (TCMSC) of a class is defined as average of class method spatial complexity of all methods of the class.

$$\text{TCMSC} = \frac{\sum_{i=1}^{m} \text{CMSC}_i}{m},$$

where m is count of methods of the class.

As the class is an encapsulation of attributes and methods, the class spatial complexity is an integration of both types of spatial complexities, and hence the Class Spatial Complexity (CSC) of a class is proposed as:

$$\text{CSC} = \text{TCASC} + \text{TCMSC}.$$

This measure of class spatial complexity depends only on intra-properties of the class. In a way, this measure helps in measurement of the understandability and cohesiveness of the class from the point of view of cognitive abilities. This measure does not take care of the possible use of that class in the form of objects, which ultimately interact with each other for achieving the complete functionality of the object-oriented software. The spatial complexity generated because of the various objects is measured in form of object spatial complexity.

5.3. *Significance of class spatial complexity*

In order to study the effect of class spatial complexity on readability and understandability of the object-oriented software, we applied this concept on 15 different object-oriented software projects (in C++) of undergraduate and postgraduate students of computer engineering. The students projects considered for measurement were chosen to be not pure data-intensive just because of simple reason that data intensive object-oriented software consisted of very less use of computation functions, and their whole functionality revolved around insertion/deletion/retrieval of data only and did not demand much interaction among classes and objects. The length of the software considered varied from 398 lines of code (LOC) to 3257 LOC. The class spatial complexity of these 15 projects was collected for all classes present in each of these object-oriented software. Total number of classes present in all these projects was 464 as some of the projects consisted more than 80 classes. Hence it is not possible to list values of CSC of all classes here. An average value of the CSC of all these 15 projects was also computed and is being listed in Table 1 and the corresponding plot with LOC is shown in Fig. 1. From Fig. 1, it can be easily noticed that average CSC values have no correlation with LOC and thus, LOC cannot be used as a measure to predict the understandability of classes of object-oriented software, etc., but average CSC can be helpful for these purposes as discussed below. The concept of class spatial complexity defined here for the first time in the literature may be used in the following ways.

(1) The value of average class spatial complexity gives a hint about the understandability of a class. Higher value of class spatial complexity of a particular class means more cognitive effort needed to understand the purpose and functionality of the class. In order to verify this intuition, we tried to do the reverse engineering of these 15 projects up to design level and aim was to generate the class diagram of all of these projects. A class diagram shows set of classes, interfaces, collaborations and their relationship[15] and

Table 1. CSC and Rev-Engg Time of 15 projects.

S.N	Lines of Code (LOC)	No. of classes	Total CSC	Average CSC = total CSC/ No. of classes	Rev-Engg Time (in hours)	Rev-Engg Time/ Average CSC
1	398	7	387	55.29	14	0.25323
2	521	3	236	78.67	28	0.355932
3	603	6	443	73.83	28	0.379233
4	929	5	387	77.40	30	0.387597
5	994	15	765	51.00	20	0.392157
6	1085	19	1945	102.37	34	0.332134
7	1256	29	2687	92.66	40	0.431708
8	1472	39	1431	36.69	17	0.463312
9	1506	47	3044	64.77	25	0.386005
10	1811	83	4077	49.12	23	0.468236
11	2154	32	1287	40.22	18	0.447552
12	2398	59	3578	60.64	26	0.428731
13	2526	37	3056	82.59	30	0.36322
14	3055	47	4269	90.83	33	0.363317
15	3257	36	3109	86.36	30	0.347379

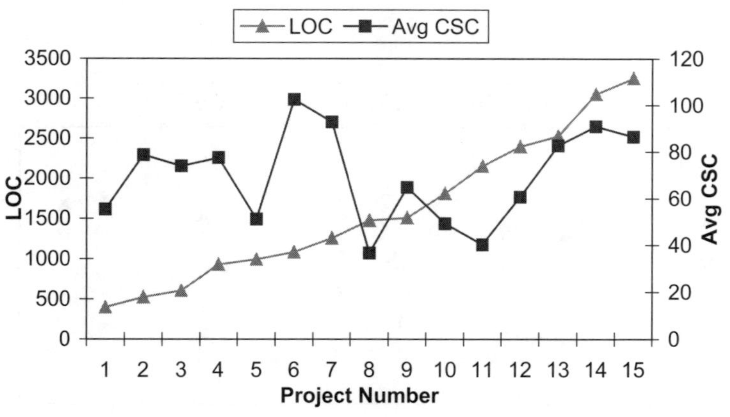

Fig. 1. Average CSC versus lines of code.

their generation from the source code required efforts to comprehend the working and semantics of every class. The time taken in generation of these class diagrams for each of the projects was noted and is shown in Table 1 itself. This time is denoted as Rev-Engg Time and is directly dependent on understandability of the classes. If more time is needed for generating class diagram of a particular project, it clearly indicates that the classes of that project require more cognitive abilities and thus are more difficult to understand. A graph of average CSC value and Rev-Engg Time of these 15 projects is plotted in Fig. 2 and their correlation has been found to be approximately 0.9, which clearly proves our intuition that the software having higher value of average CSC are likely to be more difficult to understand. It can be easily observed from Fig. 2 and Table 1 that reverse engineering time of project numbers 8 and 11 has been much lesser than all other projects (except project number 1, which is too small (1/5th of these projects)). The reverse engineering time of these two projects is much lesser than many of the projects having higher value of LOC, e.g., project numbers 12-15. On the other hand, reverse engineering time of project numbers 6 and 7 highest among all projects. If we look at the corresponding values

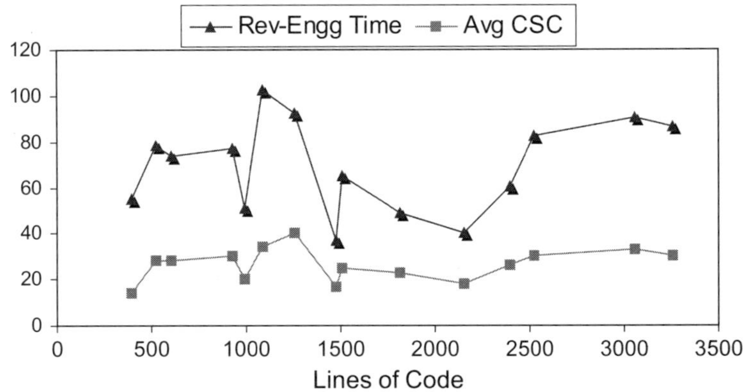

Fig. 2. Plot of Rev-Engg Time and Avg CSC.

of average CSC, it can be easily observed that the average CSC values of project numbers 8 and 11 are the lowest and that of project numbers 6 and 7 are the highest among all the projects. Thus, average CSC value can used for measuring understandability of classes.

(2) The readability of the classes can be improved by identifying those classes, which have got the higher value of class spatial complexity as compared to other classes. In that case, those classes can be modified so as to improve the readability. Table 2 shows data about those 10 classes, which have got higher value of CSC as compared to the rest of the classes. Their length in LOC is also shown in Table 2, which was counted as total number of lines of member declarations and definitions of the class.

These 10 classes belonged to different projects. When these 10 classes were carefully analyzed, it was found that almost all of these had some design defects. Many of these were having unnecessary grouping of data members and functions, and thus, clearly lacked in cohesion and were good candidates for splitting.[16,17] Because of the unnecessary grouping, many of data members in these classes were being used far from their definition by many

Table 2. Classes having highest 10 values of CSC.

Class no.	LOC	CSC	CSC/LOC
1	126	487	3.87
2	86	411	4.78
3	143	383	2.68
4	89	342	3.84
5	94	319	3.39
6	76	297	3.91
7	107	295	2.76
8	56	276	4.93
9	68	266	3.91

corresponding disjoint member functions, resulting into a sharp increase of CSC value. Redesign of these classes by proper splitting etc. was needed to improve their readability.

(3) It is quite obvious that lengthy classes will be more difficult to understand. Hence, class spatial complexity is likely to increase with increase in lines of code, because more lines of code are likely to increase the distance between usages of attributes or between definitions of the methods from their declarations. But the length of the class cannot be easily controlled to a great extent. The bigger classes will definitely have more number of lines of code, and hence, will require more effort to understand. If we want to compare understandability of two classes, the understandability of the classes needs to be measured within constraints of the length. One smaller class may have lesser understandability than a bigger class. Thus, the percentage increase in class spatial complexity with respect to class size (in lines of code) can be another important factor to find out the level of difficulty of understanding. This ratio can be used to compare the understandability of two classes of different length, keeping aside the length factor. This ratio has been computed for the 10 classes having highest values of CSC and is shown in Table 2. Most of the values for these 10 classes are more than 3. As mentioned in point 2 above, most of these classes had some design defect, and thus, possessed poor understandability. On the other hand, the smallest value of this ratio for some of the classes was of the order of 0.2–0.4. Table 3 shows the CSC values of those 10 classes, which have got the smallest values of this ratio. The corresponding ratio has also been computed and is shown in Table 3. While studying their source code, almost all of these classes were found to be very well-designed and systematically coded.

While finding these classes, classes having LOC more than 30 were considered only, as we found that classes laving LOC in the range of 10–30 were too primitive and do not really need any comparison of difficulty. The CSC value for these 10 classes are

Table 3. Classes having lowest 10 values of CSC.

Class no.	LOC	CSC	CSC/LOC
48	36	0.75	48
59	32	0.54	59
63	65	1.03	63
65	23	0.35	65
69	76	1.10	69
77	46	0.60	77
80	37	0.46	80
81	65	0.80	81
84	25	0.30	84

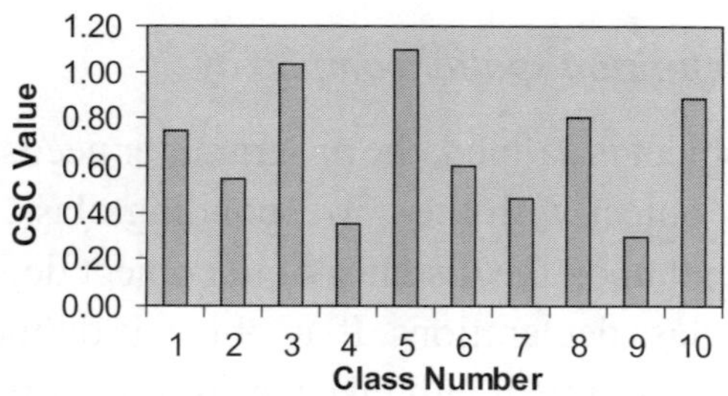

Fig. 3. 10 classes having Min CSC.

plotted in Fig. 3, which clearly shows that value of CSC is lowest for class number 9 of Table 3, and hence, is one of the very easily understandable class, although it has LOC value as 84, which is much larger than many other classes of the 15 projects.

(4) Certain guidelines can be derived about the acceptable ranges of the CSC and ratio of CSC with LOC. These ranges can be used by the object-oriented software managers to judge the understandability of the classes of the software. But this will require a lot

more empirical data and use of statistical techniques, which can be a good direction for future work.

6. Object Spatial Complexity

The object-oriented software works with help of objects and their interactions. The different methods of the class are called through objects in a specific sequence so as to obtain the proper results from the software. No researcher has tried to measure the spatial complexity of objects and their interaction. For the first time in the literature, we define the object spatial complexity to be of two types — object definition spatial complexity and object-member usage spatial complexity.

6.1. *Object definition spatial complexity*

As soon as an object is defined, the programmer needs to establish the relation of this object with the corresponding class. This cognitive effort will depend upon the distance of the object definition from the corresponding class declarations. If an object is defined immediately after its class declaration, it will take almost no effort to comprehend the purpose of the object, as the details of the corresponding class will be present in the working memory of the person. On the other hand, if an object is defined much before/after the class declaration, then more spatial abilities are needed for understanding the orientation of that object. Thus, the Object Definition Spatial Complexity of an object (ODSC) as the distances of definition of the object from the corresponding class declaration. If the object is defined in the same source-code file where the corresponding class has been declared, the distance can be calculated as above, but if the object-oriented software is written using multiple source-code files, and the object is defined in a different file than the file containing class declaration, then the effort are dependent on many files, as already discussed. In that case, distance for that particular object is defined as:

$$\begin{aligned}\text{Distance} = {} & (\text{Distance of object definition from top of current file}) \\ & + (\text{Distance of declaration of the corresponding class} \\ & \quad \text{from top of file containing class}) \\ & + (0.1 * (\text{total lines of code of remaining files})/2).\end{aligned}$$

6.2. *Object-member usage spatial complexity*

Once the objects are defined, they keep on calling various members (methods mostly, but attributes also may be referred sometimes). Whenever any member of the method is called through an object, the programmer needs to recollect details about that member of the class. For this purpose, he/she has to establish a connection between the call and corresponding definition. Higher the distance between call and usage of object-members, more are the cognitive effort needed to comprehend the processing logic.[11] If an object-member is being called immediately after its definition, the understanding will be easier as no searching for that member is to be done, and the details of that module are present in the working memory of the human being. On the other hand, if an object-member is called after a long distance from its definition, spatial abilities needed will be much more. Thus, the Object-Member Usage Spatial Complexity of a member through a particular object (OMUSC) is defined as average of distances (in terms of lines of code) between the call of that member through the object and definition of the member in the corresponding class, i.e.,

$$\text{OMUSC} = \frac{\sum_{i=1}^{n} \text{Distance}_i}{n},$$

where n represents count of calls/use of that member through that object and Distance_i is equal to the absolute difference in number of lines between the method definition and the corresponding call/use through that object. This measure is totally different than the class method spatial complexity, which measure the spatial abilities needed to understand the significance of the class, and that measure does not know anything about the usage of that class in solving a particular problem with help of other classes. In some sense, it can be said that from spatial ability point of view, class method spatial complexity measures the cohesiveness of the class and object-member usage spatial complexity measures the coupling of that class. The OMUSC measure concentrates on the usage of the classes through objects, which do interact with other processing blocks (such as main) and other classes (in which the object of another class may have been defined). Just like previous cases, in case of multiple files coming into picture for measurement of this distance, the distance is defined as:

$$\begin{aligned}\text{Distance} = {} & (\text{Distance of call from top of file containing call}) \\ & + (\text{Distance of definition of the member from top of} \\ & \quad \text{the file containing definition}) \\ & + (0.1 * (\text{total lines of code of remaining files})/2).\end{aligned}$$

Total Object-Member Usage Spatial Complexity (TOMUSC) of an object is defined as average of object-member usage spatial complexity of all members being used of that method.

$$\text{TOMUSC} = \frac{\sum_{i=1}^{k} \text{OMUSC}_i}{k},$$

where k is count of object-members being called through that object.

Based on the above formulas, the Object Spatial Complexity of an object is defined as:

$$\text{OSC} = \text{ODSC} + \text{TOMUSC}.$$

This measure of object spatial complexity depends on inter-usage of the classes with in the routines or other classes of the object-oriented software. It may be noted that this measure inherently takes care of effect of inheritance and polymorphism towards understandability of the software. Through inheritance, if a member of any super class is used in any of its derived classes, then its distance is computed from the corresponding declaration in the super class itself and thus the distance value inherently includes the effect of inheritance. Similarly if there has been some method/function overloading to implement polymorphism, then each of the overloaded method/function is considered independently while computing OSC, as each of them has got a separate definition and use, which can be differentiated through count/type of parameters, etc. while using. Thus, the contribution of use of polymorphism and inheritance towards spatial complexity gets automatically covered through above-mentioned definition of ODSC and TOMUSC.

6.3. *Significance of object spatial complexity*

This concept of object-spatial complexity can be very useful in many ways. As compared to CSC where classes and their functionality were focused always, OSC concentrates on objects and their interaction. Various method functions of different objects are called in a particular sequence to obtain the proper results. Thus if the effectiveness of OSC is to be measured, one needs to understand their interaction of these objects and sequence of their calling. The values of OSC for all objects (total 550 objects) of the 15 projects considered above were computed. For every project, the averages of OSCs of all objects in that project were taken and are listed in Table 4. In order to verify the results, activity of perfective maintenance was applied on all these 15 projects, so that uniformity in maintenance activity could be achieved.[14] A target of improving the efficiency of these projects by around 10% was decided. In case of CSC, the generation of class diagrams was a better alternative, as the aim was to comprehend the classes and their functionality. But in case of OSC, the object-interaction and sequencing of calls to various members of the objects needs to be concentrated, which needs to be understood well, if one wants to perform some maintenance activity and perfective maintenance in the form of improving the efficiency was

Table 4. OSC and maintenance time for 15 projects.

S.N	Lines of Code (LOC)	No. of objects	Total OSC	Average OSC = total OSC/No. of objects	Maint-Time (in hours)	Maint-Time/ Average OSC
1	398	8	1773	221.63	25	0.11
2	521	3	702	234.00	35	0.15
3	603	7	1983	283.29	39	0.14
4	929	9	3047	338.56	44	0.13
5	994	23	4235	184.13	30	0.16
6	1085	13	3061	235.46	42	0.18
7	1256	36	6122	170.06	28	0.16
8	1472	60	7554	125.90	23	0.18
9	1506	33	7424	224.97	40	0.18
10	1811	71	12370	174.23	33	0.19
11	2154	40	6025	150.63	27	0.18
12	2398	65	17368	267.20	40	0.15
13	2526	57	9631	168.96	33	0.20
14	3055	60	18952	315.87	48	0.15
15	3257	65	13183	202.82	37	0.18

used by us as it also ensured the uniformity of the work along with the thorough understanding of the logic, which may not be possible with other maintenance activities such as corrective maintenance and adaptive maintenance.[18] The perfective maintenance time for these 15 projects was noted and is also listed in Table 4.

A graph between LOC and Maint-Time of these 15 projects has been drawn as shown in Fig. 4, which clearly gives an indication that maintenance time does not directly depend on LOC and thus, the understandability of the object-oriented software is influenced by some other factors.

Our results have indicated that OSC can be a useful metric for measuring these type of cognitive abilities needed to comprehend

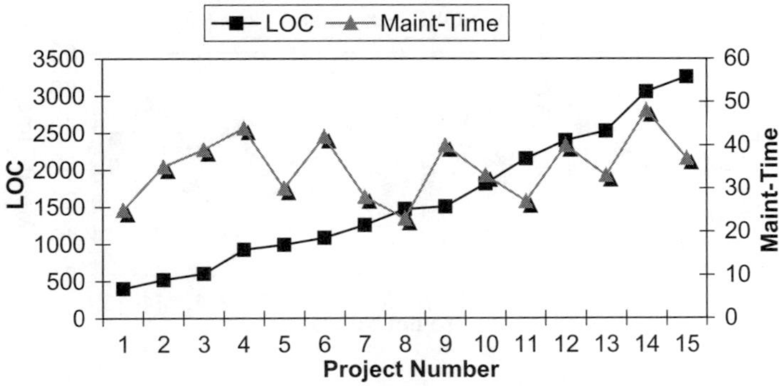

Fig. 4. Plot of Maint-Time and LOC.

the processing logic of the object-oriented software. Some of the important uses of OSC are discussed below.

(1) The object spatial complexity measure can be used to measure the understanding of processing logic through objects and their interaction, which in turn reflects effective utilization of the objects towards final solution. As discussed above, classes normally are not used directly, but through objects only. Lower value of object spatial complexity indicates that the class has been utilized through objects in close proximity of the class declaration, and hence, the understanding the use of that class towards total software working will be much easier than a class having larger value of object spatial complexity. Perfective maintenance of the object-oriented software requires understanding of all objects of various classes. A plot of average OSC and maintenance time of all of the 15 projects has been drawn in Fig. 5.

A strong correlation between average OSC and Maint-Time can be clearly observed from this figure. The correlation between these 2 parameters has been found as 0.84. On the other hand, correlation between LOC and Maint-Time was found to be 0.26 only. This strengthens our belief that OSC metric can be used to measure the understanding of the processing logic of the final solution implemented with the help of objects and their interaction.

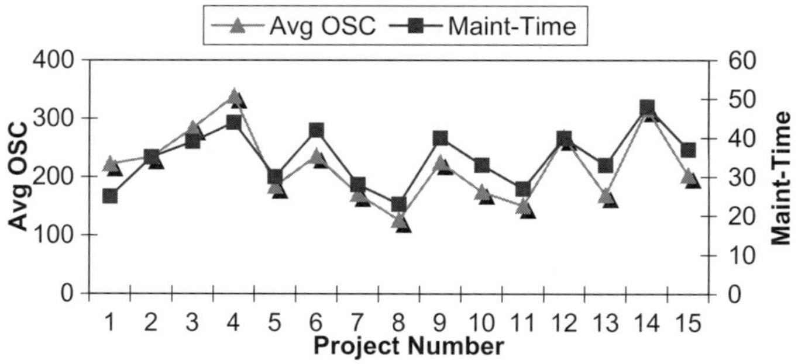

Fig. 5. Plot of Avg OSC and Maint-Time.

(2) The total object-member usage spatial complexity (TOMUSC) measure may be useful sometimes in judging the appropriateness of the attributes of the class. The higher value of TOMUSC can be possible because of distant usage of either method-members or attribute-members. If the higher value of the method-members is coming because of attribute-members usage, it clearly gives a hint about possible wrong choice/usage of the attributes of the corresponding class. For example, if the attributes are declared as public, and are used by the methods other than of the class itself, this value of TOMUSC will be high and will be helpful in pointing out this discrepancy. In the study of these 15 projects and 550 objects, we were able to identify certain attributes in four different projects, which actually should not have been part of those classes.

(3) These spatial complexity (object as well as class spatial complexity) values can help in judging the understandability of source code of object-oriented software, which contributes towards measuring the maintainability of the software.[19] Till now, the source code versus comment ratio has been used to compute the understandability of the source code while measuring.[19] But now these values of class and object spatial complexity can be used to measure the understandability of the source code, which, in turn is used to measure the software maintainability of object-oriented software.

7. Future Work

This chapter has presented two object-oriented spatial complexity metrics. Our results have been found to have a strong correlation with reverse engineering time and perfective maintenance time. But their correlation needs to be studied for lengthier projects having LOC 5000 to 100 000. The proposed measures may be much more useful, if some acceptable range of CSC and OSC could be established to prove the corresponding understandability. We have observed some intuitive figures, which could be used as acceptable values, but a statistical study of more empirical data is needed to verify our intuition and this data needs to cover the corrective maintenance as well. There is a tremendous scope of future work of recording corrective maintenance data of 10–20 projects, and then finding the correlation between this recorded data and above proposed object-oriented spatial metrics. Effect of templates, preprocessor directives, and this pointer on spatial complexity is another direction to work upon. We feel that concept of spatial complexity can play a very important role in developing maintainable software, which is most desirable in the software industry at present, as maintenance cost of some of the software has been reported to be 70–75% of the total cost.

8. Conclusion

We have proposed two spatial complexity measures for object-oriented software in this chapter, which can be very useful in judging the understandability of the object-oriented software. Class spatial complexity concentrates on the attributes and methods of the class and measures the effort required to comprehend the purpose and functionality of the classes. The object spatial complexity concentrates on usage of the corresponding class in the working of overall software. The values of the class spatial complexity and object spatial complexity can be useful in many ways, which have been pointed out. Empirical data has been collected for these metrics and results have been validated. CSC has been found to be very useful indicator of readability of classes and OSC was able to measure the understandability of interaction among objects leading to the processing logic of object-oriented software. Lower value of class spatial complexity indicates lower cognitive effort needed for understandability of the class and lower value of object spatial complexity gives a hint about effective design and utilization of objects towards final solution.

References

1. T. DeMarco, *Controlling Software Projects* (Yourdon Press, Englewood Cliffs, New Jersey, 1982).
2. G. K. Gill and C. F. Kemerer, Cyclomatic complexity density and software maintenance productivity, *IEEE Transactions on Software Engineering* **17** (1991) 1284–1288.
3. N. E. Fenton and S. Fleeger, *Software Metrics — A Rigorous and Practical Approach* (Thomson International Press, 2002).
4. T. J. McCabe, A complexity measure, *IEEE Transactions on Software Engineering* **SE-2** (1976) 308–319.
5. M. H. Halstead, *Elements of Software Science* (North Holland, New York, 1977).
6. Y. Singh and P. Bhatia, Module weakness: A new measure, *ACM SIGSOFT* **23** (1998) 81–82.
7. K. K. Aggarwal, Y. Singh and J. K. Chhabra, Computing program weakness using module coupling, *ACM SIGSOFT* **27** (2002) 63–66.
8. J. F. Peters and W. Pedrycz, *Software Engineering: An Engineering Approach* (John Wiley & Sons, 2000).
9. K. Laitnen, Estimating understandability of software documents, *ACM SIGSOFT* **21** (1996) 81–92.
10. A. Baddeley, *Human Memory: Theory and Practice*, revised edn. (Hove Psychology Press, 1997).
11. C. R. Douce, P. J. Layzell and J. Buckley, Spatial measures of software complexity, Technical Report, Information Technology Research Institute, University of Brighton, UK (1999).
12. R. Brooks, Towards a theory of the comprehension of computer programs, *International Journal of Man–Machine Studies* **18** (1983) 543–554.
13. J. K. Chhabra, K. K. Aggarwal and Y. Singh, Code and data spatial complexity: Two important software understandability measures, *Information and Software Technology* **45** (2003) 539–546.

14. S. D. Conte, H. E. Dunsmore and V. Y. Shen, *Software Engineering Metrics and Models* (Cummings Pub., Inc. USA, 1986).
15. G. Booch, J. Rumbaugh and I. Jacobson, *The Unified Modeling Language User Guide* (Pearson Education, 2002).
16. S. R. Chidamber and C. F. Kemerer, A metrics suite for object-oriented design, *IEEE Transactions on Software Engineering* **20** (1994) 476–493.
17. L. C. Briand, J. W. Daly and J. K. Wust, A unified framework for cohesion measurement of object-oriented systems, *Empirical Software Engineering Journal* **3** (1998) 65–117.
18. J. K. Chhabra and Y. Singh, Software Maintenance: Write software that is easy to maintain, *Information Technology Magazine*, May 2001, pp. 63–66.
19. K. K. Aggarwal, Y. Singh and J. K. Chhabra, Maintainability of object-oriented software, *International Journal of Management and Systems* (2003), to appear.

CHAPTER 9

A Quality Engineering Approach to Human Factors in Design-Review Process for Software Reliability Improvement

Shigeru Yamada* and Ryotaro Matsuda†

Department of Social Systems Engineering,
Faculty of Engineering, Tottori University,
Tottori-shi, 680-8552 Japan
**yamada@sse.tottori-u.ac.jp*
†98t7055@sse.tottori-u.ac.jp

1. Introduction

Software faults introduced by human errors in development activities of complicated and diversified software systems have occurred a lot of system failures of modern computer systems. Since these faults concern with mutual relations among human factors in such software development projects, it is difficult to prevent from software failures beforehand in the software production control. Additionally, most of these faults are detected and corrected after software failure occurrences during the testing phase.

If we can make the mutual relations among human factors[1–3] clear, then the problem for software reliability improvement is expected to be solved. So far, several studies have been carried out to

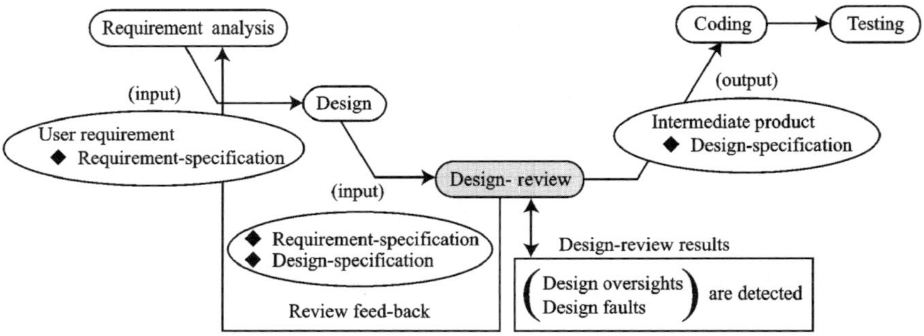

Fig. 1. Inputs and outputs in the software design process.

investigate the relationships among software reliability and human factors by performing software development experiments and providing fundamental frameworks for understanding the mutual relations among various human factors.[4,5]

In this paper, we focus on a software design-review process which is more effective than the other processes for elimination and prevention of software faults (see Fig. 1). Then, we adopt a quality engineering approach for analyzing the relationships among the quality of the design-review activities, i.e., software reliability, and human factors to clarify the fault-introduction process in the design-review process.

We conduct a design-review experiment of graduate and undergraduate students as subjects. First, we discuss human factors categorized in inhabitors and inducers in the design-review process, and set up controllable human factors in the design-review experiment. Especially, we lay out the human factors on an orthogonal array based on the method of design of experiment.[6] Second, in order to select human factors which affect the quality of the design-review, we perform a software design-review experiment reflecting an actual design process based on the method of design of experiment. For analyzing the experimental results, we adopt a quality engineering approach, i.e., Taguchi-method. That is, applying the orthogonal array $L_{18}(2^1 \times 3^7)$ to the human factor experiment, we carry out the analysis of variance by using the data of signal-to-noise ratio (defined as SNR)[7] which can evaluate the stability of quality characteristics, discuss effective human factors, and obtain the optimal levels for the selected inhabitors and inducers.

2. Design-Review and Human Factors

2.1. *Design-reviews*

The inputs and outputs for the design-review process are shown in Fig. 1. The design-review process is located in the intermediate process between design and coding phases, and have software requirement-specifications as inputs and software design-specifications as outputs. In this process, software reliability is improved by detecting software faults effectively.[8]

2.2. *Human factors*

The attributes of software designers and design process environment are mutually related for the design-review process (see Fig. 1). Then, influential human factors for the design-specification as outputs are classified into two kinds of attributes in the following[9–11] (see Fig. 2):

(i) Attributes of the design reviewers (Inhabitors)

Attributes of the design reviewers are those of software engineers who are responsible for design-review work. For example,

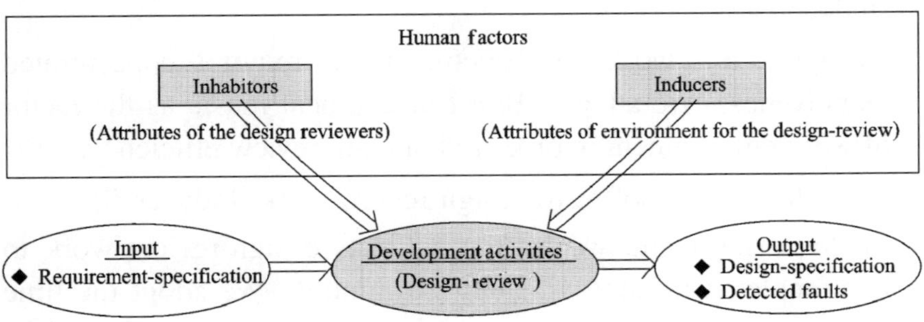

Fig. 2. A human factor model including the inhabitors and inducers.

they are the degree of understanding of software requirement-specifications and software design-methods, the aptitude of programmers, the experience and capability of software design, the volition of achievement of software design, etc. Most of them are psychological human factors which are considered to contribute directly to the quality of software design-specification.

(ii) Attributes of environment for the design-review (Inducers)

In terms of design-review work, many kinds of influential factors are considered such as the education of software design-methods, the kind of software design methodologies, the physical environmental factors in software design work, e.g., temperature, humidity, noise, etc. All of these influential factors may affect indirectly to the quality of software design-specification.

3. Design-Review Experiment

3.1. *Human factors in the experiment*

In order to find out the relationships among the reliability of software design-specification and its influential human factors, we have performed the design experiment by selecting five human factors as shown in Table 1 as control factors which are concerned in the review work.

- BGM of classical music in the review work environment (Inducer *A*)

  Design-review work for detecting faults requires concentrated attentiveness. We adopt a BGM of classical music as the factor of work environment in order to maintain review efficiency.

- Time duration of software design-review work (Inducer *B*)

  In this experiment, we set the subjects design-review work to be completed in approximately 20 minutes. We adopt the time duration of software design-review work with three levels such as 20 minutes, 30 minutes and 40 minutes as the factor of work time.

Table 1. Controllable factors in the design-review experiment.

	Control factor	Level 1	Level 2	Level 3
$A^{\dagger\dagger}$	BGM of classical music in the review work environment	A_1: yes	A_2: no	—
$B^{\dagger\dagger}$	Time duration of software design-review work (minute)	B_1: 20 min	B_2: 30 min	B_3: 40 min
$C^{\dagger}$	Degree of understanding of the design-method (*R*-Net Technique)	C_1: high	C_2: common	C_3: low
$D^{\dagger}$	Degree of understanding of requirement-specification	D_1: high	D_2: common	D_3: low
$E^{\dagger\dagger}$	Check list (indicating the matters that require attention in review work)	E_1: detailed	E_2: common	E_3: nothing

†Inhabitors, ††Inducers.

- Check list (Inducer *E*)

  We prepare the check list (CL) which indicates the matters to be noticed in review work. This factor has the following three levels: Detailed CL, common CL, and without CL.

- Degree of understanding of the design-method (Inhabitor *C*)

  Inhabitor *C* of two inhabitors is the degree of understanding of the design-method of *R*-Net (requirements network). Based on the preliminary tests on the ability to understand the *R*-Net technique, the subjects are divided into the following three groups: High, common, and low ability group.

- Degree of understanding of requirement-specification (Inhabitor D)

  Inhabitor D of two inhabitors is the degree of understanding of requirement-specification. In the similar case as Inhabitor C, based on the preliminary tests on the ability of geometry, are divided into the following three groups: High, common, and low ability group.

3.2. *Summary of experiment*

In this experiment, we conduct an experiment to clarify the relationships among human factors affecting software reliability and the reliability of design-review work by assuming a human factor model consisting of inhabitors and inducers as shown in Fig. 2. The actual experiment has been performed by 18 subjects based on the same design-specification of a triangle program which receives three integers representing the sides of a triangle and classifies the kind of triangle such sides form.[12] We measured the 18 subjects' capability of both the degrees of understanding of design-method and requirement-specification by the preliminary tests before the design of experiment. Further, we seeded some faults in the design-specification intentionally. Then, we have executed such a design-review experiment in which the 18 subjects detect the seeded faults.

We have performed the experiment by using the five control factors with three levels as shown in Table 1, which are assigned to the orthogonal-array $L_{18}(2^1 \times 3^7)$ of the design of experiment as shown in Table 3.

4. Analysis of Experimental Results

4.1. *Definition of SNR*

We define the efficiency of design-review, i.e., the reliability, as the degree that the design reviewers can accurately detect correct and incorrect design parts for the design-specification containing seeded faults. There exists the following relationship among the total number of design parts, n, the number of correct design parts, n_0, and the number of incorrect design parts containing seeded faults, n_1:

$$n = n_0 + n_1 \,. \tag{1}$$

Therefore, the design parts are classified as shown in Table 2 by using the following notations:

n_{00} = the number of correct design parts detected accurately as correct design parts,
n_{01} = the number of correct design parts detected by mistake as incorrect design parts,
n_{10} = the number of incorrect design parts detected by mistake as correct design parts,
n_{11} = the number of incorrect design parts detected accurately as incorrect design parts,

where two kinds of error rate are defined by:

$$p = \frac{n_{01}}{n_0} \,, \tag{2}$$

$$q = \frac{n_{10}}{n_1} \,. \tag{3}$$

Considering the two kinds of error rate, p and q, we can derive the standard error rate, p_0,[7] as:

$$p_0 = \frac{1}{1 + \sqrt{\left(\frac{1}{p} - 1\right)\left(\frac{1}{q} - 1\right)}} \,. \tag{4}$$

Then, the signal-to-noise ratio based on Eq. (4) is defined by (see Ref. 7):

$$\eta_0 = -10 \log_{10} \left\{ \frac{1}{(1 - 2p_0)^2} - 1 \right\} . \tag{5}$$

Table 2. Input and output tables for two kinds of error.

(i) Observed values

Input \ Output	0 (True)	1 (False)	Total
0 (true)	n_{00}	n_{01}	n_0
1 (false)	n_{10}	n_{11}	n_1
Total	r_0	r_1	n

(ii) Error rates

Input \ Output	0 (True)	1 (False)	Total
0 (true)	$1 - p$	p	1
1 (false)	q	$1 - q$	1
Total	$1 - p + q$	$1 - q + p$	2

The standard error rate, p_0, can be obtained from transforming Eq. (5) by using the signal-to-noise ratio of each control factor as:

$$p_0 = \frac{1}{2}\left\{1 - \frac{1}{\sqrt{10^{(-\eta_0/10)} + 1}}\right\}. \tag{6}$$

4.2. *Orthogonal-array* $L_{18}(2^1 \times 3^7)$

The method of experimental design based on orthogonal-arrays is a special one that requires only a small number of experimental trials to help us discover the main factor effects. On traditional researches,[5,9] the design of experiment has been conducted by using orthogonal-array $L_{12}(2^{11})$. However, since the orthogonal-array $L_{12}(2^{11})$ has two levels for grasp of factorial effect to the human factors experiment, the middle effect between two levels cannot be measured. Thus, in order to measure it, we adopt the orthogonal-array $L_{18}(2^1 \times 3^7)$ can lay out one factor with two levels (1, 2) and seven factors with three levels (1, 2, 3) as shown in Table 3, and dispense with $2^1 \times 3^7$ trials by executing experimental independent 18 experimental trials each other. For example, as for the experimental trial no. 10, we executed the design-review work under the conditions A_2, B_1, C_1, D_3, and E_3, and obtained the computed SNR as 2.099 (dB) from the observed values $n_{00} = 110$, $n_{01} = 1$, $n_{10} = 11$, and $n_{11} = 7$.

Additionally, the interaction between two factors can be estimated without sacrificing any factor. And that between any pair of human factors are confounded partially with the effect of remaining factors. Therefore, we have evaluated large main effects as high reproducible human factors because the selected optimal levels of the relatively large effect factor has larger effect than that of the relatively small one.

Considering such circumstances, we can obtain the optimal levels for the selected inhabitors and inducers efficiently by using the orthogonal-array $L_{18}(2^1 \times 3^7)$.

Table 3. The orthogonal array $L_{18}(2^1 \times 3^7)$ with assigned human factors.

Experiment no.	Control factors								Observed values				
	A	B	C	D	E	e (error)	e (error)	e (error)	n_{00}	n_{01}	n_{10}	n_{11}	SNR (dB)
1	1	1	1	1	1	1	1	1	110	1	2	16	8.404
2	1	1	2	2	2	2	2	2	108	3	10	8	−0.515
3	1	1	3	3	3	3	3	3	109	2	16	2	−6.050
4	1	2	1	1	2	2	3	3	111	0	2	16	10.008
5	1	2	2	2	3	3	1	1	107	4	4	14	2.889
6	1	2	3	3	1	1	2	2	104	7	11	7	−4.559
7	1	3	1	2	1	3	2	3	111	0	4	14	8.104
8	1	3	2	3	2	1	3	1	106	5	8	10	−0.780
9	1	3	3	1	3	2	1	2	110	1	11	7	2.099
10	2	1	1	3	3	2	2	1	110	1	11	7	2.099
11	2	1	2	1	1	3	3	2	106	5	4	14	2.260
12	2	1	3	2	2	1	1	3	105	6	12	6	−4.894
13	2	2	1	2	3	1	3	2	105	6	10	8	−2.991
14	2	2	2	3	1	2	1	3	108	3	15	3	−5.784
15	2	2	3	1	2	3	2	1	105	6	10	8	−2.991
16	2	3	1	3	2	3	1	2	109	2	2	16	6.751
17	2	3	2	1	3	1	2	3	107	4	4	14	2.889
18	2	3	3	2	1	2	3	1	103	8	9	9	−3.309

5. Investigation of Analysis Results

5.1. *Analysis of experimental results*

The experimental results of observed values of design parts discussed in Sec. 4.1 in the software design-specification are shown in Table 3. The data of the SNR calculated by Eq. (5) are also shown in Table 3.

5.2. *Analysis of variance*

The result of analysis of variance for observed correct and incorrect design parts is shown in Table 4 by using the data of SNR as shown in Table 3. In Table 4, f, S, V, F_0, and ρ represent the degree of freedom, the sum of squares, the unbiased variance, the unbiased variance ratio,

Table 4. The result of analysis of variance by using the SNR.

Factor	f	S	V	F_0	ρ (%)
A	1	36.324	36.324	10.578*	7.4
B	2	33.286	16.643	4.847*	5.9
C	2	229.230	114.615	33.377**	49.8
D	2	86.957	43.479	12.661**	17.9
E	2	3.760	1.880○		
$A \times B$	2	33.570	16.785	4.888*	6.0
e	6	23.710	3.952○	—	
e'	8	27.470	3.434	—	13.0
T	17	446.837	—	—	100.0

○: pooled.
*: 5% level of significant.
**: 1% level of significant.

and the contribution ratio, respectively, for performing the analysis of variance. In order to obtain the precise analysis results, the factor of check list (Factor E) is pooled in the factor of error (Factor e). Then, we performed the analysis of variance based on the factor of new pooled error (Factor e').

In the results, the effect of control factors such as BGM (Factor A), the time duration of design-review work (Factor B), the degree of understanding of software design-method (Factor C), and the degree of understanding of requirement-specification (Factor D) are recognized in the design-review experiment.

5.3. *Discussion*

As a result of experimental analysis, the effective control factors such as the BGM of classical music to review work environment (Factor A), the time duration of design-review work (Factor B), the degree of understanding of software design-method (Factor C), and the degree of understanding of requirement-specification (Factor D) were recognized. Especially, Factors A and B are mutually interacted.

Then, we can find that both our experience from actual software development[9] and the experiment result above of design-review are equivalent. Table 5 shows the comparisons of SNR's and standard error rates. The improvement ratio of the reliability of design-review is calculated as 20.909 dB (33.1% measured in the standard error rate in Eq. (4)) by using the SNR based on the optimal condition (A_1, B_3, C_1, D_1) of control factors such as Factors A, B, C, and D of which effects are recognized in Fig. 3.

Therefore, it is expected that quantitative improvement of the reliability of design-review can be controlled by using these control factors.

Table 5. The comparison of SNR and standard error rates.

	Optimal conditions	Worst conditions	Deviation
Signal-to-noise ratio (dB)	10.801	−10.108	Δ 20.909
Confidence interval	±3.186		
Standard error rates (%)	2.0	35.1	Δ 33.1

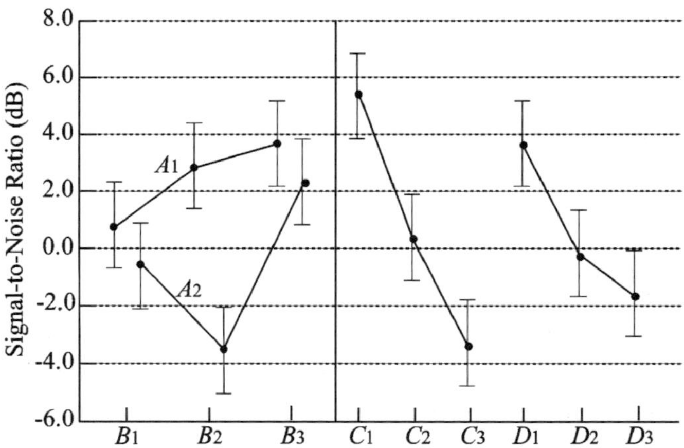

Fig. 3. Estimation of significant factors.

6. Approval of Experimental Results

Table 6 shows the optimal and worst levels of control factors of design-review discussed in Chapter 5. Considering the circumstances, we conduct an additional experiment to approve the experimental results by using the SNR.

6.1. *Additional experiment*

We focus on the effect of faults detected under the optimal condition of design-review work. Similarly to the design of experiment discussed in Chapter 3, the design-specification is for the triangle program reviewed by 18 subjects. We measured their capability of both the degree of understanding of design-method and the degree of understanding requirement-specification by the preliminary tests before the design of additional experiment.

We also seeded some faults in the design-specification intentionally. Then, we have executed the same design-review experiment discussed in Sec. 3.2 under the same review condition (the optimal levels for the selected inhabitors). Additionally, we have approved

Table 6. The optimal and worst levels of design-review.

	Control factor	Level: Optimal	Level: Worst
Inducer A	BGM of classical music to review work environment	A_1: yes	A_2: no
Inducer B	Time duration of design-review work (minute)	B_3: 40 min	B_2: 30 min
Inhabitor C	Degree of understanding of design-method (R-Net Technique)	C_1: high	C_3: low
Inhabitor D	Degree of understanding of requirement-specification	D_1: high	D_3: low

Table 7. The SNR's in the optimal levels for the selected inducers.

No.	Observed values n_{00}	n_{01}	n_{10}	n_{11}	SNR (dB)	Standard error rates	Observed values Factor C	Factor D
1	109	2	3	15	5.613	0.027	high	common
2	111	0	5	13	7.460	0.040	common	high
3	108	3	2	16	3.943	0.078	high	low
4	107	4	4	14	2.889	0.094	high	low
5	111	0	2	16	10.008	0.023	high	high
6	109	2	3	15	5.613	0.057	low	high
7	107	4	4	14	2.889	0.094	common	low
8	107	4	4	14	2.889	0.094	low	common
9	111	0	2	16	10.008	0.023	high	high
10	109	2	4	14	4.729	0.068	low	high
11	107	4	3	15	3.825	0.080	common	common
12	107	4	6	12	1.344	0.120	low	common
13	101	10	8	10	−3.385	0.220	low	low
14	105	6	3	15	2.707	0.097	common	low
15	107	4	3	16	3.825	0.080	common	common
16	111	0	4	14	8.104	0.035	common	high
17	111	0	5	13	7.460	0.040	high	common
18	98	13	9	9	−3.369	0.025	low	low

that the selected inhabitors divided by the preliminary tests are consistent with the optimal levels of two inducers.

The experimental results of the observed values of correct and incorrect design parts and the preliminary tests are shown in Table 7 with the data of SNR calculated by Eq. (5).

6.2. *Comparison of factorial effects in the optimal inducer condition*

Figure 4 shows the optimal levels of control factors of design-review based on the additional experiment. If both inhabitors are high conditions, the effect of detecting faults is improved. Additionally, Table 8

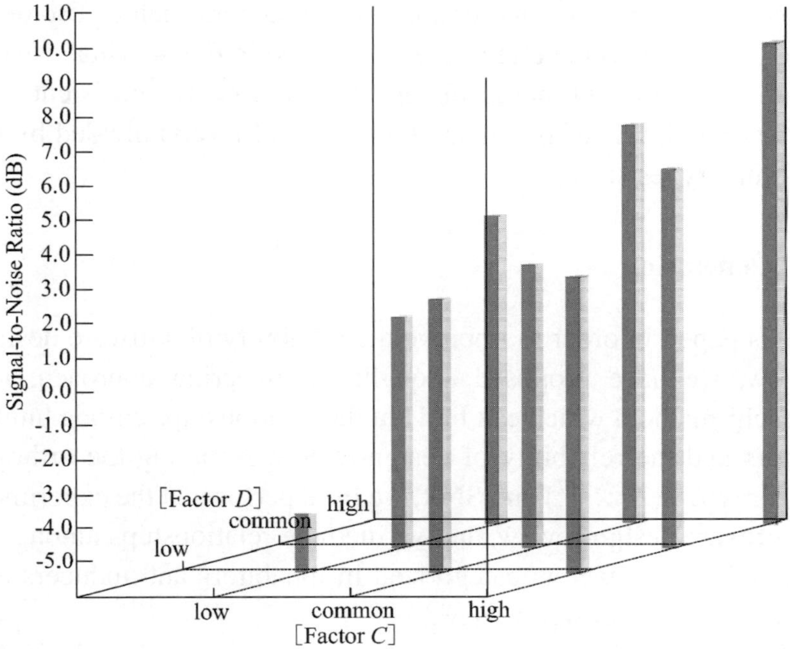

Factor *C*: the degree of understanding of design-metohd
Factor *D*: the degree of understanding of requirement specification

Fig. 4. The comparison of factorial effects.

Table 8. The comparison of SNR's and standard error rates between the optimal levels for the selected inducers.

	Factor *C* and factor *D*		
	High	Low	Deviation
Signal-to-noise Ratio (dB)	10.008	−3.510	Δ 13.518
Standard error rates (%)	2.3	22.3	Δ 20.0

shows the comparison of SNR's and standard error rates between the optimal levels for the selected inducers. The improvement ratio of the reliability of design-review is calculated as 13.518 dB (20.0% measured in the standard error rate) by using the signal-to-noise ratio based on the optimal condition of control factors such as Factor A, B, C, and D of which effects are recognized in Fig. 4. Thus, we can approve that the optimal levels of two inducers are consistent with the optimal levels of two inhabitors (optimal levels) divided by the preliminary tests.

7. Conclusion

In this paper, in order to improve the reliability of software design-review, we have proposed a quality engineering approach, i.e., Taguchi-method, which can find out the relationships among human factors and the reliability of design-review. Applying the orthogonal array $L_{18}(2^1 \times 3^7)$ and SNR, we have performed the experiment of software design-review, and verified the relationships among the selected human factors categorized in inhabitors and inducers and the reliability of design-review.

It has been shown that the result of experimental analysis discussed in this paper is consistent with previous studies[5,9] of software design-review process. Additionally, we have shown that the BGM (Factor A) is efficient for detecting faults by design-review work which requires concentrated attentiveness. However, it has been recognized for the selected human faults (Factor A, B, C, and D) that the optimal levels occur at the ends in the result of experimental analysis.

Further studies on human factors on the software design process by using this approach are needed to support the findings in this paper.

Additionally, we have approved that the selected inhabitors divided by preliminary tests the are consistent with the optimal levels of two inducers.

Acknowledgments

The authors would like to thank the graduate and undergraduate students of the Department of Social Systems Engineering, Tottori University, for their help as subjects in the experiments. This work was supported in part by the Grant-in-Aid for the Scientific Research (C)(2) from the Ministry of Education, Culture, Sports, Science and Technology of Japan under Grant No. 15510129.

References

1. V. R. Basili and R. W. Reiter, Jr, An investigation of human factors in software development, *IEEE Computer Magazine* **12** (1979) 21–38.
2. B. Curtis (ed.), *Tutorial: Human Factors in Software Development* (IEEE Computer Society Press, Los Alamitos, CA, 1985).
3. T. Nakajo and H. Kume, A case history analysis of software error cause-effect relationships, *IEEE Trans. Software Engineering* **17** (1991) 830–838.
4. K. Esaki and M. Takahashi, Adaptation of quality engineering to analyzing human factors in software design, *J. Quality Engineering Forum* **4** (1996) 47–54 (in Japanese).
5. K. Esaki and M. Takahashi, A software design review on the relationship between human factors and software errors classified by seriousness, *J. Quality Engineering Forum* **5** (1997) 30–37 (in Japanese).
6. G. Taguchi, *A Method of Design of Experiment*, 2nd edn. (Maruzen, Tokyo, 1976) (in Japanese).
7. G. Taguchi (ed.), *Signal-to-Noise Raito for Quality Evaluation* (Japanese Standards Association, Tokyo, 1998) (in Japanese).
8. S. Yamada, *Software Reliability Models: Fundamentals and Applications* (JUSE Press, Tokyo, 1994) (in Japanese).
9. K. Esaki, S. Yamada and M. Takahashi, A quality engineering analysis of human factors affecting software reliability in software design review process, *Trans. IEICE Japan* **J84–A** (2001) 218–228 (in Japanese).
10. S. Yamada, T. Kageyama, M. Kimura and M. Takahashi, An analysis of human errors and factors in code-review-process for reliable software development, *Trans. IEICE Japan* **J81–A** (1998) 1238–1246 (in Japanese).
11. S. Yamada and R. Matsuda, A quality engineering evaluation for human factors affecting software reliability in design review process, *J. Japan Industrial Management Association* **54** (2003) 71–79 (in Japanese).
12. I. Miyamoto, *Software Engineering — Current Status and Perspectives* (TBS Publishing, Tokyo, 1982) (in Japanese).

CHAPTER 10

Tree-Based Software Quality Classification Using Genetic Programming

Taghi M. Khoshgoftaar*, Yi Liu and Naeem Seliya

Empirical Software Engineering Laboratory,
Department of Computer Science and Engineering,
Florida Atlantic University, Boca Raton, FL 33431, USA
**taghi@cse.fau.edu*

1. Introduction

The knowledge of the likely problematic areas of a software system is very useful for improving its overall quality. Based on such information, a more focussed software testing and inspection plan can be devised. More specifically, the limited resources allocated for software quality and reliability improvement can be expended in a cost-effective manner. Some of the practical software quality and reliability improvement techniques include, rigorous code design and code reviews, extensive software testing, and skill-based placement of personnel. In software development practice, the amount of project resources allocated for software quality improvement is usually a small fraction of the total budget, thus, asserting the importance of a cost-effective software quality improvement for allowing greater return on investment.

*Corresponding author.

Software measurements, such as software product and process metrics, have been shown to be excellent indicators of software quality.[1,2] Based on such metrics, software quality classification (SQC) models can be built to predict the risk-based class membership of a software (program) module.[3] For example, program modules can be predicted as either fault-prone (*fp*) or not fault-prone (*nfp*). With the aid of a SQC model, the software quality team can target the available resources to improve the modules predicted as *fp*, thus, allowing for a better resource utilization. A SQC model is built based on a training data set, which consists of a set of program modules with known values for their software metrics and membership to the *fp* or *nfp* classes. Subsequently, the efficacy of the trained model is evaluated by predicting the class-membership of the modules (with known values of their software metrics) in a test data set. A further discussion on the problem of SQC is presented in the next section.

Among the existing techniques for SQC, such as logistic regression,[1] discriminant analysis,[4] and artificial neural networks,[5] we feel that the decision tree-based modeling approach has practical attractiveness due to its white-box and comprehensible classification model which can be directly interpreted by observing the tree-structure.[2,6] The software quality team can directly observe from the decision tree (DT) which software metrics (and their threshold values) are more useful to predict the quality of their system. A commonly used DT-based SQC model is a binary tree with query (internal) and leaf (terminal) nodes. A query node is a logical equation (with an independent variable(s), a software metric in our case) which returns either true or false, and can be viewed as a classifier that partitions a set of modules into two subsets. A leaf node is a terminal which assigns a class label, such as *fp* or *nfp*, to all the modules that belong to that leaf node. Additional discussion on DT-based classification is presented in the next section.

A software measurement-based SQC model is affected by the characteristics of the software system and the training data set.[7] Many classification techniques (e.g., logistic regression, and discriminant analysis) are based on an underlying assumption of the form and structure that the resulting SQC model will take. In contrast, genetic programming (GP)-based prediction models are more suited for automatically extracting the underlying relationship between the software metrics and the software quality by mimicking the natural evolution process. In our previous initial attempt we investigated GP for predicting the class-membership of modules into the *fp* and *nfp* classes.[8] In this study, we present for the first time a GP-based decision tree modeling approach for the SQC problem as applied to a real-world industrial system.

The practical essence of a DT-based classification model is its simplicity and comprehensibility. Hence, in addition to achieving maximum classification accuracy, obtaining a DT with a comprehensible size is also very important.[9] Classification accuracy is often measured in terms of the misclassification error rates, while simplicity of decision trees is often expressed in terms of the number of nodes. Hence, an optimal DT is one that has low misclassification error rates and has a (relatively) few number of nodes. However, accuracy and tree size are generally speaking, conflicting objectives for DT-based modeling, i.e., very good accuracy may be associated with a very large tree, while poor accuracy may be associated with a very small tree. Safavian and Landgrebe[10] have pointed out that the simultaneous optimization of both accuracy and efficiency for a DT is difficult. Moreover, decision trees are also susceptible to the presence of outliers and noise in the training data set, leading to misrepresentation of the underlying relationship between the independent variables (software metrics) and the dependent variable (software quality).

Genetic programming is a logical solution to the problems that require a multi-objective optimization, primarily because it is based on the process of natural evolution which involves the simultaneous optimization of several factors. As a component of evolutionary computation techniques,[11–14] GP does not assume any general form for the problem solution. It performs a global stochastic search across the space of computable functions to discover the form and the parameters of the prediction model. In standard GP, each individual (or model) in the population is an S-expression tree, which is a symbolic regression tree consisting of functions and terminals.

Very few studies have investigated GP-based decision tree models.[13,15–17] Several of the previous works related to GP-based classification models have focused on the standard GP process, which requires that the function and terminal sets have the *closure property*.[13] This property implies that all the functions in the function set must accept any of the values and data types defined in the terminal set as arguments. However, since each decision tree has at least two different types of nodes, i.e., query nodes and leaf nodes, the closure property requirement of standard GP does not guarantee the generation of a valid individual(s). Strongly Typed Genetic Programming (STGP) has been used to alleviate the closure property requirement,[18] by allowing each function to define the different kinds of data types it can accept. For example, in our study of building SQC models, a function in the function set can only be in query nodes, while the terminal variables such as *fp* and *nfp* can only be in the leaf nodes. In our study we use STGP to build the GP-based decision trees.

In a recent study, Bot and Langdon[15] demonstrated the calibration of decision trees using STGP. In the context of multiple case studies, multivariate decision trees (i.e., each query node may consist of more than one independent variable) were built such that the fitness function incorporated classification accuracy as well as a penalty for large trees. The overall error rate (percentage of observations that are misclassified for the given data set) was used to evaluate the classification accuracy[15] of the decision trees. However, such an approach is not suitable for the SQC problem because from a software engineering point of view, the costs of misclassifying a *fp* and *nfp* module are invariably different. In our study, a weighted cost of misclassification is considered during modeling to address the influence of the costs of the two types of misclassifications.

In this chapter, we present in the context of the SQC problem a simplified GP-based multi-objective optimization method for automatically building univariate decision trees (each query nodes consists of only one independent variable) that have a high classification accuracy rate and a relatively small tree size. Consequently, two fitness functions are used for optimization purposes: the average weighted cost of misclassification and the tree size. In addition, the obtained classification performances of the GP-based decision trees are compared with those obtained by standard GP which was investigated in our previous study.[8]

In comparison to other DT-based methods (such as C4.5) which can usually only minimize the classification accuracy, GP-based decision trees are more flexible and can allow optimization of performance objectives other than accuracy. More specifically, in addition to the two objectives used in this chapter, an analyst can also include and optimize other fitness functions. Moreover, GP provides a practical solution for building models in the presence of conflicting objectives, a commonly observed issue in software development practice. Existing non-GP decision tree methods are not suited for optimizing objectives other than accuracy and tree size (by varying certain parameters). However, our future work will compare the proposed approach with other decision tree methods.

The remainder of this chapter continues with a discussion on SQC modeling, with a focus on decision trees. This is followed by a discussion on GP and multi-objective optimization with GP. The remainder of the chapter presents the modeling methodology, fitness functions, the empirical case study and its results, and a summary.

2. Software Quality Classification

The importance and benefits of software quality classification models can be clearly seen in related research works.[1,4,6,19] A commonly used SQC approach is to predict software modules as being either *fp* or *nfp*. Such models are built based on software measurement attributes, such as product and process metrics. In order to categorize the modules of the training data set into the *fp* and *nfp* groups, a software quality factor, such as expected number of faults or lines of code churn, is used. More specifically, the software quality management team determines a threshold value of the quality factor in order to segregate the two classes.

The application of a two-group SQC model involves building (training) a model based on software metrics and quality factor data from previously developed system releases or similar projects. Subsequently, the predicted classes for the modules of a currently under-development system release or similar project can be determined by applying the trained model. The software quality team can then target software quality improvement toward modules predicted as *fp*. Depending on when (during the development process) such a classification model is to be applied, the appropriate software metrics are used to train the model. For example, design-level metrics are used to estimate the software quality during the implementation phase, providing a guidance for placing experienced programmers to implement the likely problematic modules.

A SQC model is usually not perfect, i.e., it will have some misclassifications. In the context of the SQC models in our study, a Type I error occurs when a *nfp* module is misclassified as *fp*, whereas a Type II error occurs when a *fp* module is misclassified as *nfp*. From a software engineering point of view, the cost of a Type II error is more severe since it entails a missed opportunity for improving a poor quality module, leading to corrective efforts during system operations. In contrast, the cost of a Type I error is relatively lower since it entails unproductive inspections (prior to deployment) of a module that is already of good quality. Therefore, it is important to incorporate the misclassification cost disparity during SQC modeling.

In our previous studies with SQC models, we have observed an inverse relationship between the Type I and Type II error rates for a given classification technique and its modeling parameters. More specifically, as the Type II error rate decreases the Type I error rate increases, and vice versa. This relationship is important in obtaining the preferred balance (between the error rates), which may be dictated by the software application domain and the quality improvement goals of the project. The usefulness of a SQC model is affected by the attained balance between the two error rates. For example, a medical or safety-critical software system may prefer a model with a very low Type II error rate, regardless of its Type I error rate. On the other hand, a high-assurance software system with limited quality improvement resources may prefer approximately equal misclassification error rates.[6]

As effective data mining tools,[10] DT-based classification models represent rules underlying the training data with hierarchical or sequential structures that recursively partition the data space. Given an object or observation to classify, a DT is traversed along a path (via internal nodes) from its root node to a leaf node in which the estimated class of the object is assigned.[20] For a graphical representation of a decision tree, please refer to Fig. 1.

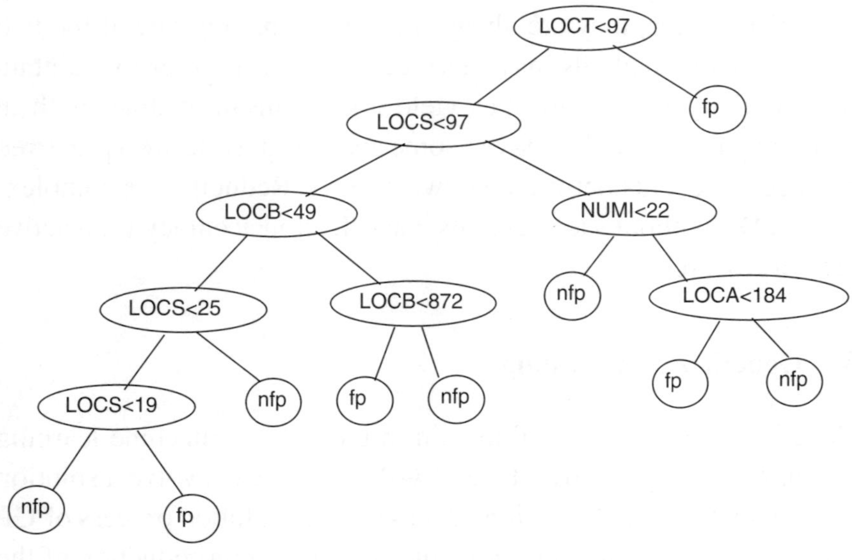

Fig. 1. Decision tree based on number of faults.

The complexity of decision trees is usually measured in terms of the number of nodes in the tree. Therefore, the process of simplifying a DT is concerned with reducing the number of nodes, without any significant reduction in the classification accuracy.[9,10] From the model comprehensibility point of view, a smaller tree is preferred. However, a tree with a few number of nodes may not have a good classification accuracy. A "right" sized decision tree is such that it yields better results in terms of accuracy and simplicity, and is less sensitive to the noise in the training data set.[10] Breslow and Aha discussed that one of the reasons analysts prefer simple trees is because they "provide insight and understanding into the predictive structure of data".[9]

Decision trees are very prone to the overfitting problem,[21] and often a DT model becomes very large to accommodate data noise and outliers. Another situation that can lead to a large DT is "a mismatch of representational biases".[22] In such a case, some tree representations are hard to express the underlying concepts concisely while others can greatly reduce the model complexity. Since there is usually a conflicting trade-off between the accuracy and the simplicity of decision trees, they are always built favoring any one of the two criteria. Some methods may sacrifice accuracy in order to maintain the simplicity, while some may yield small gains in accuracy with an extremely large tree. Moreover, some analysts may desire a preferred balance of benefits between the two aspects. Reducing the complexity of a DT without sacrificing its classification accuracy is an active area of research.[9]

3. Genetic Programming

Genetic programming is a domain in the field of machine learning systems.[13] A unique advantage of GP is that it can evolve a solution automatically from the training data set. The evolution process of GP imitates the Darwinian principle of survival and reproduction of the fittest individuals. Each individual in a GP is a S-expression. A fitness value of a individual (or model in our case) indicates its quality with respect to the problem domain. Hence, it gives a probability of who can be selected for mating and reproducing for the next generation. In this section, we discuss GP-related issues pertaining to decision tree-based SQC models: however, for additional details regarding GP and its features, please refer to Ref. 23.

Standard GP, i.e., *S*-expression tree, requires that the function and terminal sets have the closure property. To address the closure property requirement of GP, the use of STGP has been proposed.[18] It introduces additional criteria for genetic operations, i.e., crossover, mutation, and reproduction.[18,24] More specifically, each function, variable, and terminal are specified by certain types in STGP. Given a precise description of the data types for the arguments of each function and terminal, STGP will only produce individuals according to the constraint that *"the arguments of all functions are of the correct type"*.[18] The remaining of the section discusses STGP issues related to GP-based decision trees as applied to the SQC problem.

(1) Constraint: Different types of functions and terminals may not crossover or mutate under certain constraints. The leaf node can only be a function which returns the class of a module. The internal nodes can only be a simple logical equation which returns either true or false. Hence, only constants and independent variables can appear in the logical equations, and they cannot exist in leaf nodes. Moreover, the function which returns the class of a module cannot be used in the internal nodes. The root node can then be either a leaf node or a logical equation.
(2) Crossover: Further limitations are applied to the genetic operation. We defined that the type of subtree is the type of its root node. When two subtrees crossover, they are required to have the same type so that an appropriate decision tree can be generated.
(3) Mutation: If a subtree is selected to mutate, the replaced subtree also must have the same type or at least a "similar" type, i.e., a subtree that can be replaced with a different type provided that

the new tree is a permissible decision tree. For example, the leaf node is a function which returns the class of a module: however, it can be replaced by a new subtree whose root is a logical equation.

4. Multi-Objective Optimization

In a real-world system, it is often desired that several objectives or performance criteria be simultaneously satisfied. For example, during software development a software quality initiative has to be addressed with respect to the amount to resources available and the time-to-release schedule. Consequently, a multi-objective optimization is desired by analyzing the conflicting trade-off (if any) between the different criteria. A multi-objective optimization solution always strives for a set of Pareto-optima, which represents feasible solutions such that no criterion can be improved without deteriorating some other criterion. Therefore, the critical analysis component is obtaining the best solution among the set Pareto-optima.

A Pareto-optimal set is defined as a mathematical expression[25,26]: Let $\mathbf{x}$ be a vector of the independent variables and $f_j(\mathbf{x})$ be an objective function. A vector $\mathbf{x}^*$ is Pareto-optimal if and only if there is no vector $\mathbf{x}$ that exists with the characteristics:

$$
\begin{aligned}
f_j(\mathbf{x}) &\leq f_j(\mathbf{x}^*), && \text{for all } j \in \{1, \ldots, m\}, \text{ and} \\
f_j(\mathbf{x}) &< f_j(\mathbf{x}^*), && \text{for at least one } j \in \{1, \ldots, m\}.
\end{aligned}
$$

The most frequently used methods for generating Pareto-optima are based on the idea of replacing the multi-objective optimal problem by a parameterized scalar problem. Typically, by varying the value of each parameter for each objective, it is possible to generate all or parts of the Pareto-optimal set. The method used in this chapter is based on the Constraint Oriented Transformation,[25] in which an objective is initially chosen to be minimized and the other objectives are transformed into constraints.

For example, $\min(f_1(\mathbf{x}))$, with $f_j(\mathbf{x}) \leq \bar{y}_j$, where $j = 2, \ldots, m$. $f_1(\mathbf{x})$ is the main objective, generally the most important one for the given system; $f_2, \ldots, f_m$ are the secondary objectives, and $\bar{y}_j$ is the jth constraint defined for the system. Therefore, given a set of selected/allowable values, different solutions are obtained. By varying the limits of the allowable values, a Pareto-optima set will be generated.

In this chapter, instead of transforming $f_2, \ldots, f_m$ into constraints, we minimize the objectives one at time according to their importance as defined by the analyst. More specifically, we first minimize the most important objective, then minimize the second one, then the third one, and so on. That is for $\mathbf{x} \in X$: $\min(f_1(\mathbf{x}))$, $\min(f_2(\mathbf{x})), \ldots, \min(f_m(\mathbf{x}))$. In our study, the most important objective is to minimize the average weighted cost of misclassification, as discussed in the next section. Subsequently, the secondary objective is to simplify the decision trees by controlling their size. Once the first objective is minimized, it is not sacrificed during the minimization of the second objective.

5. Building GP-Based Decision Trees

5.1. *Fitness function for classification*

In the context of software quality modeling, the cost of misclassifying a *fp* module is invariably greater than that of a *nfp* module. During the process of SQC modeling, the disparate misclassification costs should be incorporated. However, since the actual costs of misclassifications are unknown at the time of modeling, we use a weighted approach for determining the relative costs of a Type I and Type II error. Therefore, a SQC model that achieves the lowest average weighted cost of misclassification will be selected for the first objective, as shown later in this section.

We define C_I as the (modeling) cost for a Type I error and C_{II} as the cost for a Type II error. Since $C_{II} > C_I$, we express C_{II} in terms of C_I, i.e., the misclassification costs are normalized with respect to C_I. For convenience we denote the normalized C_{II} as c, which can also be viewed as the cost ratio, C_{II}/C_I. For a given individual or DT, if a *nfp* module is misclassified as *fp*, a penalty of C_I is added to the fitness of the individual; and if a *fp* module is misclassified as *nfp*, a penalty of C_{II} is added to the fitness of the model. Moreover, if the absolute value of the predicted number of faults (or lines of code churn) is extremely large, then a penalty of C_{III} is added to the fitness. For simplicity, the value of C_{III} is empirically set to the value of C_{II}.

The measurement of fitness includes the raw fitness and the number of hits, which is defined as the number of correct classifications. The raw fitness is the weighted cost of misclassification, and is given by:

$$\text{fitness} = C_I * N_I + C_{II} * N_{II} + C_{III} * N_{III}, \tag{1}$$

where N_I and N_{II} are the numbers of Type I and Type II errors respectively, and N_{III} is the number of modules which have an extremely large predicted value of the quality factor.

5.2. *Fitness function for tree size*

Obtaining a comprehensible model is an important feature for decision trees. We define the second fitness function as the number of tree nodes, the definition of which is the same as for traditional decision trees. The smaller the tree, the better is its fitness. However, in order to prevent GP from losing major diversity during the early generations of a run, we limit the number of nodes to a minimum of three nodes, i.e., a root and two children.

5.3. *Selection process of GP model*

In our GP modeling process, a lower value of the raw fitness indicates a better individual and a greater probability of survival. The individual selection process during a GP run is summarized by: (1) Group the individuals according to the first fitness function. Thus, each group will have the same value for the first fitness function; (2) Select one group based on the first fitness function; (3) Check whether only one member is in this group or not. If yes, go to (5), otherwise continue; (4) Select one individual in this group using the value for the second fitness function; (5) The selection process ends. One individual is selected to crossover, mutate or reproduce.

6. System Description

The case study (denoted as WLTS) involved data collection efforts from initial releases of two large Windows©-based embedded system applications used primarily for customizing the configuration of wireless telecommunications products. The two embedded applications, written in C++, provide similar functionalities and contained common source code. The systems are analyzed simultaneously, and the primary difference between them is the type of wireless product they support. The software metrics obtained reflected aspects of source files. Hence, a module in this case study comprised of a source file.

The two applications consisted of over 1400 source code files and contained more than 27 million lines of code each. Software inspection metrics were obtained by observing the configuration management systems of the applications. The problem reporting system tracked and recorded problem statuses. Information such as, how many times a source file was inspected prior to system tests, were logged in its database. Several software metrics were collected to record information such as, fault severity, inspection time, and major/minor errors. However, only a few primitive metrics were selected for modeling purposes since they provided the most relevant and concise information pertaining to the project, as per discussions with the management team.

Software measurement and quality data was collected for the source files. However, upon the removal of data points with missing observations, 1211 modules remained. These modules are used in our case study. Over 66% of modules (809) were observed to have no faults, and the remaining 402 modules had at least 1 or more faults.

The software metrics used to build the models included four product metrics and one process (inspection) metric. The product metrics used are statement metrics for the source files. They primarily indicated the number of lines of source code prior to the coding phase (i.e., auto-generated source code) and just before system tests.

The five software metrics are: NUMI, the number of times the source file was inspected prior to system test; LOCB, the number of lines of code for the source file prior to the coding phase; LOCT, the number of lines of code for the source file prior to system test; LOCA, the number of lines of comments for the source file prior to the coding phase, and LOCS, the number of lines of comments for the source file prior to system test. The first metric, i.e., NUMI, is a process metric, while the other four are product metrics. The respective lines of code prior to the coding phase represent auto-generated source code. Available data collection tools determined the number and types of software metrics collected. Hence, another software project may collect and utilize a different set of software metrics.[19,27]

The quality factors recorded include: (1) the number of faults discovered during system test, and (2) the number of lines of code churn during system test, where code churn for a module is defined as the summation of the number of source code lines added, deleted or modified. Therefore, two types of GP-based decision trees are calibrated in this study. A random and impartial data splitting of the 1211 modules was performed such that the *fit* data set consisted of 807 modules while the *test* data set consisted of 404 modules. The fit data set is used to train the DT model, whereas the test data set is used to evaluate its predictive capability for new (not used for training) modules.

The first SQC model classifies the modules as *fp* and *nfp* based on the number of faults as the quality factor. According to our discussions with the software development team, a software module with two or more faults was considered *fp*, and *nfp* otherwise. Subsequently, the fit data set consisted of 632 *nfp* modules and 175 *fp* modules, while the test data set consisted of 317 *nfp* modules and 87 *fp* modules. The second SQC model classifies the modules as *fp* and *nfp* based on the number of lines of code churn as the quality factor. In the case of the second model, a software module was considered as *fp* if it had four or more lines of code churn, and *nfp* otherwise. Consequently, for the second model the fit data set consisted of 618 *nfp* modules and 189 *fp* modules, while the test data set consisted of 308 *nfp* modules and 96 *fp* modules.

The model selection strategy used for the SQC model for this case study consists of obtaining a classification model such that the Type I and Type II error rates are approximately equal, with the latter being the lowest among competing models. This strategy is based on the software quality improvement goals of the management team.

7. Results and Analysis

According to the model calibration process described earlier and the two fitness functions, the respective GP-based decision tree models are built. The modeling tool used is the "lilgp 1.01" which is developed by Zongker and Punch of Michigan State University.[a] Several different values of the modeling parameter, c, were considered for the first fitness function. However, our empirical studies for both types of classification models determined that only the c values of 3, 3.5, 4, 4.5, and 5 yielded classification models that were of interest. Decision trees based on the other values of c (such as 0.5, 1, 2, 5.5, ...) were not suitable according to the model selection strategy for WLTS. The decision trees built using the proposed approach are univariate binary decision trees, i.e., only one predictor is associated with each internal node.

The GP parameters for each value of c considered for the software system include: a population size of 1000; a maximum of

---

[a]Downloaded from http://garage.cse.msu.edu/software/software-index.html.

200 generations; a crossover rate of 0.6; a reproduction rate of 0.1; a mutation rate of 0.3; fitness-proportional selection; a function set of {*if*, C1, C2, <}, where C1 and C2 represent the *nfp* and *fp* classes; a terminal set of {*independent variables*, *constants*}; and a termination criterion of not exceeding the maximum number of generations. Optimizing these parameters was not considered, however, is part of our future research. We performed 20 runs (using the fit data set) for each value of *c*, and during each run the five best individuals (decision trees) were recorded. Therefore for each value of *c*, we obtained 100 candidate models. Among the 100 models recorded for a given *c*, we select the best/preferred decision tree according to the model selection strategy stated in the previous section.

7.1. *Decision trees based on number of faults*

The preferred DT model for each of the five *c* values are presented in Table 1. The table includes the quality-of-fit (i.e., based on the fit data set) performances of the respective models in terms of their Type I, Type II, and Overall misclassification error rates. The Overall error rate simply indicates the total proportion of modules that are misclassified. The average error rates are indicated in the last row. The table also indicates the number of nodes in the respective DT. We observe that the tree size is very similar for the different values of *c*, and the preferred model is the one with $c = 4.5$. A general trade-off between the Type I and Type II error rates is observed with respect to *c*, i.e., as *c* is increased, the Type I error rate tends to increase, while the Type II error rate tends to decrease, and vice versa.

The results of the models when applied to the test data set are also summarized in Table 1. In the context of the predictive capability (for test data set) of the GP-based decision trees, the average Type I error rate is 22.08%, the average Type II error rate is 16.55% and the average Overall error rate is 20.89%. These values are very similar to the respective quality-of-fit values, indicating that the GP models are not prone to overfitting. In the case of the DT for each value of *c*, upon the deletion of the introns which do not impact the final result, and the

Table 1. Decision trees based on number of faults.

Tree model		Fit data set		
c	Nodes	Type I (%)	Type II (%)	Overall (%)
3	17	24.53	22.73	24.16
3.5	17	18.51	22.16	19.33
4	17	21.20	17.61	20.45
4.5	17	20.89	12.50	19.08
5	16	24.84	13.64	22.43
Average	17	21.99	17.73	21.09
Tree model		**Test data set**		
c	Nodes	Type I (%)	Type II (%)	Overall (%)
3	17	24.53	22.73	24.16
3.5	17	18.51	22.16	19.33
4	17	21.20	17.61	20.45
4.5	17	20.89	12.50	19.08
5	16	24.84	13.64	22.43
Average	17	22.08	16.55	20.89

branches that will never be traversed, the size of the best DT contains only eight internal nodes and nine leaf nodes. The best DT model for classification based on number of faults is shown in Fig. 1. The figure shows that all five software metrics are to be used for prediction of *fp* and *nfp* modules. Each nonleaf node is associated with one software metric and a threshold value. For a given nonleaf node, all the modules that have a value lower than the displayed threshold will be sent to the left child, and to the right child otherwise.

7.2. *Decision trees based on code churn*

The respective decision tree model for each value of *c* are presented in Table 2. The table includes the quality-of-fit performances of the

Table 2. Decision trees based on code churn.

Tree model		Fit data set		
c	Nodes	Type I (%)	Type II (%)	Overall (%)
3	19	19.42	22.75	20.20
3.5	19	17.48	22.75	18.71
4	29	17.48	23.28	18.84
4.5	13	19.09	22.75	19.95
5	23	26.38	14.29	23.54
Average	21	19.97	21.16	20.25
Tree model		**Test data set**		
c	Nodes	Type I (%)	Type II (%)	Overall (%)
3	19	16.88	20.83	17.82
3.5	19	16.56	18.75	17.08
4	29	17.21	21.88	18.32
4.5	13	18.18	19.79	18.56
5	23	23.38	10.42	20.30
Average	21	18.44	18.33	18.42

models in terms of their error rates and their averages across the five models. Upon observing the number of nodes for each DT, we note though the average is 21 nodes there is considerable variability in the tree size with respect to c. Subsequently, in addition to the preferred balance between the Type I and Type II error rates, the tree size is also considered while selecting the preferred model. Hence, the best model is the DT when $c = 3.5$.

The results of the trained models when applied to the test data set are also summarized in Table 2. In the context of the predictive capability of the decision trees, the average Type I error rate is 18.44%, the average Type II error rate is 18.33% and the average Overall error rate is 18.42%. These values are better than the respective quality-of-fit

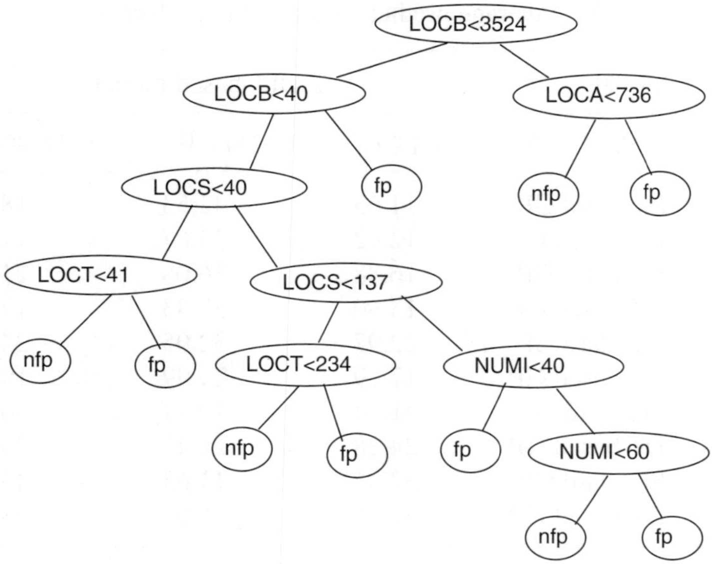

Fig. 2. Decision tree based on code churn.

values, indicating that the GP models do not suffer from overfitting. The size of the preferred decision tree is 19 nodes with nine internal nodes and 10 leaf nodes. The best DT model for classification based on code churn is shown in Fig. 2. The figure shows that only four software metrics are to be used for prediction, with NUMI not being used.

7.3. *Standard GP versus GP-based decision trees*

The models based on standard GP and GP-based decision trees are evaluated based on their average performances among the 20 runs. For a given *c* value, the average performances of classification models indicates the average among the 100 models that were recorded for that value of *c*. We opted to compare the average performances because each run in GP can yield a different model. Therefore, comparing the two techniques across all the runs is more justifiable. The techniques are compared for models based on number of faults and for those based on code churn.

Table 3 presents the respective average error rates (for the test data set) of standard GP and GP-based decision trees for SQC models

Table 3. Average predictive (test data) performances.

Model		Faults-based models		
c	Algorithm	Type I (%)	Type II (%)	Overall (%)
3	Standard GP	11.63	42.53	18.33
	DT-based GP	12.92	34.67	17.63
3.5	Standard GP	16.88	36.08	21.04
	DT-based GP	13.60	33.43	17.90
4	Standard GP	22.97	32.05	24.94
	DT-based GP	17.19	29.99	19.97
4.5	Standard GP	41.14	19.41	36.64
	DT-based GP	24.28	22.45	23.89
5	Standard GP	52.91	11.63	43.96
	DT-based GP	31.86	16.68	28.57
Model		Code churn-based models		
c	Algorithm	Type I (%)	Type II (%)	Overall (%)
3	Standard GP	18.87	30.48	21.59
	DT-based GP	16.98	24.10	18.65
3.5	Standard GP	22.43	26.60	23.41
	DT-based GP	19.43	20.75	19.74
4	Standard GP	34.61	18.18	30.76
	DT-based GP	19.68	21.46	20.10
4.5	Standard GP	43.51	12.21	36.18
	DT-based GP	20.37	20.32	20.36
5	Standard GP	49.74	8.52	40.08
	DT-based GP	26.11	17.14	24.01

based on number of faults and models based on code churn. The techniques are compared using the test data set because in software quality modeling we are interested in the predictive capability of a SQC model. The relative performances of the preferred models based on the two techniques can be observed in Table 4. Recall, for the WLTS case study the preferred model is one that has the best balance

Table 4. Best predictive (test data) performances.

Faults-based tree models			
Algorithm	Type I (%)	Type II (%)	Overall (%)
Standard GP	27.76	24.14	26.97
DT-based GP	21.45	16.09	18.56
Code churn-based trees models			
Algorithm	Type I (%)	Type II (%)	Overall (%)
Standard GP	18.83	21.88	19.54
DT-based GP	16.56	18.75	17.07

between the Type I and Type II error rates, with Type II being as low as possible.

The two tables indicate that the GP-based DT technique is better at obtaining the preferred classification model than those based on standard GP. When comparing the predictive performances of the preferred SQC models (Table 4), we observe that both the Type I and Type II error rates are lower for the GP-based decision tree technique. Moreover, the Overall error rates of the GP-based decision trees are always lower than those of the standard GP models (see Tables 3 and 4). Therefore, in addition to achieving better classification, the proposed technique yields univariate binary decision trees, which are more comprehensible than the symbolic (S-expression) regression trees obtained by standard GP.

8. Conclusion

An automated approach for building simplified GP-based decision tree models in the context of the SQC problem is presented. Software quality prediction prior to testing or software operations is very effective in reducing the corrective maintenance problems. Optimizing the classification accuracy and tree size of decision trees is an important and difficult issue. STGP is utilized to build optimal multi-objective decision trees. Decision trees are built for software data collected from two embedded systems. The models predict the class membership of software modules as either *fp* or *nfp*, depending on the type of quality factor used: number of faults or code churn. It is shown that the GP-based decision tree modeling approach achieves better results than SQC models calibrated using standard GP. Moreover, the proposed model achieved good optimization with respect to the tree size and the model selection strategy of balancing the two error rates.

The GP-based decision tree modeling procedure presented, provides flexibility in building optimal classification models in the presence of multiple constraints which is very common in software development. More specifically, in addition to accuracy and tree size, additional performance objectives may be added according to the needs of the software project. Traditional DT-based approaches are generally not suited for optimizing objectives other than accuracy. However, in those methods the tree size can be varied through a parameter.

A software management team can utilize the multi-objective optimization approach presented in this chapter for building goal-oriented software quality estimation models. For example, in addition to classification accuracy and tree size, the GP-based approach presented can be modified to include an objective which ranks the program modules within each class from the most faulty to the least faulty. Similarly, other practical goal-oriented objectives can be included in the optimization process.

Future work will involve studies in which the performance of GP-based and non-GP-based decision trees will be compared. In addition, the performance of the proposed GP-based decision tree approach will be compared with other decision tree techniques based on genetic algorithms or genetic programming.

Acknowledgment

This work was supported in part by the National Science Foundation Grant CCR-9970893.

References

1. N. F. Schneidewind, Investigation of logistic regression as a discriminant of software quality, *Proceedings: 7th International Software Metrics Symposium* London, UK (2001), pp. 328–337.
2. R. Takahashi, Y. Muraoka and Y. Nakamura, Building software quality classification trees: Approach, experimentation, evaluation, *Proceedings: 8th International Symposium on Software Reliability Engineering*, Albuquerque, NM, USA (1997), pp. 222–233.
3. K. E. Imam, S. Benlarbi, N. Goel and S. N. Rai, Comparing case-based reasoning classifiers for predicting high-risk software components, *Journal of Systems and Software* **55** (2001) 301–320.
4. N. Ohlsson, M. Zhao and M. Helander, Application of multivariate analysis for software fault prediction, *Software Quality Journal* **7** (1998) 51–66.
5. R. Kumar, S. Rai and J. L. Trahan, Neural-network techniques for software-quality evaluation, *Proceedings of the Annual Reliability and Maintainability Symposium*, Anaheim, CA, USA (1998), pp. 155–161.
6. T. M. Khoshgoftaar, X. Yuan and E. B. Allen, Balancing misclassification rates in classification tree models of software quality, *Empirical Software Engineering Journal* **5** (2000) 313–330.
7. M. Shepperd and G. Kadoda, Comparing software prediction techniques using simulation, *IEEE Transactions on Software Engineering* **27** (2001) 1014–1022.
8. Y. Liu and T. M. Khoshgoftaar, Genetic programming model for software quality prediction, *Proceedings: 6th International High Assurance Systems Engineering Symposium*, Boca Raton, Florida, USA (2001), pp. 127–136.
9. L. A. Breslow and D. W. Aha, Simplifying decision trees: A survey, *The Knowledge Engineering Review* **1** (1997) 1–40.
10. S. R. Safavian and D. Landgrebe, A survey of decision tree classifier methodology, *IEEE Transactions on Systems, Man and Cybernetics* **21** (1991) 76–85.
11. P. Angeline and K. E. Kinner, *Advances in Genetic Programming*, Vol. II (MIT Press, Cambridge, 1996).
12. H. Iba, H. de Garis and T. Sato, Genetic programming using a minimum description length principle, *Advances in Genetic Programming* **12** (1996) 265–284.
13. J. R. Koza, *Genetic Programming*, Vol. I (MIT Press, New York, 1992).

14. W. B. Langdon, *Data Structures and Genetic Programming, Advances in Genetic Programming 2* (MIT Press, Cambridge, 1996).
15. M. C. J. Bot and W. B. Langdon, Application of genetic programming to induction of linear classification trees, *GECCO: Proceedings of the Third European Conference on Genetic Programming* (2000), pp. 247–258.
16. N. I. Nikolev and V. Slavov, Inductive genetic programming with decision trees, *Proceedings: 9th European Conference on Machine Learning*, Prague, Czech Republic (1997), pp. 183–190.
17. M. Shirasaka, Q. Zhao, O. Hammami, K. Kuroda and K. Saito, Automatic design of binary decision trees based on genetic programming, *Proceedings: 2nd Asia-Pacific Conference on Simulated Evolution and Learning (SEAL'98)*, ed. C. Newton, Canberra, Australia (1998), pp. 24–27.
18. D. J. Montana, Strongly typed genetic programming, *Evolutionary Computation* **23** (1995) 199–230.
19. M. Reformat, W. Pedrycz and N. J. Pizzi, Software quality analysis with the use of computational intelligence, *Proceedings: IEEE International Conference on Fuzzy Systems*, Vol. 2, Honolulu, HI, USA (2002), pp. 1156–1161.
20. S. K. Murthy, Automatic construction of decision tree from data: A multidisciplinary survey, *Data Mining and Knowledge Discovery* **2** (1998) 345–389.
21. J. R. Quinlan, Induction of decision trees, *Machine Learning* **1** (1986) 81–106.
22. C. Schaffer, Deconstructing the digit recognition problem, *Proceedings of the Ninth International Conference on Machine Learning* **1** (1992) 394–399.
23. W. Banzhaf, P. Nordin, R. E. Keller and F. D. Francone, *Genetic Programming: An Introduction on the Automatic Evolution of Computer Programs and Its Application* (PWS Publishing Company, New York, 1998).
24. T. D. Haynes, D. A. Schoenefeld and R. L. Wainwright, Type inheritance in strongly typed genetic programming, *Advances in Genetic Programming* **2** (1996) 359–376.
25. H. Eschenauer, J. Koski and A. Osyczka, *Multicriteria Design Optimization* (Springer-Verlag, Berlin, Heidelberg, Germany, 1971).
26. W. Stadler, *Preference Optimality and Application of Pareto-Optimality* (Springer-Verlag, Berlin, Germany, 1975).
27. T. M. Khoshgoftaar, E. B. Allen, W. D. Jones and J. P. Hudepohl, Data mining of software development databases, *Software Quality Journal* **9** (2001) 161–176.

CHAPTER 11

An Approach to Quantifying Process Cost and Quality

Greg Twaites

General Dynamics Advanced Information Systems,
8800 Queen Avenue South, Bloomington, MN 55431, USA

Constance Hoffman

8201 E. McDowell Rd., MS H177, Scottsdale, AZ 85257, USA

1. Introduction

Designing and implementing an effective product development process can be difficult and expensive. Once an organization has crafted a process, it can be extremely difficult to quantify the processes cost and effectiveness. Often, process evaluation will be an *ad-hoc* combination of subjective and objective metrics. While subjective elements, such as ease of following the process, are valid considerations, more objective metrics are desirable. This chapter proposes metrics that measure (1) the effort associated with following a process and (2) the quality of the product generated by the process and the quality of the process itself.

2. Measuring Effort

Effort data measures how much time/money is spent performing various activities associated with designing and building a product.

Collecting effort data on all labor produces a consistent measure of the activity types associated with project design and execution. The data can aid in better understanding the design process and help identify areas for improvement. While intended primarily for measuring development activities, this collection activity can apply to all personnel involved with any aspect of a project and can assist in achieving process maturity ratings.

Many organizations track the amount of effort spent in various development phases, e.g., requirements, design, implementation, and integration and test. This is usually achieved via a project Work Breakdown Structure (WBS) delineating standard development phases. While WBS-level data represent a good start, they do not give enough visibility into the day-to-day allocation of effort.

Table 1 shows typical effort data collected for each phase of a project. Cursory data analysis indicates a reasonable distribution of effort across the development phases. A valid conclusion would be that the project is in reasonable shape.

Table 2 shows a finer granularity of the effort data collected for activities within each phase of a project. Analysis of these data show that this project is (1) expending very little effort on reviewing products, and (2) spending increasing amounts of time performing rework. One conclusion from these data might be that the lack of reviews is causing errors to be missed, resulting in more rework. This higher resolution breakdown portrays a different situation and indicates that further analysis may be required.

Table 1. Coarse measurement of project effort.

Phase	Effort (hours)	% of Total
Requirements	1500	15
Design	2500	25
Implementation	3500	35
Integration and test	2500	25

Table 2. Finer resolution of project effort.

Phase	Effort (hours)	Effort (hours)		% of Total
Requirements	1500	Generation	1200	15
		Peer review	0	
		Rework	300	
		Test	0	
Design	2500	Generation	1300	25
		Peer review	200	
		Rework	1000	
		Test	0	
Implementation	3500	Generation	1500	35
		Peer review	500	
		Rework	1000	
		Test	500	
Integration and test	2500	Generation	200	25
		Peer review	0	
		Rework	1500	
		Test	800	

Appendix B of this chapter presents a more detailed example of using effort codes to manage a project.

3. Effort Codes

Effort data can be collected several ways. At a lower level, staff can keep a separate log of daily activities, automated software can monitor activities, activities can be based on specific work products, and so on. At a higher level, companies often use an automated time-tracking system for payroll purposes. In addition, organizations performing work for the federal government have strict time-reporting rules imposed upon them. In these cases, the infrastructure needed to support the collection of effort metrics is already in place and can be modified to accommodate effort data collection. This chapter assumes that a suitable time reporting system is available for collecting effort data.

This chapter proposes tracking effort with *Effort Codes*. These define the various activities in which an employee participates as part of his daily routine and are recorded on the employee's timesheet. Individual effort codes are described in the following paragraphs.

The effort code is a combination of one to three characters used to record how the employee allocated his effort while completing a task. As shown in Table 3, the first letter represents the Activity, the second letter represents the Task, and the third letter designates the engineer's role or discipline while performing the task.

One, two, or all three letters can be used depending on the organization's needs and the project's structure. For example, if the WBS or charge number adequately reflects the task being worked, the Activity code alone is sufficient.

3.1. *Activity code*

The activity code identifies the primary activity performed for work associated with the corresponding charge number on the time sheet. These codes are designed to be general enough to apply to most work products. The defined activity codes, along with the allowable task codes, are shown in Table 4. Appendix A of this chapter provides a more detailed description of activity codes (see Refs. 1 and 2).

- Generate (G) — initial work product creation where a work product may be a subsection of an overall product.

Table 3. Effor code.

1st letter	2nd letter	3rd letter
Activity code	Task code	Discipline code

Table 4. Activity codes.

Activity code	Description	Allowable task code
G	Generate product	M, Q, U, I, H, B, S, R, O, A, F, C, D
E	Evaluate product	M, Q, U, I, C, D, H, B, S, R, O, A, N, T, E, P, F
R	Rework problems	M, Q, U, I, H, B, S, R, O, A, F, P, D, C
S	Support/ management/ administration	N, T, E, P, F
T	Test	M, Q, U, I, F, H

- Evaluate (E) — formal and informal work product review. This includes preparing for and participating in inspections and collaborations. Evaluate is not intended to capture informal task review that occurs during generation.
- Rework (R) — correcting work products based on inspection results or reported defects. Reducing rework is a primary goal.
- Support (S) — performing an administrative task.
- Test (T) — executing integration and system tests.

3.2. *Task code*

This code identifies the task associated with the activity selected for the activity code (1st character). As shown in Table 4, not all task codes can be associated with a particular activity code. Task codes are summarized in Table 5.

Appendix A of this chapter provides a detailed mapping of the relationship between specific activity and task codes.

3.3. *Discipline code*

In some organizations, employees are organized by discipline, e.g., system, software, electrical, mechanical, etc., and belong to a specific

Table 5. Task codes.

Task code	Description	Allowable activity codes
A	Architecture definition/documentation	G, E, R
B	Bill of information and documentation activities	G, E, R
C	Code/build/breadboard	G, E, R
D	Detail design and analysis	G, E, R
E	Metrics and process development	E, S
F	Fixtures/tools/facilities	G, E, R, S, T
H	Health and safety	G, E, R, T
I	Integration activities	G, E, R, T
M	Materials and parts	G, E, R, T
N	Administrative activities	E, S
O	Interface control definition/requirements	G, E, R
P	Planning (task, project, etc.)	E, R, S
Q	Qualification/verification/validation activity	G, E, R, T
R	Requirements — Units, modules, SW, systems	G, E, R
S	Trade studies	G, E, R
T	Training and mentoring	E, S
U	Unit/assembly/product/system	G, E, R, T

department. In these matrixed organizations, expenses are often charged to an employee's department. It can be beneficial to track the disciplines performing each activity and task and charge the appropriate department. The discipline code identifies the engineer's primary responsibility on a particular project. For most individuals, this will be their area of expertise and will be the same for multiple charge numbers. However, it is possible that someone might be performing as a system engineer on one task and a software engineer on another. The defined discipline codes are listed in Table 6. These codes apply to all allowable combinations of activities and tasks.

Table 6. Discipline codes.

Discipline code	Description
H	Hardware — Includes electrical engineering and design, all types and disciplines, electrical technicians
S	Software — Includes software engineering, software subcontracting, software technicians, COTS software activities
Y	Systems — Includes systems engineering, system I & T, system analysis, configuration management
M	Mechanical — Includes mechanical engineering, mechanical design, PWB design, mechanical technicians
F	Manufacturing — Includes manufacturing, process engineering

In many cases, an engineer's discipline can be extracted from his personnel or payroll record and using the discipline code may be redundant. However, it can be beneficial when there is a need to track cross-disciplinary charging. In this case, the discipline code can be used.

Example 1.

A software engineer spends time developing software to support a simulation for systems engineering. This effort is counted as systems engineering rather than software engineering (charge GCY, not GCS) and the systems engineering organization is billed for the software engineer's effort.

Example 2.

A system engineer spends time reviewing software requirements. This effort is counted as software engineering rather than systems engineering (charge ERS, not ERY), and the software engineering organization is billed for the system engineer's effort.

4. Implementation

4.1. *Introducing effort codes to an organization*

The effort codes described in this chapter can be a bit daunting. To help ease the introduction and use of effort codes into day-to-day activities, it is useful to provide a map for applying effort codes. Figure 1 shows an example of an Integrated Product and Process Development (IPPD) diagram for an organization with effort codes mapped onto the associated IPPD elements. Each box represents a task performed by phase with the associated effort codes shown in brackets.

4.2. *Collection effort codes via timesheet*

One of the easiest ways to automate effort data collection is to leverage an existing infrastructure, such as a time entry system. Table 7 shows a sample time sheet showing use of multiple activity codes. In this example, all three effort codes (EC) characters are used.

4.2.1. *Monday*

The employee spends four hours creating/developing software requirements (GRS), 2.5 hours in a planning meeting for the next software increment (SPS), 1.5 hours reviewing system requirements (ERY), and one hour fixing defects in the system requirements (RRY). Even though the employee spent the entire day on one project, his efforts were devoted to different aspects of the project.

4.2.2. *Tuesday*

The employee spends six hours fixing defects in the system requirements (RRY) and three hours in a management review of the project (SNY).

4.2.3. *Wednesday*

The employee spends four hours working on software requirements for project 1234 (GRS) and four hours working on software

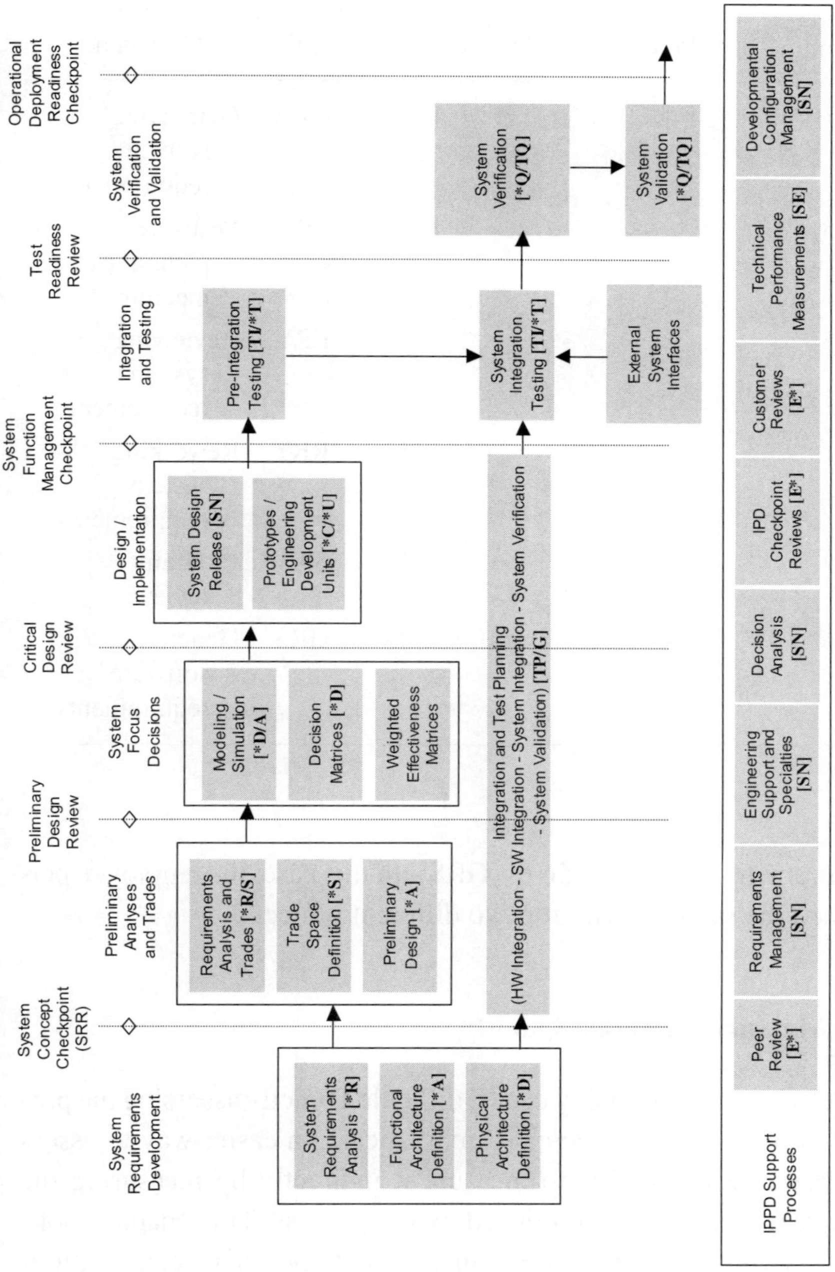

Fig. 1. Example IPPD diagram with effort code mapping.

Table 7. Sample time entry sheet with effort codes.

Mon	Tue	Wed	Thu	Fri	Project #	EC	Comment
4		4			1234	GRS	Generating software requirements
2.5					1234	SPS	Software planning meeting
1.5					1234	ERY	Reviewing system requirements
1	6				1234	RRY	Reworking system requirements
	3				1234	SNY	Program reviews
		4			5678	GRS	Generating software requirements

requirements for project 5678 (GRS). In this case, the employee performed the same activity for two different projects.

5. Measuring Process Quality

Measuring process quality can be difficult. Directly assessing the process is often not possible or is impractical. An easier way to assess process effectiveness is to measure it indirectly by measuring the quality of the products produced by the process. This chapter looks at measuring the quality of the intermediate products, e.g., requirements, designs, implementations, etc., by means of defects detected during product evaluations. These quality measures are stored and correlated with the effort data.

5.1. *Collecting quality data via evaluations*

Most mature development organizations have defined a process for evaluating work products to eliminate defects. Peer Reviews, Structured Walkthroughs, and Fagin Inspections are typical evaluation techniques.

Inspections in this chapter are the same as the SEI Level 3 KPA "Peer Reviews". It is assumed that metrics are collected during inspection.[3]

Collaborations are the informal way work products are reviewed. No metrics are required to be collected during collaborations, but design decisions may be recorded in meeting minutes. Both collaborations and inspections help reduce the number of defects in a work product.

An online database of information about each evaluation is maintained. The database supports queries to analyze its contents. Additionally, the raw data collected from each inspection may be viewed. The following types of information can be obtained from the database:

- Number of inspections performed.
- Inspection quality (as a mathematical value obtained by applying a specific formula to the metrics data collected during the inspection process).
- Average length of each inspection.
- Average number of defects found in each inspection.
- Average number of pages/drawings/lines-of-code/etc inspected per hour.
- Types of defects found (according to a pre-defined set of defect types and classifications defined within the evaluation process of the organization).
- Phase detected and phase injected.

Access to this information should be straightforward. Every project in the organization should have access to the evaluation database and a wide set of robust queries should be available. Each user can obtain a report of the total number of inspections, the inspection quality, the average inspection length, the average number of pages inspected per hour, and much more. By comparing their project's metrics against the company's metrics, the project leader can gain insight into their process performance.

6. Analysis Metrics

Initially, effort code data can provide metrics on the level of engineering participation, the cost of quality, the cost of noncompliance, and the distribution of engineering labor across all activity code categories. These metrics can be created by project, department, and organization. Data is not made available at the individual level (except lack of participation) to avoid creating personal bias in reporting activity codes. Individual projects correlate the distribution of engineering labor to the project phase. Initially, these metrics are collected and reviewed monthly. As a better understanding of the design process is gained, these metric reports may be updated more frequently. Changes to metrics to be reported should be reflected in the organization's Metrics Manual, and it should always be consulted for current information.

Figure 2 shows Percent of Engineering Staff Reporting Activity Code Data by Department. This is useful while establishing the effort code collection program. While not tied directly to the effort for any specific activity, the participation level indicates the depth of the data. For example, lack of participation may indicate tasks being performed that are not covered by Effort Codes. The codes may need to be expanded to account for tasks unique to the company.

Figure 3 indicates the amount of effort spent reviewing products as they are developed. This is often referred to as the *Cost of Quality*. This metric should be used in conjunction with other metrics

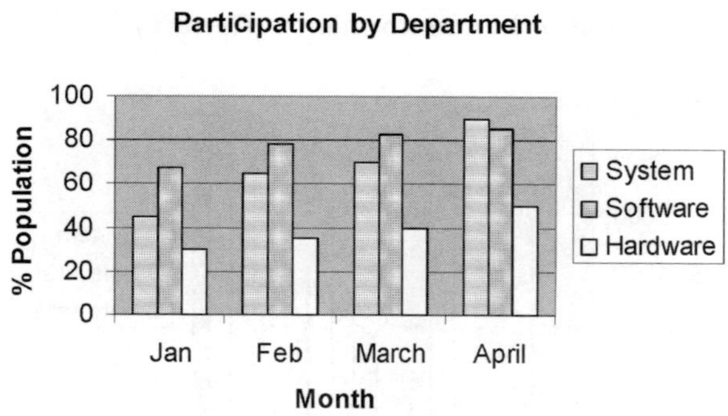

Fig. 2. Percent of engineering staff reporting activity code data by department.

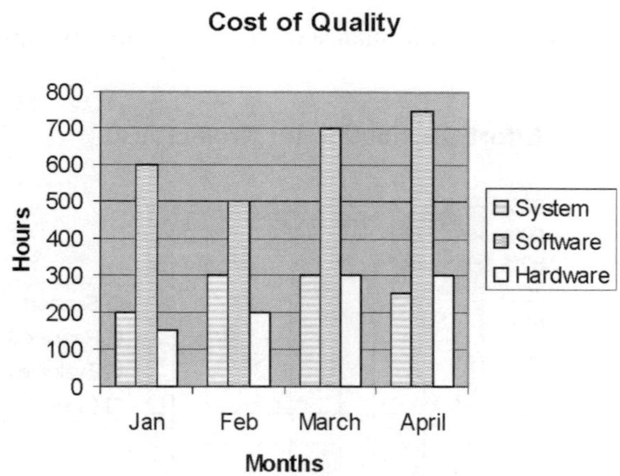

Fig. 3. Amount of effort expended on reviews (cost of quality).

to determine whether time spent reviewing the work products is in fact producing better quality.

Figure 4 shows the amount of effort spent fixing problems that have been discovered. This measure is often referred to as the *Cost of Nonconformance*.

Figure 5 shows the distribution of effort across multiple activities for a single project. Shown in percentages, this chart is a quick visual indicator of how a project is allocating its resources. In this example, we can see that in March the amount of rework effort increased dramatically and the amount of new creation decreased markedly.

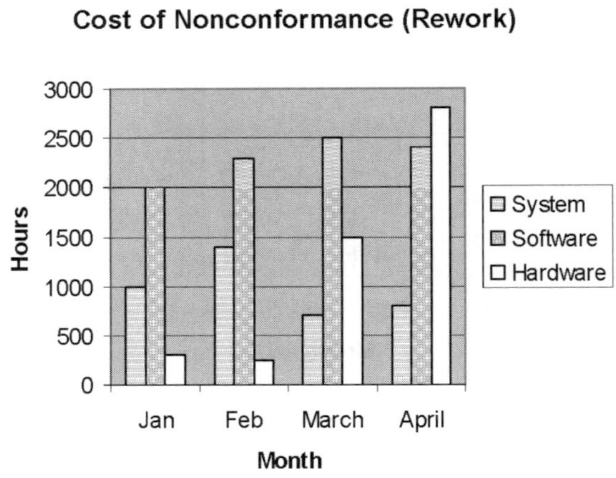

Fig. 4. Cost of nonconformance (defect correction) by discipline.

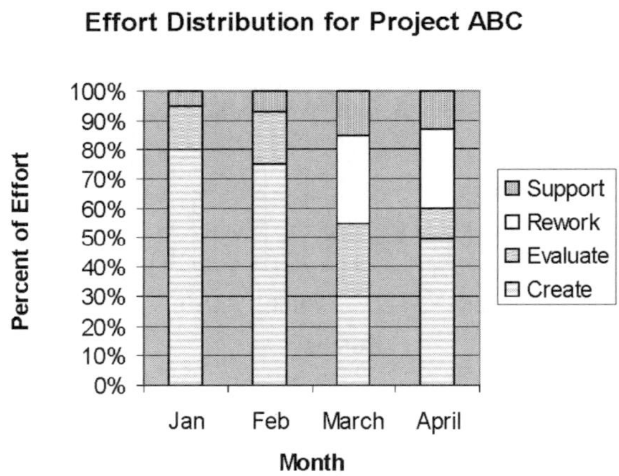

Fig. 5. Distribution of engineering activities for a project.

This is a red flag that something may be wrong and requires further investigation.

Figure 6 shows the distribution of effort for multiple projects across the organization. Shown in percentages, this chart is a visual indicator of how the organization's projects compare to each other. This chart is typical of a report that the organization or process management would review. For example, when used in conjunction with

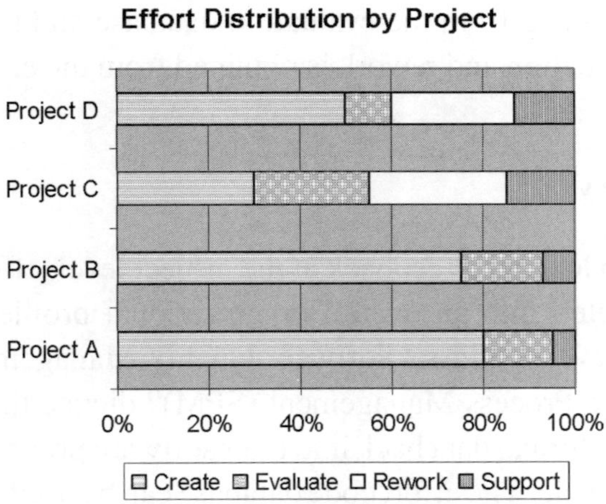

Fig. 6. Project distribution of engineering labor by month and phase.

customer-reported defects metrics, management could gain insight into which projects may have enhanced their process for better results. These lessons learned could then be shared and the company's process improved.

Effort and quality data can be correlated to produce data representation as illustrated in Table 8. For this measure, the number of

Table 8. Correlating effort and quality.

	Requirements	Design	Implementation
Hours to evaluate	100	500	500
Number of defects detected	15	80	75
Hours to rework	50	170	100
Defect detected/evaluation hour	0.15	0.16	0.15
Rework hour/detected defect	3.33	2.125	1.33

defects is obtained from the evaluation database and the number of hours for evaluation and rework is obtained from the effort database.

7. Summary

In addition to localized feedback at the project level, effort code data are consolidated into an overall organizational profile. In the SEI CMM, the Level 4 KPAs "Software Quality Management (SQM)" and "Software Process Management (SPM)" dictate that an organization feed historical data back into their software process to improve the software cycle. The effort code database can be used to meet these Level 4 KPAs.[4]

Many companies use software planning tools to predict the amount of time needed to complete a project. This tool helps the organization to size internal efforts or to bid on outside contracts. These costing tools usually require that prior organizational information be gathered and entered for consideration. Since effort code data contains concrete measurements of the organization's past effort, the information can be used to predict future behavior.

Applying effort codes to the time reporting system obtains a consistent measure of project execution activities. Measurements quantify the development and management effort for a project as it is being worked, and this information can help predict effort required for future work on the same project, similar current projects, or future projects. The information can help predict cost and schedule from actual organization historical data. Prior to the availability of Effort Data, predictions were based on less exact methods.

References

1. Activity based metrics, General Dynamics Advanced Information Systems, internal publication.
2. Concurrent engineering burden codes for communication systems division, General Dynamics Decision Systems, internal publication.

3. Software metrics manual, General Dynamics Defense Systems, internal publication (1998).
4. G. Twaites and M. Sibilla, Software engineering in an SEI level-5 organization, *International Journal of Reliability, Quality and Safety Engineering* **9** (2002) 347–365.

Appendix A: Effort Code Tables

This appendix presents a proposed set of effort codes. While these are not intended as an exhaustive set of every possible activity, they comprise a fairly complete initial effort. Each organization will tailor the code set to their specific processes.

Table 9 is an overview of the effort codes. Tables 10–14 describe each activity code and the associated allowable task codes, as indicated below.

Table 9. Overview of engineering effort codes.

Activity code	Description	Task code	Description	Discipline code	Description
E	**E**valuate/inspect/review (M, Q, U, I, C, D, H, B, S, R, O, A, N, T, E, P, F)	A	**A**rchitecture definition/documentation	H	**H**ardware
G	**G**enerate/create/design (M, Q, U, I, H, B, S, R, O, A, F) problem diagnosis/redesign/	B	**B**ill of information and documentation activities	M	**M**echanical
R	**R**ework/retest/repair (M, Q, U, I, H, B, S, R, O, A, F, P, D, C)	C	**C**onstruct/code/build/breadboard	S	**S**oftware

Table 9. (*Continued*)

Activity code	Description	Task code	Description	Discipline code	Description
S	**S**upport/ management/ administration (N, T, E, P, F)	D	**D**etail design and analysis	Y	S**y**stems
T	**T**est (M, Q, U, I, F, H)	E	M**e**trics and process development	O	**O**ther
		F	**F**ixtures/tools/ facilities		
		H	**H**ealth and safety		
		I	**I**ntegration activities		
		M	**M**aterials and parts		
		N	Admi**n**istrative activities		
		O	Interface c**o**ntrol definition/ requirements		
		P	**P**lanning (task, project, etc.)		
		Q	**Q**ualification/ verification/ validation activity		
		R	**R**equirements — Units, modules, SW, systems		
		S	Trade **S**tudies		
		T	**T**raining and mentoring		
		U	**U**nit/assembly/ product/system		

Table 10. Generate [G] activity code and associated task codes.

Task code	Description	Examples and details
A	**A**rchitecture definition/ documentation	Design or creation of hardware, software, or system architecture and associated documentation
B	**B**ill of information and documentation activities	The creation/design of the BOI and any of its items/objects except architecture (A), interface (I), qualification (Q), unit/assembly/product/system (U), or requirements (R) documentation. Includes SW build documentation, Integration plans and procedures, assembly and fabrication procedures, and CM Tasks
C	**C**onstruct/code/ build/breadboard	Construction, code, build, or breadboard of a hardware, software, or system item (or portion thereof) and its associated documentation, except items in I, H, or F
D	**D**etailed design and analysis	Design, creation, simulation, or analysis of a hardware, software, or system item (or portion thereof) and its associated documentation, except items in I, H, or F
F	**F**ixtures/tools/ facilities	Design, creation, or generation of any fixtures, tooling, facilities, test equipment, test software, etc.
H	**H**ealth and safety	Design, creation, or generation of any product or service health or safety feature
I	**I**ntegration activities	Design, simulation, and/or analysis of hardware or software interfaces. Creation, generation, or design of integration plans and procedures
M	**M**aterials and parts	Design of parts or materials

Table 10. (*Continued*)

Task code	Description	Examples and details
O	Interface c**o**ntrol definition/ requirements	Design or creation of all interface control documentation
Q	**Q**ualification/ verification/ validation activity	Design or creation of hardware, software, or system qualification/verification/ validation processes, plans, or procedures
R	**R**equirements — Units, modules, SW, systems	Design or creation of hardware, software, or system requirements and associated documentation
S	Trade **S**tudies	All activity associated with the completion of a trade study
U	**U**nit/assembly/ product/system test activities	Design, creation, or generation of test plans, test cases, and test procedures

Table 11. Evaluate [E] activity code and associated task codes.

Task code	Description	Examples and details
A	**A**rchitecture definition/ documentation	Evaluation activity including nonprocedural test efforts, evaluations, inspections, peer reviews, and internal design reviews of hardware, software, or system architecture and associated documentation

Table 11. (*Continued*)

Task code	Description	Examples and details
B	**B**ill of information and documentation activities	Evaluation activity including inspections, peer reviews, and internal design reviews of the BOI and any of its items/objects except architecture, interface, or requirements documentation. Includes SW build documentation, Integration plans and procedures, Test plans and procedures, Assembly and fabrication procedures, and CM tasks
C	**C**onstruct/code/ build/breadboard	Evaluation activity including nonprocedural test efforts, evaluations, inspections, peer reviews, and internal design reviews of construction or coding of any level of a design not in U or I
D	**D**etail design and analysis	Evaluation activity including nonprocedural test efforts, evaluations, inspections, peer reviews, and internal design reviews of detail design, simulation, modeling, or analysis of a hardware or software item not in U or I
F	**F**ixtures/tools/ facilities	Evaluation activity including nonprocedural test efforts, evaluations, inspections, peer reviews, and internal design reviews of any fixtures, tooling, facilities, test equipment, test software, etc.
H	**H**ealth and safety	Evaluation activity including nonprocedural test efforts, evaluations, inspections, peer reviews, and internal design reviews of any product or service health or safety feature

Table 11. (*Continued*)

Task code	Description	Examples and details
I	**I**ntegration activities	Evaluation activity including nonprocedural test efforts, evaluations, inspections, peer reviews, and internal design reviews of hardware or software interfaces
M	**M**aterials and parts	Evaluation activity including nonprocedural test efforts, evaluations, inspections, peer reviews, and internal design reviews of parts or materials
O	Interface c**o**ntrol definition/ requirements	Evaluation activity including nonprocedural test efforts, evaluations, inspections, peer reviews, and internal design reviews of all interface control documentation
P	**P**lanning (task, project, etc.)	Evaluation activity including nonprocedural test efforts, evaluations, inspections, peer reviews, and internal design reviews of project plans, task plans, proposal plans, bidding and estimating, and all schedules
Q	**Q**ualification/ verification/ validation activity	Evaluation activity including nonprocedural test efforts, evaluations, inspections, peer reviews, and internal design reviews of hardware, software, or system qualification/verification/validation processes, plans, or procedures
R	**R**equirements — Units, modules, software, systems	Evaluation activity including nonprocedural test efforts, evaluations, inspections, peer reviews, and internal design reviews of hardware, software, or system requirements and their associated documentation

Table 11. (*Continued*)

Task code	Description	Examples and details
S	Trade **S**tudies	Evaluation activity including nonprocedural test efforts, evaluations, inspections, peer reviews, and internal design reviews of a trade study
U	**U**nit/assembly/ product/system	Evaluation activity including nonprocedural test efforts, evaluations, inspections, peer reviews, and internal design reviews of a hardware, software, or system item and its associated documentation, except items in I, H, or F

Table 12. Rework [R] activity code and associated task codes.

Task code	Description	Examples and details
A	**A**rchitecture definition/ documentation	Redesign and documentation of hardware, software, or system architecture
B	**B**ill of information and documentation activities	Correction of drawings, specifications, test plans/procedures, instructions, manuals, etc., except interface definition documentation. Includes SW build documentation, integration plans and procedures, test plans and procedures, assembly and fabrication procedures, and CM tasks
C	**C**onstruct/code/ build/breadboard	Problem diagnosis, redesign, rework, retest, or repair of any implementation level of a design not included in U or I

Table 12. (*Continued*)

Task code	Description	Examples and details
D	**D**etail design and analysis	Problem diagnosis or redesign of the detail design, simulation, modeling, or analysis of a system, hardware, or software item not included in U or I
F	**F**ixtures/tools/ facilities	Diagnosis, documentation correction, redesign, retest caused by defects associated with fixtures/tooling/facilities
H	**H**ealth and safety	Diagnosis and repair of an electrical shock hazard
I	**I**ntegration activities	Diagnosis and correction of integration-related defects in hardware, software, or systems
M	**M**aterials and parts	Diagnosis, repair, retest due to defects in parts or materials, including purchased software modules
O	Interface c**o**ntrol definition/ requirements	Correction of drawings, specifications, test plans/procedures, instructions, manuals, etc., associated with interface definition
P	**P**lanning (task, project, etc.)	This includes all replanning activities associated with project planning, task planning, proposal planning, bidding and estimating, and all scheduling activities
Q	**Q**ualification/ verification/ validation activity	Diagnosis and correction of errors in qualification/verification/validation processes or methods
R	**R**equirements — Units, modules, SW, systems	Diagnosis and correction of errors in requirements derivation or flowdown
S	Trade **S**tudies	Correction of errors or omissions in a trade study

Table 12. (*Continued*)

Task code	Description	Examples and details
U	**U**nit/assembly/ product/system	Diagnosis, redesign, retest of a hardware, software (at the unit level), or system item that is defective

Table 13. Support [S] activity code and associated task codes.

Task code	Description	Examples and details
E	**Me**trics and process development	This includes all activities associated with developing and reviewing processes; reporting, collecting, and analyzing metric data
F	**F**ixtures/tools/ facilities	This includes all activities related to fixtures, tools, or facilities that do not relate to either the design or test of these items
N	Admi**n**istrative activities	This includes all support/management/ administrative activities not included in T, E, P, or F; e.g., checking timesheets, project reviews, resolving personnel issues, security issues, building evacuation drills, communications meetings, etc.
P	**P**lanning (task, project, etc.)	This includes all activities associated with project planning, task planning, proposal planning, bidding and estimating, and all scheduling activities
T	**T**raining and mentoring	This includes all activities associated with training and/or mentoring that you provide or receive. This includes time spent in these activities and time for preparation in support of these activities. In general, conference attendance and authoring of papers falls in this category

Table 14. Test [T] activity code and associated task codes.

Task code	Description	Examples and details
F	**F**ixtures/tools/ facilities	All procedural test of hardware and/or software fixtures, tools, test equipment, development environments, facilities, and test or development software. These items are used in the development or production of products. These items are not offered for sale, but may or may not be deliverable to a customer depending on the contract terms
H	**H**ealth and safety	All procedural testing to evaluate the health and/or safety features, effects, or characteristics of an item
I	**I**ntegration activities	All procedural test of hardware and/or software to support the integration or interfacing of two or more items
M	**M**aterials and parts	All test of parts or materials to support a project, department, division, group, sector, corporate, industry standard, etc., to a released procedure
Q	**Q**ualification/ verification/ validation activity	All procedural test of hardware and/or software for certification of performance to requirements and/or suitability for its intended application. This includes all product qualification testing, type acceptance, certification to a standard, and reliability testing used for any deliverable item
U	**U**nit/assembly/ product/system	All procedural test of hardware and/or software not included in "Q" or "I" above; e.g., alignment; characterization; execution of SW unit test; workmanship verification; testing that is not part of a deliverable item qualification, verification, or validation

- Table 10 — Task Codes associated with the [G]enerate Activity code.
- Table 11 — Task Codes associated with the [E]valuate Activity code.
- Table 12 — Task Codes associated with the [R]ework Activity code.
- Table 13 — Task Codes associated with the [S]upport Activity code.
- Table 14 — Task Codes associated with the [T]est Activity code.

A.1. *Additional task codes*

A possible extension to the normal development tasks described in is to quantify activities involved with Non-Developmental Items (NDI). NDI are products that are purchased or provided to the organization and not developed within the organization. Table 15 lists NDI-related tasks.

Table 15. Non-Developmental Item (NDI) task.

Task code	Description
V	Evaluation — This phase includes evaluation and test of the NDI products, and selection of the NDI product. This phase concludes with procurement of the NDI product
X	Assessment — This phase includes inspection of the products, installation, and test. This phase concludes with a Software Assessment Review
Y	Survey — This phase includes identification of requirements and performing a product survey. Phase completion concludes with a defined set of suitable NDI product candidates
Z	Qualification — This phase includes general evaluation and code review, followed by detailed evaluation and test. This phase concludes with complete certification of the NDI product

Table 16. Simplified effort codes software and system.

Activity code		Task code		Discipline code	
G	Generate/create	P	Plan	S	Software
E	Evaluate/examine/inspect	R	Requirements	Y	Systems
R	Rework/fix	A	Architecture design	H	Hardware
T	Test execution	D	Detailed design	M	Mechanical
S	Manage/support	C	Construct/implement		
		U	Unit test		
		I	Integration		
		Q	Qualification		

A.2. *Simplified effort codes*

The previous tables describe a fairly comprehensive set of Activity, Task, and Discipline codes. While this level of detail may be suitable for some very mature organizations, it may be overkill for many organizations. Table 16 lists a subset of the previous codes that captures many of the most commonly used codes.

Appendix B: Example of Effort Code Usage

This appendix presents an example of using effort codes to assist in managing a software development project.

B.1. *Program description*

This program involved developing a custom product for a specific customer. The customer was the driving force for the requirements and the schedule. The project was comprised of three increments. Each increment consisted of Requirement, Architecture, Design, Code and Test phases. This example examines the use of effort codes by the software team during the Increment 1. The Increment 1 software effort began after system engineering provided the overall system architecture and a set of system requirements allocated to software.

B.2. *Schedule*

The software schedule for Increment I is shown in Fig. 7.

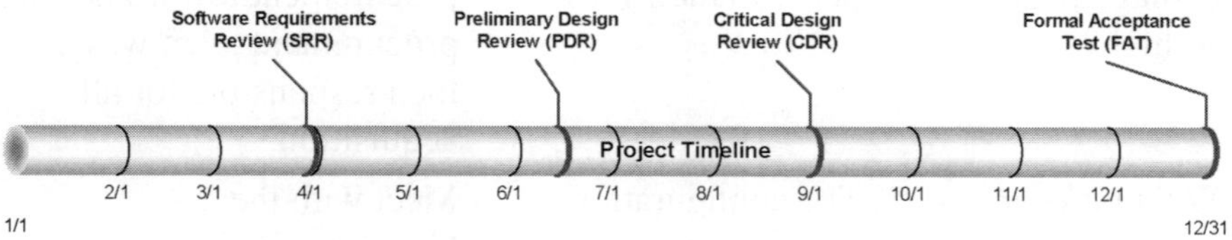

Fig. 7. Increment 1 software shedule.

Month 1. Effect codes reported.

Activity	Task	Discipline	%[a]	Description
S	P	S	20	Support Planning for Software
S	N	S	25	Support Administrative Activities for Software
G	B	S	20	Generate Bill of Information and Documentation Activities for Software
G	R	S	25	Generate Requirements for Software
G	F	S	10	Generate Fixtures/Tools/Facilities for Software

[a]Effort code metrics are shown here as percentages. Alternatively they could be reported as total hours.

Month 1. Effort code analysis.

Issues	Conclusions	Actions
Why was significant amount of planning still occurring? Why are the management tasks so high?	The development team was spending time identifying and acquisitioning resources needed to perform their jobs	Program Manager and Procurement Manager define a process by which the project submits all acquisition requests to the procurement team. The procurement team was then responsible for all acquisition
What tasks are being performed as Bill of Information and Documentation Activities?	The configuration management system had not been set up for the project and the development team was spending considerable time supporting the CM team	Meet with the Configuration Management department to identify the project's requirements. The CM department configured the project's system and committed resources to support the project

Month 2. Effort codes reported.

Activity	Task	Discipline	%	Description
G	R	Y	20	Generate Requirements for Systems
E	R	Y	25	Evaluate Requirements for Systems
G	R	S	25	Generate Requirements for Software
E	R	S	15	Evaluate Requirements for Software
K	R	S	10	Rework Requirements for Software
S	N	S	5	Support Administrative Activities for Software

Month 2. Effort code analysis.

Issues	Conclusions	Actions
Why are software engineers spending time in the Systems Engineering discipline? Why was there considerable amount of rework for the requirements?	The customer had not signed off on the system requirements. The software team was perpetually reacting to the volatile system requirements. In addition, the software team was spending time assisting the short-staffed systems engineering team	Program Manager, System Engineering Lead and Customer liaison reached an agreement on the requirements to be fully addressed at the SRR. The identified system requirements were to be baselined by the end of the week
Less time was spent on planning and management issues than in month 1. Verify that the procurement and configuration management teams are providing the necessary support	Procurement was providing full support. The required resources have been ordered. The configuration management team had established the proper environment for the project and the software team was in the process of being integrated with the configuration management system	Requested that the procurement department provide weekly status on all procurements

Month 3. Effort codes reported.

Activity	Task	Discipline	%	Description
G	R	S	55	Generate Requirements for Software
E	R	S	10	Evaluate Requirements for Software

Month 3. (*Continued*)

Activity	Task	Discipline	%	Description
G	O	S	20	Generate Interface Control Definition/Requirements for Software
S	T	S	15	Support Training and Mentoring for Software

Month 3. Effort code analysis.

Issues	Conclusions	Actions
With PDR only 2 months away, why had no architecture tasks been performed?	The entire software team was working hard to meet the SRS. However, the project's staffing profile was not adequate for the amount of work that needed to be done	Hired additional engineers who have experience in the type of work required by the project
Why had the test team not been performing any tasks? The test plan should have been started long before now	The test team leader currently only has one tester who has not worked as a tester previously. The test team lead has been training/mentoring the junior engineer and has not had sufficient time to address the testing. Mentoring has been completed	Suggested that the junior test engineer develop a training document for new test engineers in an effort to minimize future monopolization of test leader's time as more test engineers are hired

Month 4. Effort codes reported.

Activity	Task	Discipline	%	Description
K	R	S	10	Rework Requirements for Software
G	A	S	45	Generate Architecture Definition/Documentation for Software
E	A	S	10	Evaluate Architecture Definition/Documentation for Software
S	E	S	20	Support Metrics and Process Development for Software
E	T	S	15	Generate Trade Studies for Software

Month 4. Effort code analysis.

Issues	Conclusions	Actions
Why was time being used on Trade Studies when they had already been performed during the proposal work?	The team had charged their time to Trade Studies task (ETS) when they were reading the trade studies submitted with the proposal even though they were doing so in order to create the architecture documents	Provided an abridged version of the organization's effort code description to provide a better alignment with the project

Month 4. (*Continued*)

Issues	Conclusions	Actions
No inspections of the architecture were performed. CDR is one month away and the team cannot wait to the very end to review the architecture	Learned that the team was waiting until a document was almost 100% complete before inspecting it	Brought in the inspection trainer to provide a short refresher course in which it was stressed that part of a document, e.g., a chapter can be inspected instead of waiting for the entire document to be completed
Why was so much time being spent on metrics?	The subsystem leads were being bogged down trying to provide weekly metrics for management	Program Management, Organizational Management and the Software Engineering Lead reached a compromise in which the Software Engineering Lead would report status of milestones weekly and the subsystem leads would report their metrics monthly

Month 5. Effort codes reported.

Activity	Task	Discipline	%	Description
G	A	S	15	Generate Architecture Definition/Documentation for Software
E	A	S	20	Evaluate Architecture Definition/Documentation for Software
K	A	S	15	Rework Architecture Definition/Documentation for Software

Month 5. (*Continued*)

Activity	Task	Discipline	%	Description
G	Q	S	25	Generate Qualification/ Verification/Validation Activity for Software
E	T	S	5	Evaluate Trade Studies for Software
G	O	S	20	Generate Interface Control Definition/Requirements for Software

Month 5. Effort code analysis.

Issues	Conclusions	Actions
What was the reason for the significant amount of architecture evaluation and rework? CDR was two weeks away and the documentation should be have been baselined	Two of the subsystem leads were in disagreement on a couple of critical architecture issues. They were using the inspections as a forum to discuss the issues and try to persuade the other subsystem leads	Software Lead met with the subsystem leads to instill the need to work as a team. Software Lead to attend all architecture and design meetings to provide the final decision when conflicts arise
Why was the interface documentation being worked sporadically (e.g., in month 3 and then again in month 5)? Why were no interface documentation reviews being performed?	The engineers were working on the interface documentation as a side task when they had "free" time	Designated an engineer as the interface lead. Tasked the engineer to define all external interfaces first and then address internal interfaces

Month 6. Effort codes reported.

Activity	Task	Discipline	%	Description
G	A	S	5	Generate Architecture Definition/ Documentation for Software
E	A	S	5	Evaluate Architecture Definition/Documentation for Software
K	A	S	10	Rework Architecture Definition/Documentation for Software
G	O	S	15	Generate Interface Control Definition/Requirements for Software
E	O	S	5	Evaluate Interface Control Definition/Requirements for Software
K	O	S	5	Rework Interface Control Definition/Requirements for Software
G	Q	S	20	Generate Qualification/Verification/ Validation Activity for Software
E	Q	S	8	Evaluate Qualification/Verification/ Validation Activity for Software
K	Q	S	2	Rework Qualification/Verification/ Validation Activity for Software
G	D	S	25	Generate Detail Design and Analysis for Software

Month 6. Effort code analysis.

Issues	Conclusions	Actions
Verify the architecture rework was a result of the PDR	The rework associated for the architecture tasks was associated with the PDR	None required

Month 6. (*Continued*)

Issues	Conclusions	Actions
Verify the status of the interface documentation	The engineer assigned to develop the interface document was having difficulty getting the necessary information for the external interfaces	Held a meeting with the customer liaison and the interface engineer to establish points of contact. Customer liaison would personally contact each POC to insure a readily flow of information to the interface engineer
Verify that the detail design inspections are being planned in increments	There was a considerable amount of isolation in the subsystems' detail design resulting in duplication of effort	Each subsystem lead was instructed to be present in all design meetings in order to establish cohesion and to identify overlap
Ensure that the subsystem leads were planning on having the engineers start implementation once their respective detail designs were inspected	Verified	None required

Month 7. Effort codes reported.

Activity	Task	Discipline	%	Description
G	D	S	25	Generate Detail Design and Analysis for Software
E	D	S	10	Evaluate Detail Design and Analysis for Software

Month 7. (*Continued*)

Activity	Task	Discipline	%	Description
K	D	S	5	Rework Detail Design and Analysis for Software
G	O	S	20	Generate Interface Control Definition/Requirements for Software
S	E	S	5	Support Metrics and Process Development for Software
S	N	S	10	Support Administrative Activities for Software
G	C	S	25	Generate Construct/Code/Build/ Breadboard for Software

Month 7. Effort code analysis.

Issues	Conclusions	Actions
Verify the interface engineer is nearing completion on external interfaces and has begun defining the internal interfaces	The interface engineer has completed the external interfaces but forgot to have that portion inspected. The internal interfaces are being discussed	None required
Determine why there had not been any unit test activities. Unit test should be initially identified during the detail design phase and continue through coding	The engineers planned to address integration during the next month's activities	None required

Month 7. (*Continued*)

Issues	Conclusions	Actions
Find out what the test team had been doing	The test team had been designing and coding test tools they need for system testing. Clarified with the test team that they should charge their time to Integration tasks to distinguish itself from the development of the delivered product	Test team asked to get involved with the development team to identify any test tools the developers may need during unit and integration testing and to determine if the system test tools could be modified to provide the needed capabilities

Month 8. Effort codes reported.

Activity	Task	Discipline	%	Description
G	D	S	10	Generate Detail Design and Analysis for Software
E	D	S	5	Evaluate Detail Design and Analysis for Software
K	D	S	5	Rework Detail Design and Analysis for Software
G	O	S	15	Generate Interface Control Definition/Requirements for Software
E	O	S	5	Evaluate Interface Control Definition/Requirements for Software
K	O	S	5	Rework Interface Control Definition/Requirements for Software

Month 8. (*Continued*)

Activity	Task	Discipline	%	Description
G	C	S	25	Generate Construct/Code/Build/ Breadboard for Software
G	U	S	10	Generate Unit/Assembly/Product/ System for Software
G	I	S	20	Generate Integration Activities for Software

Month 8. Effort code analysis.

Issues	Conclusions	Actions
Verify with the test team that the qualification test tasks were not being set aside in favor of test tool development	Test lead planned to start qualification test plans upon completion of CDR	None required

Month 9. Effort codes reported.

Activity	Task	Discipline	%	Description
K	D	S	10	Rework Detail Design and Analysis for Software
G	O	S	15	Generate Interface Control Definition/Requirements for Software
E	O	S	5	Evaluate Interface Control Definition/ Requirements for Software
K	O	S	2	Rework Interface Control Definition/ Requirements for Software

Month 9. (*Continued*)

Activity	Task	Discipline	%	Description
G	I	S	15	Generate Integration Activities for Software
G	C	S	20	Generate Construct/Code/Build/ Breadboard for Software
G	Q	S	18	Generate Qualification/Verification/ Validation Activity for Software
K	A	S	10	Rework Architecture Definition/ Documentation for Software
K	R	S	5	Rework Requirements for Software

Month 9. Effort code analysis.

Issues	Conclusions	Actions
Determine the reason for the rework of requirements and architecture. The problem discovered during CDR was thought to be minor	The resolution of the problem discovered during CDR had a cascading effect on other requirements and a minor affect on the architecture	Assessed the impact of the requirements and architecture changes. The impacts are isolated to two subsystems. Software Lead met with the test team to minimize the slip in the test schedule. The test team will begin testing the nonimpacted subsystems in the lab using the system test tools allowing the impacted subsystems a little more time to implement the new changes

Month 9. (*Continued*)

Issues	Conclusions	Actions
Determine why there are no metric tasks being reported	The project lead had tasked an intern to create a metrics gathering and reporting tool. Consequently the metric reporting has been streamed line for the subsystem tack leads and thus requires minimal effort	Suggested that the project lead submit the new metrics tool to the company's process team in order to have it available to other projects

Month 10. Effort codes reported.

Activity	Task	Discipline	%	Description
G	R	S	10	Generate Requirements for Software
G	A	S	10	Generate Architecture Definition/Documentation for Software
G	C	S	15	Generate Construct/Code/Build/ Breadboard for Software
E	C	S	5	Evaluate Construct/Code/Build/ Breadboard for Software
K	C	S	2	Rework Construct/Code/Build/ Breadboard for Software
G	Q	S	10	Generate Qualification/Verification/ Validation Activity for Software
E	Q	S	5	Evaluate Qualification/Verification/ Validation Activity for Software

Month 10. (*Continued*)

Activity	Task	Discipline	%	Description
G	O	S	10	Generate Interface Control Definition/Requirements for Software
E	O	S	3	Evaluate Interface Control Definition/ Requirements for Software
K	O	S	2	Rework Interface Control Definition/ Requirements for Software
G	U	S	8	Generate Unit/Assembly/Product/ System for Software
E	U	S	3	Evaluate Unit/Assembly/Product/ System for Software
K	U	S	2	Rework Unit/Assembly/Product/ System for Software
T	U	S	10	Test Unit/Assembly/Product/System for Software
T	I	S	5	Test Integration Activities for Software

Month 10. Effort code analysis.

Issues	Conclusions	Actions
Why were requirements and architecture tasks being charges this late in the increment?	Determined that the requirements and architecture tasks were for increment II. The subsystem leads were implementing lessons learned from increment I and realized that the architecture designed for increment I may be affected by the requirements for increment II	Set up two activity summary reports so that the metrics for each increment were reported separately

Month 10. (*Continued*)

Issues	Conclusions	Actions
All other aspects appeared good. Verify that there were no design rework tasks due to the good design and not due to the lack of maintaining the design documentation	Updating the design documentation had taken a back seat to coding. The engineers were taking the approach that the design documents would be updated while their code was being tested. The interface documentation had been baselined	The team was informed that the design documentation must be updated before the code is given to the test team

Month 11. Effort codes reported.

Activity	Task	Discipline	%	Description
E	C	S	10	Evaluate Construct/Code/Build/ Breadboard for Software
K	C	S	5	Rework Construct/Code/Build/ Breadboard for Software
G	Q	S	15	Generate Qualification/Verification/ Validation Activity for Software
E	Q	S	5	Evaluate Qualification/Verification/ Validation Activity for Software
K	Q	S	2	Rework Qualification/Verification/ Validation Activity for Software
G	U	S	10	Generate Unit/Assembly/Product/ System for Software
E	U	S	2	Evaluate Unit/Assembly/Product/ System for Software

Month 11. (*Continued*)

Activity	Task	Discipline	%	Description
K	U	S	1	Rework Unit/Assembly/Product/ System for Software
T	U	S	10	Test Unit/Assembly/Product/System for Software
T	I	S	15	Test Integration Activities for Software
T	Q	S	10	Test Qualification/Verification/ Validation Activity for Software
K	D	S	15	Rework Detail Design and Analysis for Software

Month 11. Effort code analysis.

Issues	Conclusions	Actions
Determine whether the design rework was due to the previous month's instructions to update the design documentation before submitting code for system test or was the test team finding problems that trace back to the design	Reviewed the problem reporting metrics and determined that the test team had found a minor problem that traced back to the design. The main design rework was due to the engineers updating the documentation based on the findings during inspections	None required
Similarly, determine the cause for the code rework	Same	None required

Month 11. (*Continued*)

Issues	Conclusions	Actions
Nearing end of increment 1. Need to address staffing		Had several of the engineers whose code had been completed enroll in various classes. Some took process improvement courses and others took technical courses that are pertinent to their subsystem. Assign other engineers begin working increment 2 issues

Month 12. Effort codes reported.

Activity	Task	Discipline	%	Description
K	C	S	15	Rework Construct/Code/Build/ Breadboard for Software
T	I	S	25	Test Integration Activities for Software
T	Q	S	40	Test Qualification/Verification/ Validation Activity for Software
T	U	S	15	Test Unit/Assembly/Product/System for Software
K	O	S	5	Rework Interface Control Definition/Requirements for Software

Month 12. Effort code analysis.

Issues	Conclusions	Actions
Verify the significance of the rework tasks	The rework was attributed to a minor defect in the interface document. The changes were implemented and retested	None required
Verify the test team is ready for FAT	Team is ready	None required
Nearing end of increment 1. Need to address staffing		Shift remaining engineers to increment 2

CHAPTER 12

Software Process Improvement Activities Based on CMM

Takaji Fujiwara

Development Department 3, Development Division,
Fujitsu Peripherals Limited,
35 Saho, Yashiro-cho, Katoh-gun, Hyogo, 673-1447, Japan
takaji@fpe.fujitsu.com

Shigeru Yamada

Department of Social Systems Engineering,
Faculty of Engineering, Tottori University,
4-101 Minami, Koyama-cho, Tottori-shi, Tottori, 680-8552, Japan
yamada@sse.tottori-u.ac.jp

1. Introduction

Recently, the process assessment and improvement of the development organization that uses the 3rd edition of ISO9001 (abbreviated as ISO9001:2000)[1] or Capability Maturity Model (abbreviated as **CMM**)[2,3] attract the attention of the IT industries. As the background, there are problems with delivery time and quality of the software products which have become large-scaled and diversified. If the delivery period of the software product is missed, it becomes outdated. Moreover, the delivery delay will lead to immediate budget deficiencies. Further, the quality problems after delivery not only presses profits, but also makes customers lose trust in us. In the worst case, the company might go out of business.

So, in our company, the process improvement activities based on the CMM are tackled at the shift timing to the ISO9001:2000. Now, we are performing the improvement activities according to the following procedures:

(1) The objective analysis of process, and judging the degree of capability maturity level.
(2) The extraction of strong and weak points, and those improvement proposals for the process.
(3) The document generation of the best process procedures in the current level.
(4) The practical-use trial in actual projects.

This chapter briefly describes the CMM used for the software process improvement in Sec. 2. In Sec. 3, how to analyze the strong and weak points of the process, and the improvement activities based on the improvement proposals are shown. Finally, we illustrate the process improvement activities in our company in Sec. 4.

2. What's the CMM?

The CMM can be summarized as follows:

- The guide of the software organization which aims at growth of the corporate culture towards development and management of the software which was excellent for the purpose of the improvement in the level of the management ability to the process of software development.

Further, the purpose and features of the CMM can be epitomized as follows:

Purpose: In order to judge the current degree of our software process maturity and to raise the process and product quality, we can plan the strategy of the process improvement by discriminating the most important theme.

Features: It is the process maturity model for assessing indirectly the quality of the software product generated by the process. The CMM classifies the degree of process maturity of the organization into five levels, and defines **K**ey **P**rocess **A**rea (abbreviated as **KPA**) required in order that each level is attained (see Fig. 1). Moreover, the CMM can move into the degree of maturity level of the higher rank by attaining all the goals defined in each KPA. However, the CMM is not what described the process comprehensively, and is not the territory which covers the technical field and the field of the human elements.

On the other hand, the ISO9001:2000 can be summarized as follows:

- The system and process for which the organization should make the document when the organization wants to acquire the capability to satisfy the customer requirement matter and to maintain continuous improvement activities.

Then, why is the IT industry trying the process improvement using the CMM? The model called the CMM has a lot of advantages in using

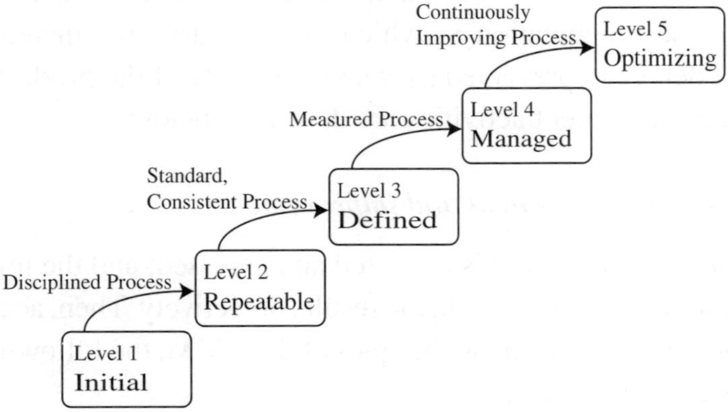

Fig. 1. Five levels of the degree of software process maturity.

it as the framework required for improvement. It helps to produce the vision: "What can be obtained in the process improvement for the organization". Moreover, the common language about the process is established and it becomes the index which determines the priority coping with software development problems. Further, measurement of the process is possible by providing the framework for performing assessment which is reliable and consistent. Generally, although subjective judgment cannot be completely excluded when performing such assessment, the basis judged objectively by the CMM is obtained. Simultaneously, the framework of this measurement provides the notional structure for improving management and development of the software product by controlling the noninconsistencies.

3. Process Improvement Activities

This section describes about the practice procedures of current assessment and the improvement, the promotion organization for process improvement and the selection method and contents of activities of improvement items as process improvement activities. "A breakaway from individual dependence" is aimed as the target of our company, that is, we do not depend on individual's skill and experience, but aim at the organization which can always hold **QCD** (Quality, Cost, Delivery) measures as the management indices. Then, we can improve the organization in which unnecessary cost increase and delivery delay are prevented by keeping quality of the products. The process improvement activities are described below.

3.1. *Current assessment and improvement*

First, the current process is analyzed and assessed, and the improvement items are extracted from the results objectively. Then, according to the process assessment techniques of the CMM, the following procedures are performed:

(1) Answer preparation to the interview contents.
(2) Creation of the interview contents.

(3) Interviews to the arbitrary personnels of the development department.
(4) Analysis of the interview results.
(5) Reporting of the assessment results.

We create interview contents described in all viewpoints to these contents of the answer, after determining the contents of the answer. They are created in order to investigate the degree of comprehension of the defined process. The questions based on the interview contents are put to the arbitrary personnels of the development department. Next, while analyzing the interview results, we generate the table in order to discriminate the strong and weak points of the development organization process. Finally, we draw up the report of assessment results (KPA profile) showing whether each KPA is achieved or not based on this table.

Next, the improvement themes based on the report of assessment results are planned out. Then, we choose the KPAs by which the great effective result is obtained from the nonattained items in the short period, and try them in the actual project. So, we can extract the problems or inefficient items of the process from the project members at on-demands, and construct the structure with which the CMM promotion team discusses the improvement policies (see Fig. 2). We

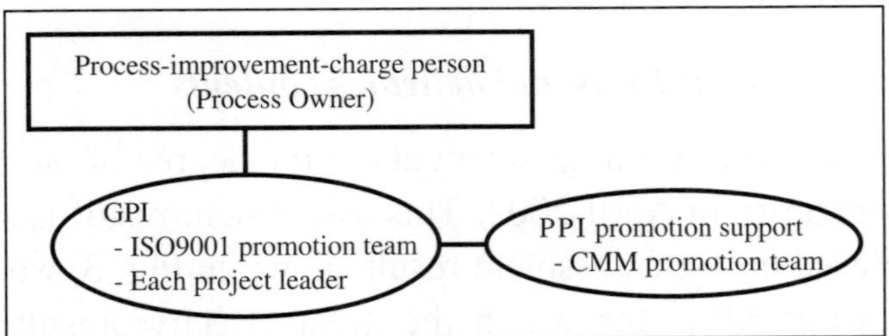

Fig. 2. The CMM promotion organization.

can powerfully promote the improvement of the current process by using this structure.

3.2. *CMM promotion organization*

As the top of the CMM promotion organization, we have stationed the process-improvement-charge person (process owner), and also established the Group of Process Improvement (abbreviated as **GPI**). As for the staff in charge of the GPI, the person in charge of the ISO9001 and each project leader were assigned. Their important roles are to inspect deviations from the ISO9001, and make the process work permeate to each project member. Further, the Project of Process Improvement (abbreviated as **PPI**) promotion support group has also been established as the CMM promotion team. The PPI members always support the GPI member. Further, their roles are to discuss the extracted process improvement items at any time, and to explain the deliberation results to all project members. Moreover, they have to participate at discussion or review of the specifications, and to inspect whether the activity procedures are satisfied or not according to the process. We can perform application and on-demand support timely for the improvement items of the project by using these structures (see Fig. 2).

3.3. *Improvement KPAs and activities contents*

We performed the self-assessment about the degree of our software process maturity in April 2002. This assessment result is shown in Fig. 3. Based on the assessment result shown in Fig. 3, we selected the following KPAs for which the great effective results can be expected:

The nonattained item: Organization Process Focus (Level 3)

The strengthening items: Software Project Planning (Level 2); Peer Reviews (Level 3)

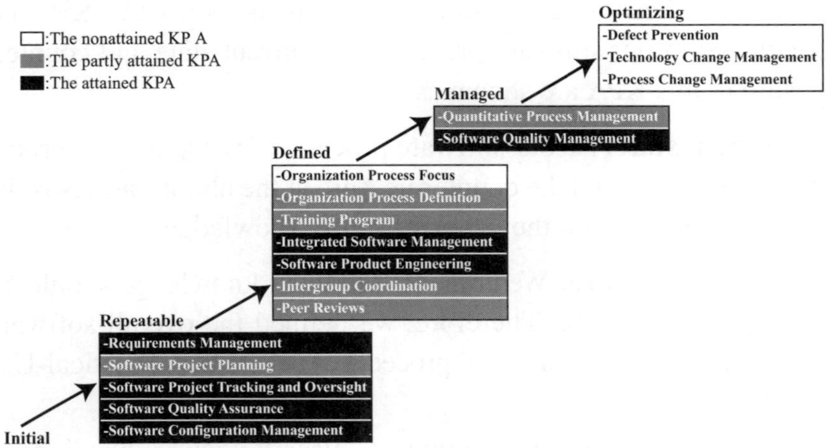

Fig. 3. The assessment results of the degree of our software process maturity.

We determined these selected KPAs as the activities contents in 2002. The selected KPAs are explained in the following.

First, as the reason for "Software Project Planning", we aimed at the perfect practice of this KPA, and planned out the maturity level 2 achievement. The contents of lack and countermeasure in our software process are shown as follows:

The lacking contents: While the development vector (policy) had been undecided or the worker's degree of comprehension to the development contents has been low, each developer's activities are started. For these reasons, the redo activities by the specification changes, and design omissions of funtions often occurred.

The countermeasure: The project manager and leader create the functional list which should be developed based on the customer's requirements. Next, based on this list, each worker creates the subdivision of development contents, i.e., "**W**ork **B**reakdown **S**tructure (abbreviated as **WBS**)", and the activity schedule table to it. Thereby, each worker's development contents become clear. Further, the severe development progress management according to the detailed development schedule is attained.

Next, we chose "Organization Process Focus", since this KPA has not attained in our software process. The current state and countermeasure to this KPA are shown as follows:

The current state: As our software process, only the normal procedures are defined, and the countermeasure at the abnormalities is the situation depending on the manager's tacit knowledge.

The countermeasure: We consider that tacit knowledge should be made open knowledge. Therefore, we defined the overall software process including the abnormal procedures in "Process Practical-Use Procedures".

Finally, the reason of selection of "Peer Reviews" was judged from the analysis result of the detected faults in the current testing phase (see Table 1). Table 1 shows the ratio of the fault classification for the CMM nonapplied projects. From Table 1, we considered that the simple mistakes in the fault classification should be detected at the review of the upstream phase. That is, we thought that about 20–40% of the faults detected in the testing phase could be beforehand inhibited by reinvestigation of the review method. So, we added the following contents to the review activities:

- The review client executes the self-review before distribution by using the review check sheets.

Table 1. The ratio of the fault classification.

	Simple mistake	Logic mistake	Specification change	Hardware fault	Others
Nonapplied Project 1	27.8	32.8	8.3	23.3	7.8
Nonapplied Project 2	38.5	47.5	9.0	0	5.0
Nonapplied Project 3	24.5	53.3	4.9	0.9	16.4

Unit: %.

- The reviewer verifies the review subject beforehand.
- The review time is limited by the maximum of two hours.[4,5]
- The review client clarifies the review purpose. Then, the review end criterion will also become clear.
- In the phase end review, the leader and person in charge of development execute the design omission check based on the development functional and WBS lists.

4. Application Results

This section shows activity results of the process improvement discussed above.

4.1. *Software project planning*

As the activity result of this KPA, we obtained the effect shown below:

- The unification of the project member's development policy to the customer requirements (i.e., the useless discussion among members was reducible).
- The WBS lists generation based on the customer requirements.
  (1) The newly developed functions and the reconstruction/reuse portions of the existing functions became clear, and the overlapping and omission of functions were exterminated.
  (2) The redone activities by specification changes decreased.
  (3) The strict development progress management could be performed because each worker's detailed scheduling became possible.

Moreover, as the effect of detailed scheduling, the communication among related groups became easy and the following phenomena were verified.

- The omission of functions extending over related groups decreased.
- Since each group can provide the high quality products, the development activities progress became good.

On the other hand, as the demerit, the determination time of the requirement specification definitions needed about twice of the conventional development, in order to unify the worker's development vector. Although this is dependent on the project size, it took the applied projects about 3–4 weeks.

4.2. *Peer reviews*

The activity results of this KPA are shown in Figs. 5–7 and Table 3. These figures represent the comparison results among the CMM nonapplied project and applied projects. The projects used for comparison are shown in Table 2. Further, the M1–M4 in Figs. 5–7 represents the common module name used in each project, and these relations can be shown in Fig. 4. That is, each project is based on the different customer requirements, and the M1–M4 means the derived module by the reconstruction for satisfying the customer requirements of each project based on the completed module.

Figure 5 shows the comparison result of the number of **L**ine **o**f **C**ode (abbreviated as LOC) reviewed per an hour. From Fig. 5, we find that the number of reviewed LOC of the CMM applied projects is small compared with that of the CMM nonapplied

Table 2. The projects used for comparison.

	Development (total) scale ($\times 10^3$ LOC)	Number of workers (Persons)	Development period (Months)
Nonapplied Project 1	26.0 (348.4)	12	8
Applied Project 1	34.0 (511.5)	10	9
Applied Project 2	30.2 (511.5)	10	8.5

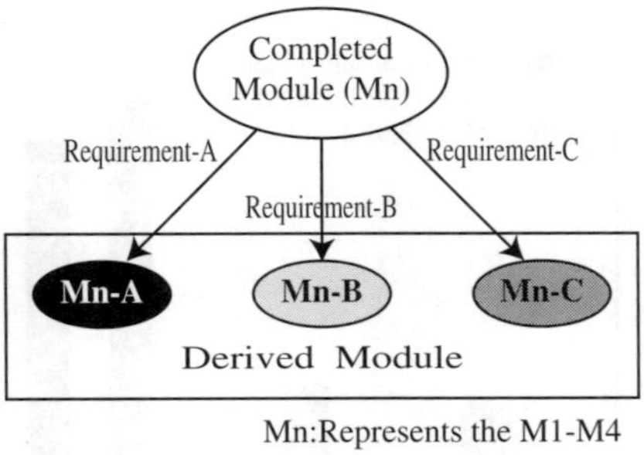

Fig. 4. The relation of M1–M4 in each project.

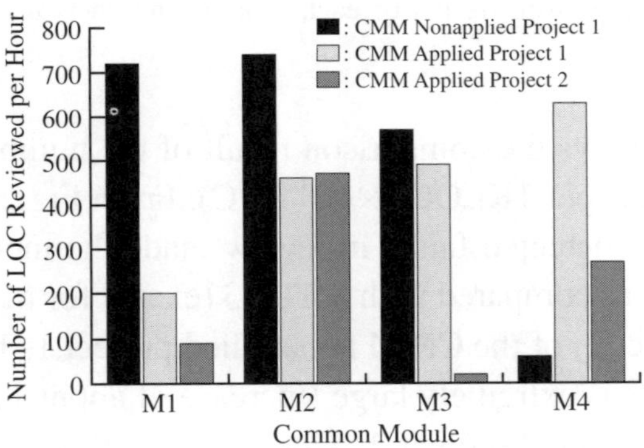

Fig. 5. The comparison results of each module (number of LOC reviewed per an hour).

project 1 in M1–M3 where the reuse ratio is low. We can consider that time was spent on all processing analysis assumed to the portion of the newly developed functions and the existing functions for reconstruction/reuse in the review of each module. On the other hand, in each CMM applied project for M4, we find that the number of reviewed LOC is large. In the case where the high reuse ratio of the existing functions, and the reconstruction portion and its influence range were clear as this module structure, we have verified the validity of using the review check sheets in the review.

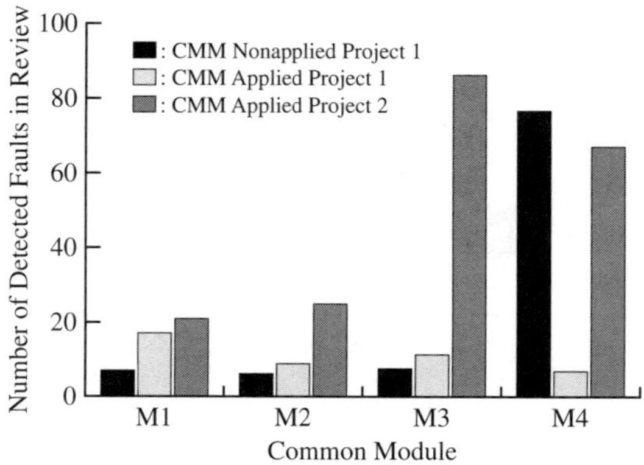

Fig. 6. The comparison results of each module (number of detected faults in review per 1KLOC).

Figure 6 shows the comparison result of the number of detected faults in review per 1KLOC ($\times 10^3$ LOC). From Fig. 6, we find that the number of detected faults in review made the increase in about 44.2–410.1% as compared with M1–M3 (except for M3 of the CMM applied project 2) of the CMM nonapplied project 1. The number of detected faults is extremely large (increase of about 1150%) for M3 of the CMM applied project 2. This reason is that a lot of faults to be detected easily were detected in order that the worker had little experience of the function. Moreover, as the reason that the number of faults is small for M4, it has high reuse ratio of the existing functions, and then the reconstruction/reuse portion and its influence range are clear. From these results, we can expect the same situation as Fig. 5.

Figure 7 shows the comparison result of the number of detected faults in the testing in terms of the newly developed functions and the existing functions for reconstruction/reuse per 1KLOC. From Fig. 7, we find that the number of detected faults in testing is very small in M1 and M4. This means that the high quality is made by the review of upstream software development phase. That is, the acvtivity policy of the reinvestigation and strengthening of "Peer Reviews" has been justified. Especially, in the case of M4, since the items extracted in the CMM applied project 1 had been improved, no fault was detected

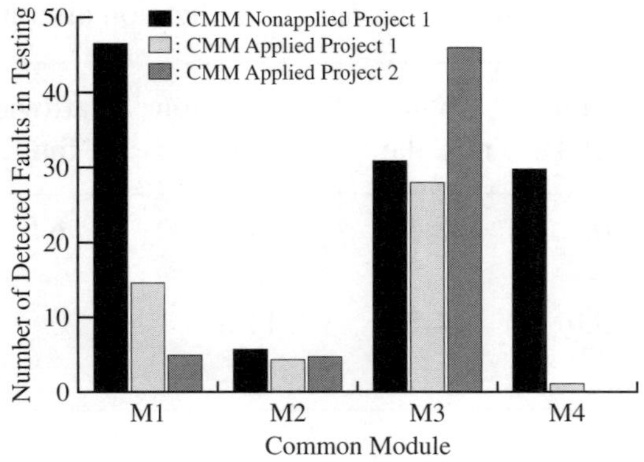

Fig. 7. The comparison results of each module (number of detected faults in testing per 1KLOC).

in the CMM applied project 2. Moreover, the number of detected faults for M2 was small, and it was almost equivalent to the CMM nonapplied project 1. The screen display module is considered to be the result which the severe review in which the customer participated had been performing conventionally about M2. Therefore, in the case of this module, the high quality was realized conventionally by the review of upstream software development phase. However, the number of detected faults for M3 was large, and it was almost equivalent to the CMM nonapplied project 1. In the case of the CMM applied project 2, the number of detected faults is large as compared with the CMM nonapplied project 1.

Finally, Table 3 shows the analysis result about the fault classification of the detected faults during the testing phase of the CMM applied projects. We have performed improvement/strengthening activities of "Peer Reviews" for the purpose of extermination of the simple mistake faults. That is, we have obtained 0% by the CMM applied project 1, and 3.9% by the CMM applied project 2 as the ratio of the faults classified into the simple mistake. These ratios mean to be decreasing rapidly as compared with those of the CMM nonapplied projects (see Table 3). Therefore, it is clear that the implementation of the review examined carefully in each upstream development phase has improved the quality of products. From now on, we need to

Table 3. The ratio of the fault classification after activity.

	Simple mistake	Logic mistake	Specification change	Hardware fault	Others
Applied Project 1	0	80.4	7.2	6.5	5.9
Applied Project 2	3.9	74.5	11.8	0	9.8

Unit: %.

advance investigation about the policy which decreases the number of faults classified into the logic mistake.

4.3. *Consideration to improvement activities*

We discuss the reduction degree of the manpower in the improvement activities which we are currently tackling. Then, we compare by the change in the manpower of each process in the CMM nonapplied and applied projects.

The relative comparison results of each manpower are shown in Fig. 8, by normalizing the manpower of the CMM nonapplied

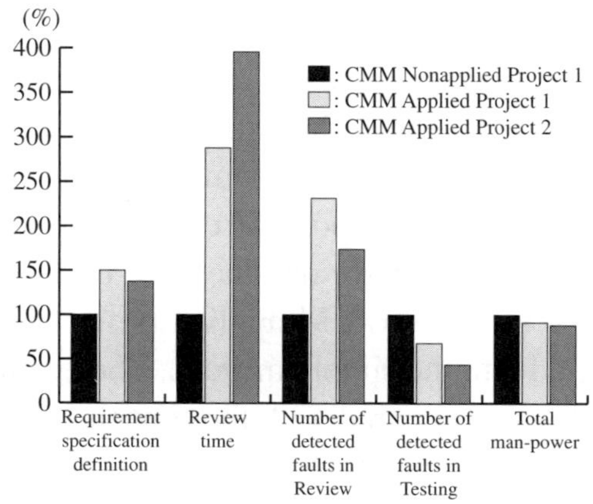

Fig. 8. The relative comparison results of each item.

project 1 into 100. We find that the review time increases by about 2.9 and 4.0 times compared with the CMM nonapplied project 1, respectively. On the other hand, we also find that the number of detected faults in testing decreases to about 67.9 and 43.5% compared with the CMM nonapplied project 1, respectively. That is, in the testing phase, when time to spend on the cause analysis, correction and validity check of the detected faults is taken into consideration, the increase in the requirement specification definition and review time is small. In each observed project, we have cut down the manpower to about 91.6 and 88.5%, respectively, keeping the quality and reliability of the software product. And we have cut down the manpower of testing phase to about 85.0 and 77.5%, respectively.

5. Concluding Remarks

We have discussed software process improvement activities by the CMM and its practical application in our company. Further, the self-assessment of our software process has been performed, and the strong and weak points of our process has been extracted based on the assessment results. Next, as the great improvement effect, the KPAs have chosen from the nonattained items, and we have performed the trial in the actual projects as the activity theme in 2002. In the upstream software development (design and coding) phase, the high quality products by the thorough-paced review can be developed, and these trial results have shown contribution to reduction of the testing effort greatly.

As the future issues, we are going to extend the standard process for the embedded software development, and have begun to investigate about the standard process for the Windows application development. The standard process for the Windows application development is planning to be applied to the new projects starting in January 2003. And the difference in the activity procedures of the Windows application and embedded software developments is due to be clarified by practical application.

Acknowledgments

The authors wish to thank Mr. Kazunori Shiotani, Senior Manager, and Mr. Koichi Sekine, Section Manager, Development Department 3, Development Division, Fujitsu Peripherals Limited for their helpful suggestions and generous supports.

This research was supported in part by the Grant-in-Aid for Scientific Research from the Ministry of Education, Culture, Sports, Science and Technology of Japan under Grant No.15510129.

References

1. S. Takabayashi, ISO9001:2000 which brings about profits — The interpretation and practical-use, Tokyo-Denki-Daigaku-Shuppan, Tokyo (2001).
2. M. C. Paulk, B. Curtis, M. B. Chrissis and C. V. Weber, Capability maturity model for software, Version 1.1, Software Engineering Institute, CMU/SEI-93-TR-24 (1993).
3. M. C. Paulk, C. V. Weber, S. M. Garcia, M. B. Chrissis and M. Bush, Key practices of the capability maturity model, Version 1.1, Software Engineering Institute, CMU/SEI-93-TR-25 (1993).
4. K. Esaki, S. Yamada and M. Takahashi, A quality engineering analysis of human factors affecting software reliability in software design review process, *Transactions of the IEICE-A* **J84-A** (2001) 218–228.
5. K. Esaki, S. Yamada and M. Takahashi, A software reliability prediction model based on software reviewing process characteristics and its evaluation, *Journal of the Society of Project Management* **3** (2001) 27–32.

CHAPTER 13

Asymptotic Properties of a Software Reliability Growth Model with Imperfect Debugging: A Martingale Approach

Winai Bodhisuwan

Department of Applied Statistics,
King Mongkut's Institute of Technology North Bangkok,
Bangkok, 10800, Thailand

Panlop Zeephongsekul

Department of Mathematics and Statistics,
RMIT University, Melbourne 3001, Australia

1. Introduction

A key objective of modern software engineering is targeted towards identifying and removing existing defects in a piece of software while preventing the injection of new ones. This is done during the testing phase of the development process where the aim is to ensure that the software reaches the required quality standard before it is released into the market place. In order to be able to effectively manage software defects, it is important that the production team be able to monitor the performance of the software. One very effective way of achieving this is to produce a stochastic model to predict when failures are likely to occur, fit it using empirical data and then use the estimated model to calculate important metrics such as rate of occurrence of failures, mean time to failure and to predict how much defects remain in the software during its operational phase. Such models are known in the software reliability literature as *software reliability growth models* (*SRGMs*).

Starting with the model in the seminal paper by Jelinski and Moranda,[1] a tremendous number of SRGMs have been proposed in the literature. For a review of some of these models, the reader is referred to Refs. 2 and 3. Recently, Chen and Singpurwalla[4] showed that practically all SRGMs that have been proposed belong to the class of *self-exciting point processes*.[5] These include the very large class of SRGMs based on the nonhomogeneous Poisson process (NHPP) (first introduced by Goel and Okumoto[6]) and SRGMs based on concatenated failure rates (to which the Jelinski–Moranda model belongs).

Closely related to self-exciting point processes is the more theoretical approach to point processes which is based on conditional intensities and likelihood functions. This approach relies on the martingale theory for counting processes (see, e.g., Refs. 7–10) and has been successfully applied in many applied statistics areas including the analysis of survival data (e.g., Ref. 11). It has the advantage of not making too many unreasonable assumptions not borne out by practical considerations and it also provides a more realistic framework with which a model builder could describe the evolution of software reliability. Papers in software reliability which utilized this approach include Refs. 12–15.

Most SRGMs assumed that debugging of a software defect is perfect and that no new defects are injected into the software in case of imperfect debugging. Generalized SRGMs based on NHPP which allow for imperfect debugging and introduction of new defects into the software modules were considered in Refs. 16–19. In this paper, one of our objectives is to present a software growth model in the *martingale framework* which incorporates imperfect debugging and the introduction of new defects into the software programs. A second objective of this paper is to look at some asymptotic properties related to the parameters of our model. Here, we utilized an approach similar to that of Ref. 13 of letting the total number of defects increases at different stages of the process instead of letting the duration of the testing time approaches infinity. This is more realistic since in practice, the testing and debugging process terminate after a finite length of time.

In the next section, we introduce the counting process associated with our model and obtain its intensity function. This is then used in Sec. 3 to obtain the likelihood function. We then give sufficient conditions under which we have consistency and asymptotic normality of the maximum likelihood estimators. Finally, some numerical examples will be presented in Sec. 4.

2. The Model

A software system comprises of program codes which may or may not contain defects. By a defect or *fault*, we are here referring to error in the codes which will lead to *failures*, or observed deviations of the output of the program from its specifications, when the program is run. A common method of identifying defects in software engineering is for the tester to run the program using a range of input selected from a family of input distributions which closely resemble the operational profile of the program. When a failure occurs, steps are taken to identify the defects causing it. This process is usually interleaved with the process of defect removal, which is the activity of making the actual changes to the codes so that failures may not occur in future. Unfortunately, this process may or may not remove the fault from the software. In some cases, imperfect removal of defects may inject new faults into the program.

In our model, we will assume that testing is done over a finite time interval $[0, T]$ so that all counting processes that arise from testing for defects will occur in that interval. Let $(\Omega, \mathcal{E}, \mathbf{P})$ be a probability space and $\mathcal{F} = \{\mathcal{F}_t, t \in [0, T]\}$ be a right continuous filtration of $\mathcal{E}$. Also, let $\mu < \infty$ be total number of defects in the software being examined. The failure times (calendar or execution) caused by the defects are

$$0 < T_1^{(0)} < T_2^{(0)} < \cdots < T_j^{(0)} \cdots \quad j \leq \mu ,$$

and the corresponding counting process will be denoted by $N_0(t)$. The smallest sub σ-algebras $\mathcal{H} = \{\mathcal{H}_t, t \in [0, T]\}$ with respect to which $N_0(t)$ is adapted is called the *internal history* of the point process. In practice, it is usually sufficient, and this we will assume throughout, to let $\mathcal{F}$ be generated by $\mathcal{H}$ and $\mathcal{F}_0$, where the latter subsumes all prior information concerning the point process and subsequent displacements.

Since not all original defects are removed, we define the following counting processes of the number of faults *successfully removed*, $N_{0S}(t)$, and those *unsuccessfully removed*, $N_{0U}(t)$, by

$$N_{0S}(t) = \sum_{n \geq 1} X_n^{(0)} I\big(T_n^{(0)} \leq t\big) ,$$

$$N_{0U}(t) = \sum_{n \geq 1} (1 - X_n^{(0)}) I\big(T_n^{(0)} \leq t\big) ,$$

where $X_n^{(0)}, n = 1, 2, \ldots$ is a sequence of *i.i.d.* (independent and identically distributed) {0,1} valued random variables $\big(X_n^{(0)} = 1$ if the defect is successfully removed; otherwise it equals $0\big)$ with $P\big(X_n^{(0)} = 1\big) = p_0$ and $I(A)$ is the indicator of the set A.

In the event that some defects were not successfully removed, new faults will be injected into the system. Due to a number of considerations, for example clustering effect, we assume in our model that the effect of a multiple injection to be identical to singular injection in terms of subsequent discovery. Let

$$T_j^{(1)} = \begin{cases} T_j^{(0)} + Y_j^{(0)} & \text{if } X_j^{(0)} = 0 , \\ \infty & \text{if } X_j^{(0)} = 1 , \end{cases} \tag{1}$$

where $Y_n^{(0)}, n = 1, 2, \ldots$ is a sequence of non-negative real valued *i.i.d.* random variables, each of which represents the time differences between the introduction of a new defect by an imperfectly removed fault to its eventual detection. This sequence is assumed to be statistically independent of the original sequence of failure times. The common density function of the random variables will be denoted by $g_0(t) := g_0(t; \psi_0)$, ψ_0 belongs to some open subset $\Psi \subset \mathbb{R}^p$ where ψ_0 is an unknown vector parameter. It is assumed to be a continuous function of t. The injected defects will result in the following failure times uncovered with further testing:

$$0 < T_{(1)}^{(1)} < T_{(2)}^{(1)} < \cdots < T_{(j)}^{(1)} \ldots , \tag{2}$$

where $T_{(j)}^{(1)}$, $j = 1, 2, \ldots$ are the *order statistics* obtained from the sequence (1). The counting process corresponding to the sequence $T_{(j)}^{(1)}$, $j = 1, 2, \ldots$ will be denoted by $N_1(t)$. Note that the second branch of Eq. (1) ensures that perfectly debugged defects are not counted by $N_1(t)$.

The above process may be repeated several times although empirical evidence (see, e.g., Ref. 17) seems to suggest that secondary defects, once identified and removed, does not inject further defects into the software. However, if the process does go beyond the first stage described, and the process is repeated up to the i stage, $i \geq 2$, then the counting processes corresponding to the number of introduced faults successfully and unsuccessfully removed in $[0, t)$ at this stage are respectively

$$\begin{aligned} N_{i-1,S}(t) &= \sum_{n \geq 1} X_n^{(i-1)} I\big(T_n^{(i-1)} \leq t\big) , \\ N_{i-1,U}(t) &= \sum_{n \geq 1} \big(1 - X_n^{(i-1)}\big) I\big(T_n^{(i-1)} \leq t\big) , \end{aligned} \tag{3}$$

where $X_n^{(i-1)}$, $n = 1, 2, \ldots$ is a sequence of *i.i.d.* $\{0, 1\}$ valued random variables with $P\big(X_n^{(i-1)} = 1\big) = p_{i-1}$. In the event of an imperfect debugging, new defects are added to the system resulting in the following failure times

$$0 < T_{(1)}^{(i)} < T_{(2)}^{(i)} < \cdots < T_{(j)}^{(i)} \ldots ,$$

where

$$T_j^{(i)} = \begin{cases} T_j^{(i-1)} + Y_j^{(i-1)} & \text{if } X_j^{(i-1)} = 0\,, \\ \infty & \text{if } X_j^{(i-1)} = 1\,. \end{cases} \tag{4}$$

As before, $Y_n^{(i-1)}, n = 1, 2, \ldots$ is a sequence of non-negative real valued *i.i.d.* random variables statistically independent of the sequence $T_n^{(i-1)}, n = 1, 2, \ldots$. The common density function of these random variables will be denoted by $g_{i-1}(t) := g_{i-1}(t; \psi_{i-1})$, $\psi_{i-1} \in \Psi \subset \mathbb{R}^p$ and is assumed to be a continuous function of t. The counting process corresponding to Eq. (4) is denoted by $N_i(t)$.

We now consider the stochastic intensities of the above processes. Let us assume that the rate of occurrence of a failure (*hazard rate*) due to each indigenous defect exists and is denoted by $\beta(t) := \beta(t; \phi)$, ϕ belongs to some open subset $\Phi \subset \mathbb{R}^q$ where ϕ is an unknown vector parameter, i.e.,

$$\begin{aligned} &P\{\text{an indigenous fault causes a failure} \\ &\quad \text{in the interval } (t, t+h) | \mathcal{H}_{t-}\} = \beta(t)h + o(h). \end{aligned}$$

Then it follows that intensity of the counting process $N_0(t)$ is given by $\beta(t)M(t-)$ where $M(t) = \mu - N_0(t)$, i.e.,[8]

$$\lim_{h \downarrow 0} \frac{1}{h} E[N_0(t+h) - N_0(t) | \mathcal{H}_{t-}] = \beta(t)M(t-)\,. \tag{5}$$

We remark here that the Jelinski–Moranda model corresponds to the case when $\beta(t) = \beta_0$, a constant. It also follows that

$$\lim_{h \downarrow 0} \frac{1}{h} E[N_{0S}(t+h) - N_{0S}(t) | \mathcal{H}_{t-}] = \beta(t) p_0 M(t-)\,, \tag{6}$$

$$\lim_{h \downarrow 0} \frac{1}{h} E[N_{0U}(t+h) - N_{0U}(t) | \mathcal{H}_{t-}] = \beta(t)(1-p_0)M(t-)\,, \tag{7}$$

and the processes defined by:

$$\begin{aligned} M_{0S}(t) &= N_{0S}(t) - p_0 \int_0^t M(t-)\beta(t)dt\,, \\ M_{0U}(t) &= N_{0U}(t) - (1-p_0) \int_0^t M(t-)\beta(t)dt\,, \end{aligned} \tag{8}$$

are square integrable martingales.[11]

Since $N_1(t)$ is derived from $N_{0U}(t)$ through random translations, its intensity process is given by:

$$\lambda_1(t) = \int_0^t g_0(t-u)dN_{0U}(u)\,, \tag{9}$$

and the process

$$M_1(t) = N_1(t) - \int_0^t \int_0^v g_0(v-u)dN_{0U}(u)dv\,, \tag{10}$$

is a square integrable martingale. We note here for future reference that using Eqs. (8) and (10),

$$\lambda_1(t) = (1-p_0) \int_0^t \int_0^v g_0(v-u)M(u-)\beta(u)dudv\,. \tag{11}$$

This follows since, given any $\mathcal{F}$-predictable function $C(t, \omega)$

$$\begin{aligned} E\left(\int_0^T CdN_1\right) &= E\left(\int_0^T \int_0^t C(t)g_0(t-u)dN_{0U}(u)dt\right) \\ &= (1-p_0)E\left(\int_0^T\int_0^t C(t)g_0(t-u)M(u-)\beta(u)dudt\right), \end{aligned}$$

hence, by Definition 2.11 in Ref. 10, Eq. (11) follows.

Similarly, at the ith stage, the intensity of $N_i(t)$ is:

$$\lambda_i(t) = \int_0^\tau g_{i-1}(t-u)dN_{i-1,U}(u)\,, \tag{12}$$

and the process

$$M_i(t) = N_i(t) - \int_0^t \int_0^v g_{i-1}(v-u)dN_{i-1,U}(u)dv\,, \qquad (13)$$

is a square integrable martingale.

3. Asymptotic Properties of Maximum Likelihood Estimators

In this section, we will consider only a *single stage* of the dual processes of defect removal and injection of new faults. The functional form of $g(\cdot)$ is assumed known but not the value of the parameter vector $\eta := (p, \mu, \psi, \phi) \in \Theta := [0, 1) \times \mathbb{N} \times \Psi \times \Phi \subset \mathbb{R}^{p+q+2}$. We shall be concerned with the asymptotic properties, namely consistency and asymptotic normality, of maximum likelihood estimators (MLE) of these unknown parameters. The counting process corresponding to Eq. (2) and its intensity function (11) will be denoted by $N(t)$ and $\lambda(t)$ respectively where

$$\lambda(t;\eta) = (1-p)\int_0^t g(t-u;\psi)M(u-)\beta(u;\phi)du\,, \quad t \in [0,T]\,,$$
$$0 \le p < 1, \quad \psi \in \Psi \subset \mathbb{R}^p, \quad \mu \in \mathbb{N}, \quad \phi \in \Phi \subset \mathbb{R}^q\,. \qquad (14)$$

The MLE of the unknown parameters are obtained by solving the system of equations

$$\frac{\partial}{\partial \eta} \ln L(\eta; T) = 0\,, \qquad (15)$$

where the likelihood function at time t, $L(\eta; t)$ is given by (c.f. Ref. 11)

$$L(\eta;t) = \exp\left[\int_0^t \ln \lambda(s;\eta)dN(s) - \int_0^t \lambda(s;\eta)ds\right]. \qquad (16)$$

In order to derive the asymptotic properties of the MLE, it is not reasonable to let the testing time $T \to \infty$ since in practice, the process of testing terminates after some finite time. However, asymptotic convergence requires us to increase the value of one of the parameters in the model. Following Ref. 13, we increase the total number of defects in the software, μ, i.e., we re-parameterize

$$\mu = \nu\gamma\,, \tag{17}$$

where γ is an unknown constant and we let ν increases over the set of natural numbers. For each ν, the counting processes corresponding to $N_0(t)$ and $N(t)$ will be denoted by $N_{0\nu}(t)$ and $N_\nu(t)$ respectively. Also, the MLE $\hat{\eta} := (\hat{p}_\nu, \hat{\gamma}_\nu, \hat{\psi}_\nu, \hat{\phi}_\nu)$ for the true values $\eta_0 := (p_0, \gamma_0, \phi_0, \psi_0)$ will be obtained by solving

$$\frac{\partial}{\partial\eta}\ln L_\nu(\eta; T) = 0\,, \tag{18}$$

where

$$L_\nu(\eta; t) = \exp\left[\int_0^t \ln\lambda_\nu(s;\eta)dN_\nu(s) - \int_0^t \lambda_\nu(s;\eta)ds\right]. \tag{19}$$

In Eq. (19), $\lambda_\nu(t;\eta)$, the intensity of $N_\nu(t)$, has the form

$$\lambda_\nu(t;\eta) = \nu(1-p)\int_0^t g(t-u;\psi)(\gamma - x_\nu(u-))\beta(u;\phi)du\,, \tag{20}$$

where

$$x_\nu(t) = \frac{N_{0\nu}(t)}{\nu}, \quad t\in[0, T]\,. \tag{21}$$

We note that for each ν, $x_\nu \in D := D(0, T]$, the Skorohod space of cadlag functions on $(0, T]$.

In the sequel, $\stackrel{P}{\to}$ will denote convergence in probability and $\stackrel{d}{\to}$ denotes convergence in distribution on the function space D.

To simplify notation, we shall express

$$\lambda_\nu(t;\eta) = \nu h(t;\eta, x_\nu)\,, \tag{22}$$

where $h(t;\eta,x_\nu) = (1-p)\int_0^t g(t-u;\psi)(\gamma - x_\nu(u-))\beta(u;\phi)du$.

We shall assume the following conditions:

(a) For all $(\psi, \phi) \in \Psi \times \Phi$

$$\sup_{t \leq T} \int_0^t g(t-u; \psi)\beta(u; \phi)du < \infty .$$

It follows easily from Condition (a) that there exist constants C_1, C_2 and D, not depending on t, such that for any $x, y \in D$ and $t \in (0, T]$,

$$\sup_{t \leq T} h(t; \eta, x) \leq C_1 + C_2 \sup_{t \leq T} |x(t)| , \tag{23}$$

and

$$|h(t; \eta, x) - h(t; \eta, y)| \leq D \sup_{s \leq t} |x(s) - y(s)| . \tag{24}$$

Using a law of large numbers due to Ref. 20, Eqs. (23) and (24) imply that the $x_\nu(t)$ defined by Eq. (21), converges uniformly on $(0, T]$ in probability to a function $x_0(t)$ as $\nu \to \infty$, where $x_0 \in D$ is the unique solution of

$$x(t) = \int_0^t h(s; \eta_0, x)ds . \tag{25}$$

(b) There exists a neighborhood Θ_0 and D_0 of η_0, x_0 respectively, such that the function $h(t; \eta, x)$ and its derivatives with respect to η of the first, second and the third order exist, are continuous functions of η and x, bounded on $[0, T] \times \Theta_0 \times D_0$. Furthermore, $h(t; \eta, x)$ is bounded away from zero on $[0, T] \times \Theta_0 \times D_0$.

(c) The matrix $\Sigma = \{\sigma_{ij}(\eta_0)\}$, where

$$\sigma_{ij}(\eta) = \int_0^T \frac{\frac{\partial}{\partial \eta_i} h(s; \eta, x_0) \frac{\partial}{\partial \eta_j} h(s; \eta, x_0)}{h(s; \eta, x_0)} ds , \tag{26}$$

is positive definite.

(d) The stochastic process x_ν defined in Eq. (21) has to satisfy for all $i, j \in \{1, 2, \dots, n\}$ as $\nu \to \infty$:

$$\sqrt{\nu} \int_0^T \frac{\frac{\partial}{\partial \eta_i} h(s; \eta_0, x_\nu) \frac{\partial}{\partial \eta_j} h(s; \eta_0, x_\nu)}{h(s; \eta_0, x_\nu)} ds$$
$$\xrightarrow{P} \int_0^T \frac{\frac{\partial}{\partial \eta_i} h(s; \eta_0, x_0) \frac{\partial}{\partial \eta_j} h(s; \eta_0, x_0)}{h(s; \eta_0, x_0)} ds$$
$$< \infty ,$$

and $$\sqrt{\nu} \int_0^T \left[\frac{\partial^2}{\partial \eta_i \partial \eta_j} \ln h(s; \eta_0, x_\nu) \right]^2 h(s; \eta_0, x_\nu) ds$$
$$\xrightarrow{P} \int_0^T \left[\frac{\partial^2}{\partial \eta_i \partial \eta_j} \ln h(s; \eta_0, x_0) \right]^2 h(s; \eta_0, x_0) ds$$
$$< \infty .$$

(e) There exist predictable processes G and H not depending on θ and neighborhoods Θ_0, D_0 of η_0 and x_0 respectively such that for all $t \in [0, T]$ and $x \in D_0$

$$\sup_{\eta \in \Theta_0} \left| \frac{\partial^3 h(t; \eta, x_\nu)}{\partial \eta_i \partial \eta_j \partial \eta_k} \right| \le G(t; x_\nu) ,$$

$$\sup_{\eta \in \Theta_0} \left| \frac{\partial^3 \ln h(t; \eta, x_\nu)}{\partial \eta_i \partial \eta_j \partial \eta_k} \right| \le H(t; x_\nu) ,$$

for all i, j, k. Moreover the processes H and G satisfy as $\nu \to \infty$:

$$\sqrt{\nu} \int_0^T H(s; x_\nu) h(s; \eta_0, x_\nu) ds \xrightarrow{P} \int_0^T H(s; x_0) h(s; \eta_0, x_0) ds$$
$$< \infty ,$$

$$\sqrt{\nu} \int_0^T H^2(s; x_\nu) h(s; \eta_0, x_\nu) ds \xrightarrow{P} \int_0^T H^2(s; x_0) h(s; \eta_0, x_0) ds$$
$$< \infty ,$$

$$\text{and} \quad \sqrt{\nu} \int_0^T G(s; x_\nu) ds \xrightarrow{P} \int_0^T G(s; x_0) ds < \infty .$$

For each sequence of models indexed by $\nu = 1, 2, \ldots,$ let us define for $\eta \in \Theta$, $t \in [0, T]$, $i, j, k \in \{1, 2, \ldots, p+q+2\}$

$$C_\nu(\eta; t) := \ln L_\nu(\eta, t)\,, \tag{27}$$

$$U_{\nu i}(\eta; t) := \frac{\partial C_\nu(\eta; t)}{\partial \eta_i}\,, \tag{28}$$

$$I_{\nu ij}(\eta; t) := -\frac{\partial^2 C_\nu(\eta; t)}{\partial \eta_i \partial \eta_j}\,, \tag{29}$$

$$R_{\nu ijk}(\eta; t) := \frac{\partial^3 C_\nu(\eta; t)}{\partial \eta_i \partial \eta_j \partial \eta_k}\,. \tag{30}$$

In the above, we let $\eta_1 = p$ and $\eta_2 = \gamma$.

The next two theorems will be proved using the ideas employed in Refs. 21 and 13 to prove their asymptotic results. The proofs mainly rely on combining Lenglart's inequality with the martingale central limit theorem in the forms expressed in Ref. 11 to show the validity of some auxiliary results, and then use these to infer the main results with the support of some classical results due to Ref. 22.

Theorem 1. *Consider the sequence of counting processes with intensities $\lambda_\nu(t; \eta)$ defined by Eq.* (20) *where $\eta = (p, \gamma, \psi, \phi)$ denotes an unknown $n = p + q +$ two-dimension parameter. Under Conditions* (*a*) *to* (*e*), *there is a unique consistent solution for the MLE of η_0, the true value of η. This solution provides a local maximum of the likelihood function* (19).

Proof. Using a Taylor series expansion about $\eta_0 = (\eta_{10}, \ldots, \eta_{n0})$, we get for any $\eta \in \Theta_0$,

$$U_{\nu i}(\eta; T) = U_{\nu i}(\eta_0; T) - \sum_{i=1}^{n} (\eta_i - \eta_{i0}) I_{\nu ij}(\eta_0; T) + \frac{1}{2} \sum_{j=1}^{n} \sum_{k=1}^{n} (\eta_j - \eta_{j0})(\eta_k - \eta_{k0}) R_{\nu ijk}(\eta^*; T)\,,$$

where η^* is on the line segment joining η and η_0.

The theorem will be proved, using the same argument as that used in Ref. 22, pages 10–13, if we can show that

$$(\sqrt{\nu})^{-1} U_{\nu i}(\eta_0; T) \xrightarrow{P} 0\,, \tag{31}$$

$$(\sqrt{\nu})^{-1} I_{\nu ij}(\eta_0; T) \xrightarrow{P} \sigma_{ij}(\eta_0)\,, \tag{32}$$

and there exists a finite constant L, not depending on η, such that

$$\lim_{\nu\to\infty} P(|(\sqrt{\nu})^{-1} R_{\nu ijk}(\eta; T)| < L) = 1\,, \tag{33}$$

for all i, j, k and $\eta \in \Theta_0$.

We can write the score function evaluated at η_0 as

$$U_{\nu i}(\eta_0; T) = \int_0^T \frac{\partial}{\partial \eta_i} \ln h(s; \cdot) dN_\nu(s) - \int_0^T \frac{\partial}{\partial \eta_i} \nu h(s; \cdot) ds\,. \tag{34}$$

Notice that Eq. (34) can also be expressed as:

$$U_{\nu i}(\eta_0; T) = \int_0^T \frac{\partial}{\partial \eta_i} \ln h(s; \cdot) dM_\nu(s)\,, \tag{35}$$

where

$$M_\nu(t) = N_\nu(t) - \int_0^t \nu h(s; \eta_0, x_\nu) ds\,, \tag{36}$$

$\nu = 1, 2, \ldots$ are square integrable martingales with the predictable variation processes

$$\langle M_\nu \rangle(t) = \int_0^T \nu h(s; \eta_0, x_\nu) ds\,. \tag{37}$$

By Condition (b), the stochastic integrals (35) are square integrable martingales, hence Lenglart's inequality (cf., Ref. 11) implies that for all $\delta, \varepsilon > 0$ we have

$$\begin{aligned}
&P\left(\sup_{t\in(0,T]}\left|(\sqrt{\nu})^{-1}U_{\nu i}(\eta_0;t)\right| > \varepsilon\right)\\
&\quad\le \frac{\delta}{\varepsilon^2} + P(\langle(\sqrt{\nu})^{-1}U_{\nu i}(\eta_0,T),(\sqrt{\nu})^{-1}U_{\nu i}(\eta_0,T)\rangle > \delta)\\
&\quad= \frac{\delta}{\varepsilon^2} + P((\sqrt{\nu})^{-2}\langle(U_{\nu i}(\eta_0,T),U_{\nu i}(\eta_0,T)\rangle > \delta)\\
&\quad= \frac{\delta}{\varepsilon^2} + P\left(\nu^{-1}\nu\int_0^T\left(\frac{\partial h(s;\cdot)/\partial\eta_i}{h(s;\cdot)}\right)^2 h(s;\eta_0,x_\nu)ds > \delta\right)\\
&\quad= \frac{\delta}{\varepsilon^2} + P\left(\int_0^T\left(\frac{\partial h(s;\cdot)/\partial\eta_i}{h(s;\cdot)}\right)^2 h(s;\eta_0,x_\nu)ds > \delta\right).
\end{aligned}$$

Since δ can be made arbitrarily small and Condition (d) implies that the integral in the last term above converges in probability to 0 as $\nu \to \infty$, we can conclude that

$$(\sqrt{\nu})^{-1}U_{\nu i}(\eta_0;T) \xrightarrow{P} 0\,,$$

which proves Eq. (31).

To prove Eq. (32), we first note that

$$\begin{aligned}
I_{\nu ij}(\eta_0,T) &= -\int_0^T \frac{\partial^2 \ln h(s;\cdot)}{\partial\eta_i\partial\eta_j}\,dN_\nu(ds) + \nu\int_0^T \frac{\partial^2 h(s;\cdot)}{\partial\eta_i\partial\eta_j}ds\\
&= -\int_0^T \frac{\partial^2 \ln h(s;\cdot)}{\partial\eta_i\partial\eta_j}(dN_\nu - \nu h(s;\cdot)ds)\\
&\quad - \int_0^T \frac{\partial^2 \ln h(s;\cdot)}{\partial\eta_i\partial\eta_j}\nu h(s;\cdot)ds + \nu\int_0^T \frac{\partial^2 h(s;\cdot)}{\partial\eta_i\partial\eta_j}ds\,.
\end{aligned}$$

Since

$$\begin{aligned}
\int_0^T \frac{\partial^2 \ln h(s;\cdot)}{\partial\eta_i\partial\eta_j}\nu h(s;\cdot)ds = \int_0^T &\left(-\frac{1}{h(s;\cdot)^2}\frac{\partial h(s;\cdot)}{\partial\eta_i}\frac{\partial h(s;\cdot)}{\partial\eta_j}\right.\\
&\left.+\frac{1}{h(s;\cdot)}\frac{\partial^2 h(s;\cdot)}{\partial\eta_i\partial\eta_j}\right)\nu h(s;\cdot)ds\,,
\end{aligned}$$

it follows that

$$I_{\nu ij}(\eta_0, T) = -\int_0^T \frac{\partial^2 \ln h(s;\cdot)}{\partial\eta_i\partial\eta_j}\, dM_\nu(s) + \nu\int_0^T \frac{\frac{\partial}{\partial\eta_i}h(s;\cdot)\frac{\partial}{\partial\eta_j}h(s;\cdot)}{h(s;\cdot)} ds\,, \tag{38}$$

which, by Condition (b), implies that $I_{\nu ij}$ is a semimartingale. Next, we further note that

$$\begin{aligned}
&\left\langle \int_0^T \frac{\partial^2 \ln h(s;\cdot)}{\partial\eta_i\partial\eta_j}\, dM_\nu(s), \int_0^T \frac{\partial^2 \ln h(s;\cdot)}{\partial\eta_i\partial\eta_j} dM_\nu(s)\right\rangle \\
&= \int_0^T \left(\frac{\partial^2 \ln h(s;\cdot)}{\partial\eta_i\partial\eta_j}\right)^2 d\langle M_\nu\rangle(s) \\
&= \nu\int_0^T \left(\frac{\partial^2 \ln h(s;\cdot)}{\partial\eta_i\partial\eta_j}\right)^2 h(s;\cdot) ds\,,
\end{aligned}$$

and, applying Lenglart's inequality, for all $\delta, \varepsilon > 0$,

$$\begin{aligned}
&P\left(\sup_{t\in(0,T]}\left|(\sqrt{\nu})^{-1}\int_0^T \frac{\partial^2 \ln h(s;\cdot)}{\partial\eta_i\partial\eta_j} dM_\nu(s)\right| > \varepsilon\right) \\
&\leq \frac{\delta}{\varepsilon^2} + P\left(\nu^{-1}\left\langle \int_0^T \left(\frac{\partial^2 \ln h(s;\cdot)}{\partial\eta_i\partial\eta_j}\right) dM_\nu(t), \right.\right. \\
&\qquad \left.\left. \int_0^T \left(\frac{\partial^2 \ln h(s;\cdot)}{\partial\eta_i\partial\eta_j}\right) dM_\nu(t)\right\rangle > \varepsilon\right) \\
&= \frac{\delta}{\varepsilon^2} + P\left(\int_0^T \left(\frac{\partial^2 \ln h(s;\cdot)}{\partial\eta_i\partial\eta_j}\right)^2 h(s;\cdot) ds > \delta\right).
\end{aligned}$$

By Condition (d), it follows that

$$(\sqrt{\nu})^{-1}\int_0^T \frac{\partial^2 \ln h(s;\cdot)}{\partial\eta_i\partial\eta_j} dM_\nu(s) \xrightarrow{P} 0\,,$$

and therefore Eq. (38) and another application of Condition (d) imply that

$$(\sqrt{\nu})^{-1} I_{\nu ij}(\eta_0; T) \xrightarrow{P} \sigma_{ij}(\eta_0)\,,$$

which prove Eq. (32).

Finally, to prove Eq. (33), we have to show that

$$\left|(\sqrt{\nu})^{-1} R_{\nu ijk}(\eta; T)\right| < L\,,$$

with probability 1 as $\nu \to \infty$ for some constant L and for all i, j, k and all $\eta \in \Theta_0$. Since

$$R_{\nu ijk}(\eta; T) = \int_0^T \frac{\partial^3 \ln h(s;\cdot)}{\partial\eta_i \partial\eta_j \partial\eta_k}\, dN_\nu(s) - \nu \int_0^T \frac{\partial^3 h(s;\cdot)}{\partial\eta_i \partial\eta_j \partial\eta_k}\, ds\,,$$

and, using Condition (e),

$$\begin{aligned}
\left|(\sqrt{\nu})^{-1} R_{\nu ijk}(\eta; T)\right| &\le (\sqrt{\nu})^{-1} \int_0^T \left|\frac{\partial^3 \ln h(s;\cdot)}{\partial\eta_i \partial\eta_j \partial\eta_k}\right| dN_\nu(s) \\
&\quad + \sqrt{\nu} \int_0^T \left|\frac{\partial^3 \ln h(s;\cdot)}{\partial\eta_i \partial\eta_j \partial\eta_k}\right| ds \\
&\le (\sqrt{\nu})^{-1} \int_0^T H(s; x_\nu) dN_\nu(s) \\
&\quad + \sqrt{\nu} \int_0^T G(s; x_\nu) ds\,. \qquad (39)
\end{aligned}$$

Applying Langlart's inequality again, for all $\delta, \varepsilon > 0$,

$$\begin{aligned}
&P\Bigg(\sup_{t\in(0,T]} \Bigg|(\sqrt{\nu})^{-1} \int_0^t H(s; x_\nu) dN_\nu(s) \\
&\qquad - (\sqrt{\nu})^{-1}\nu \int_0^t H(s; x_\nu) h(s;\cdot) ds\Bigg| > \varepsilon\Bigg) \\
&= P\left(\sup_{t\in(0,T]} \left|(\sqrt{\nu})^{-1} \int_0^t H(s; x_\nu)\, dM_\nu(s)\right| > \varepsilon\right) \\
&\le \frac{\delta}{\varepsilon^2} + P\left(\nu^{-1}\nu \int_0^T H^2(s; x_\nu) h(x;\cdot) ds > \varepsilon\right).
\end{aligned}$$

By Condition (e), the integral in the last term converges in probability to 0, therefore

$$\begin{aligned}
&\operatorname*{plim}_{\nu\to\infty} (\sqrt{\nu})^{-1} \int_0^T H(s; x_\nu) dN_\nu \\
&\quad = \operatorname*{plim}_{\nu\to\infty} \sqrt{\nu} \int_0^T H(s; x_\nu) h(s; \eta_0, x_\nu) ds \\
&\quad = \int_0^T H(s; x_0) h(s; \eta_0, x_0) ds < \infty .
\end{aligned}$$

Combining with the condition

$$\sqrt{\nu} \int_0^T G(s; x_\nu) ds \xrightarrow{P} \int_0^T G(s; x_0) ds < \infty ,$$

and referring to Eq. (39), Eq. (33) is proved. This completes the proof of Theorem 1.

Theorem 2. *Assume that Conditions* (a)–(e) *holds, and let* $\hat{\eta}_\nu \xrightarrow{P} \eta_0$ *be the consistent solutions of equations* (18). *Then*

$$\nu^{1/4}(\hat{\eta}_\nu - \eta_0) \xrightarrow{d} \mathcal{N}(0, \Sigma^{-1}) ,$$

where the matrix $\Sigma = \{\sigma_{ij}(\eta_0)\}$ *defined by Eq.* (26) *can be estimated consistently with* $(\sqrt{\nu})^{-1} I_{\nu ij}(\hat{\eta}_\nu)$.

Proof. Using a Taylor series expansion of $U_{\nu i}(\hat{\eta}_\nu; T)$ about η_0 gives

$$\begin{aligned}
0 = \nu^{-1/4} U_{\nu i}(\hat{\eta}_\nu; T) &= \nu^{-1/4} U_{\nu i}(\eta_0; T) \\
&\quad + \sum_{j=1}^{n} \nu^{1/4}(\hat{\eta}_{j\nu} - \eta_{j0}) \frac{1}{\sqrt{\nu}} I_{\nu ij}(\eta_\nu^*; T) ,
\end{aligned}$$

where $\hat{\eta}_\nu = (\hat{\eta}_{1\nu}, \ldots, \hat{\eta}_{n\nu})$ and η_ν^* is a point on the line segment between $\hat{\eta}_\nu$ and η_0. If we can prove that

$$\nu^{-1/4} U_{\nu i}(\eta_0; T) \xrightarrow{d} \mathcal{N}(0, \Sigma) , \tag{40}$$

and

$$\frac{1}{\sqrt{\nu}} I_{\nu ij}(\eta_\nu^*; T) \xrightarrow{P} \sigma_{ij}(\eta_0) , \tag{41}$$

for all i, j and for any η_ν^* such that $\eta_\nu^* \xrightarrow{P} \eta_0$ as $\nu \to \infty$, then appealing to Theorem 2.2 in Ref. 22 will complete the proof of the theorem.

We use the Martingale Central Limit Theorem to prove Eq. (40). Firstly, for all i, j, Condition (d) gives

$$\begin{aligned}
&\langle \nu^{-1/4} U_{\nu i}(\eta_0; T), \nu^{-1/4} U_{\nu j}(\eta_0; T)\rangle \\
&= \sqrt{\nu} \int_0^T \left(\frac{\partial h(s;\cdot)}{\partial \eta_i} \frac{\partial h(s;\cdot)}{\partial \eta_j} \right) \frac{1}{h(s;\eta_0, x_\nu)} ds \\
&\xrightarrow{P} \int_0^\tau \left(\frac{\partial h(s;\cdot)}{\partial \eta_i} \frac{\partial h(s;\cdot)}{\partial \eta_j} \right) \frac{1}{h(s;\eta_0, x_0)} ds \\
&= \sigma_{ij}(\eta_0)\,.
\end{aligned}$$

Secondly, for each i, $\epsilon > 0$ and $t \in (0, T]$, let us define

$$\begin{aligned}
&U_{\nu i \epsilon}(\eta; t) \\
&\quad = \nu^{-1/4} \int_0^t \frac{\partial}{\partial \eta_i} \ln h(s;\cdot) I \left(\nu^{-1/4} \frac{\partial}{\partial \eta_i} \ln h(s;\cdot) > \epsilon \right) dM_\nu(s)\,.
\end{aligned}$$

Then, dominated convergence and Condition (d) imply

$$\begin{aligned}
&\langle U_{\nu i \epsilon}(\eta_0; T), U_{\nu i \epsilon}(\eta_0; T)\rangle \\
&= \sqrt{\nu} \int_0^T \left(\frac{\frac{\partial h(s;.)}{\partial \eta_i}}{h(s;\eta_0, x_\nu)} \right)^2 h(s;\eta_0, x_\nu) I \\
&\quad \times \left(\nu^{-1/4} \left| \frac{\frac{\partial h(s;\eta_0, x_\nu)}{\partial \eta_i}}{h(s;\eta_0, x_\nu)} \right| > \epsilon \right) ds \\
&\xrightarrow{P} 0\,,
\end{aligned}$$

as $\nu \to \infty$. By the Martingale Central Limit theorem (see Appendix I, Ref. 11), Eq. (40) follows.

Finally, expanding $I_{\nu ij}(\eta^*_\nu; T)$ by a Taylor series about η_0 gives

$$\nu^{-1/2} I_{\nu ij}(\eta^*_\nu; T) = \nu^{-1/2} I_{\nu ij}(\eta_0; T) + \nu^{-1/2} \times \sum_{k=1}^{n} (\eta^*_{k\nu} - \eta_{k0}) R_{\nu ijk}(\bar{\eta}_\nu; T) , \quad (42)$$

where $\eta^*_\nu = (\eta^*_{1\nu}, \ldots, \eta^*_{n\nu})$, $\bar{\eta}_\nu$ is on the line joining η^*_ν to η_0 and $R_{\nu ijk}(\eta; t)$ is defined by Eq. (30). We have already shown in the proof of Theorem 1 that Eq. (32) holds, i.e., $(\sqrt{\nu})^{-1} I_{\nu ij}(\eta_0; T) \xrightarrow{P} \sigma_{ij}(\eta_0)$. Also, by Eq. (33), $\nu^{-1/2} R_{\nu ijk}(\bar{\eta}_\nu; T)$ is bounded above in probability. Therefore, the last term in Eq. (42) converges in probability to 0 as $\eta^*_\nu \xrightarrow{P} \eta_0$. This proves Eq. (41) which also implies, since $\hat{\eta}_\nu \xrightarrow{P} \eta_0$, that $\nu^{-1/2} I_{\nu ij}(\hat{\eta}_\nu; T)$ is a consistent estimator for Σ.

4. Simulation Results

In the previous section, we provide very general conditions under which the sequence of MLEs of parameters of our model converges towards the normal distribution. In this section, we provide some simulation results with the view to investigating how rapid this convergence is. Works by van Pul[13] on the convergence of MLEs of parameters of the Jelinska–Moranda model to the normal distribution seem to suggest that the rate of convergence to the normal distribution is rather slow.

We will assume that the original failure times follow a Jelinska–Moranda model with failure rate $\beta(t; \phi) = \beta$, a constant, and the displacements have a common distribution exponentially distributed, i.e.,

$$g(t; \psi) = \psi \exp(-\psi t) .$$

From Eq. (15), the intensity function of the failure times of the introduced errors is given by:

$$\lambda(t; \eta) = (1 - p)\phi\psi \int_0^t \exp(-\psi(t - u))(\mu - N(u-))du . \quad (43)$$

Now suppose that successive failures were observed to occur at times $0 < t_0 < t_1 < t_2 < \cdots < t_{N(T)}$. Then, using Eq. (43),

$$\lambda(t;\eta) = \begin{cases} (1-p)\phi\mu(1-\exp(-\psi t)) & \text{if } 0 < t \le t_1\,, \\ (1-p)\phi(\mu - i + 1) & \text{if } t_{i-1} < t \le t_i, \\ \quad (1-\exp(-\psi t)) & \quad 2 \le i \le N(T)\,. \end{cases} \tag{44}$$

Applying Eq. (16) and noting Eq. (44), we obtain the likelihood function

$$\begin{aligned} L(\eta;T) &= \prod_{i=1}^{N(T)} \lambda(t_i;\eta)\exp\left(-\int_0^T \lambda(t;\eta)dt\right) \\ &= [(1-p)\phi]^{N(T)} \prod_{i=1}^{N(T)} (\mu - i + 1)(1-\exp(-\psi\, t_i)) \\ &\quad \times \exp\Bigg[-(1-p)\sum_{i=1}^{N(T)+1} \phi(\mu - i + 1) \\ &\quad \times \left(t_i - t_{i-1} + \frac{\exp(-\psi t_i) - \exp(-\psi t_{i-1})}{\psi}\right)\Bigg]. \end{aligned} \tag{45}$$

In the above, we have let $t_{N(T)+1} = T$.

To find MLEs $\hat{p}$, $\hat{\phi}$, $\hat{\psi}$ and $\hat{\mu}$ we solve the following equations

$$\frac{\partial}{\partial p}\ln L(\eta;T) = 0\,, \quad \frac{\partial}{\partial \phi}\ln L(\eta;T) = 0\,,$$
$$\frac{\partial}{\partial \psi}\ln L(\eta;T) = 0\,, \quad \frac{\partial}{\partial \mu}\ln L(\eta;T) = 0\,.$$

For our simulation experiments, we set $p = 0.9$, $\psi = 1$ and $\phi = 10$ and let $\mu = 100, 500$ and 1000 respectively in order to compare the results for different starting number of indigenous faults. For the convenient display of our results, we scaled all failure times, indigenous and introduced, so that they are constrained to lie within the interval [0, 1], i.e., we set $T = 1$.

We ran 1000 simulations and obtained MLEs from the results of each simulation using the likelihood function given in Eq. (44). These simulations resulted in 1000 different sets of MLEs ($\hat{p}$, $\hat{\mu}$, $\hat{\psi}$, $\hat{\phi}$) for each of the three starting numbers of indigenous faults. The histograms and normal probability plots of these MLEs are displayed in Figs. 1–12. The results are mixed insofar that while the normal approximation appears good for some parameters, it is poor for others. This seems to confirm van Pul's[13] observation that the rate of convergence using the asymptotic theory developed is not uniformly fast for all parameters.

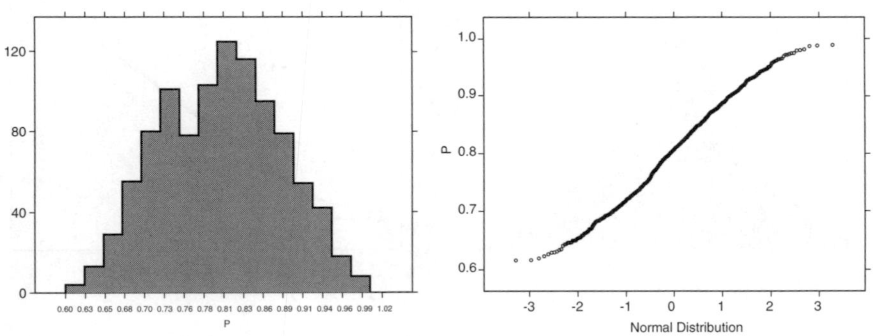

Fig. 1. Histogram and normal probability plot of $\hat{p}$ for 100 simulation runs.

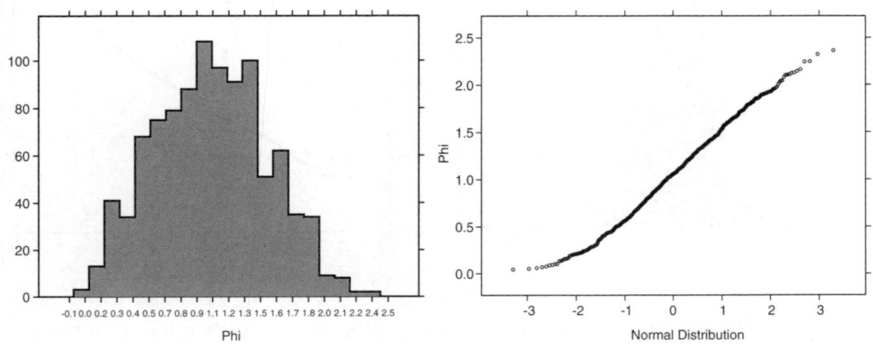

Fig. 2. Histogram and normal probability plot of $\hat{\phi}$ for 100 simulation runs.

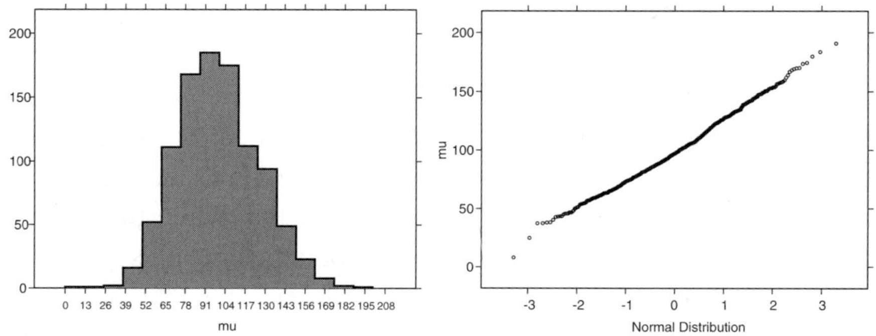

Fig. 3. Histogram and normal probability plot of $\hat{\mu}$ for 100 simulation runs.

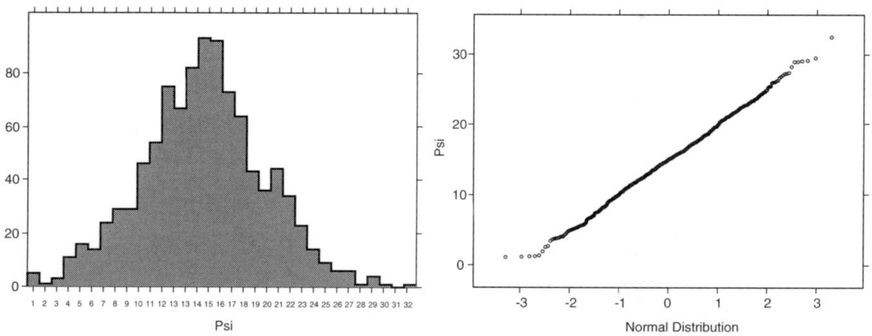

Fig. 4. Histogram and normal probability plot of $\hat{\psi}$ for 100 simulation runs.

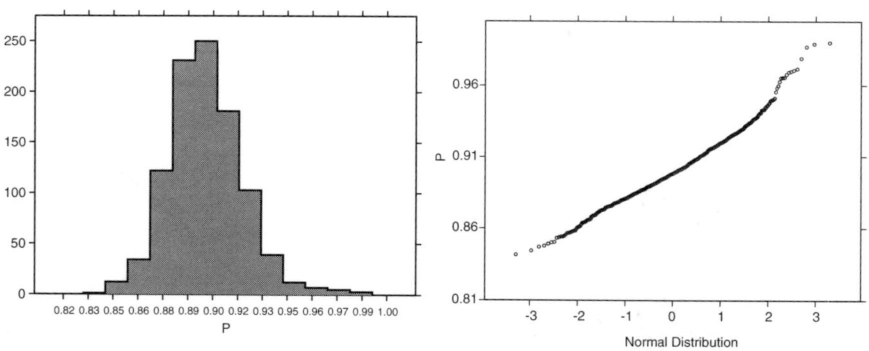

Fig. 5. Histogram and normal probability plot of $\hat{p}$ for 500 simulation runs.

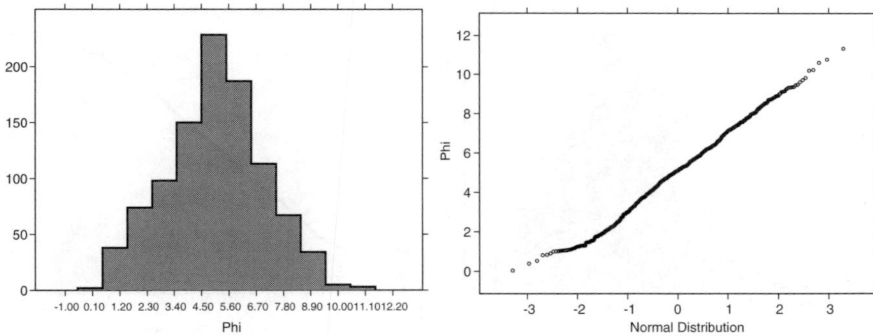

Fig. 6. Histogram and normal probability plot of $\hat{\phi}$ for 500 simulation runs.

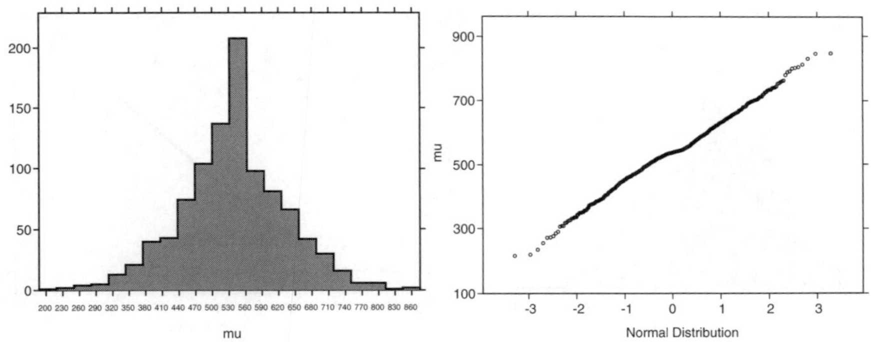

Fig. 7. Histogram and normal probability plot of $\hat{\mu}$ for 500 simulation runs.

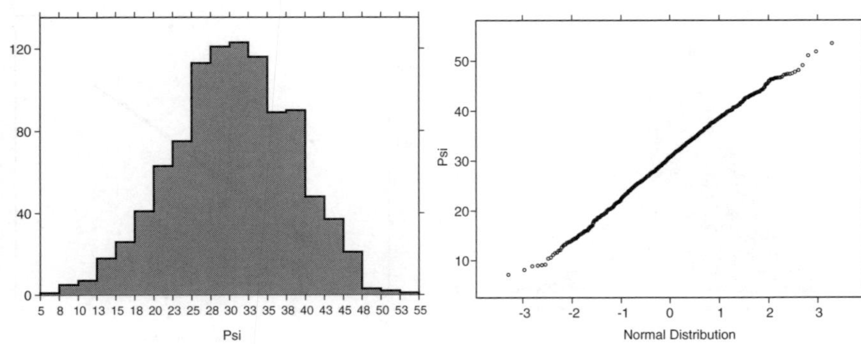

Fig. 8. Histogram and normal probability plot of $\hat{\psi}$ for 500 simulation runs.

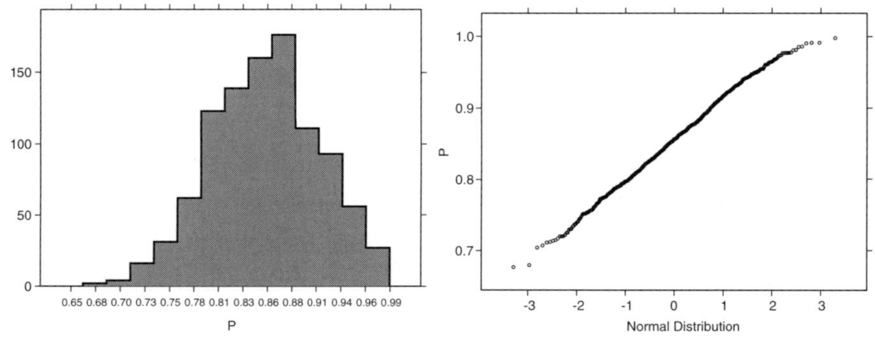

Fig. 9. Histogram and normal probability plot of $\hat{p}$ for 1000 simulation runs.

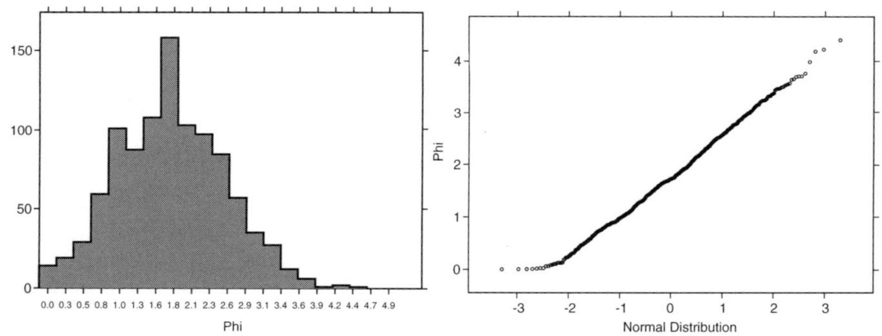

Fig. 10. Histogram and normal probability plot of $\hat{\phi}$ for 1000 simulation runs.

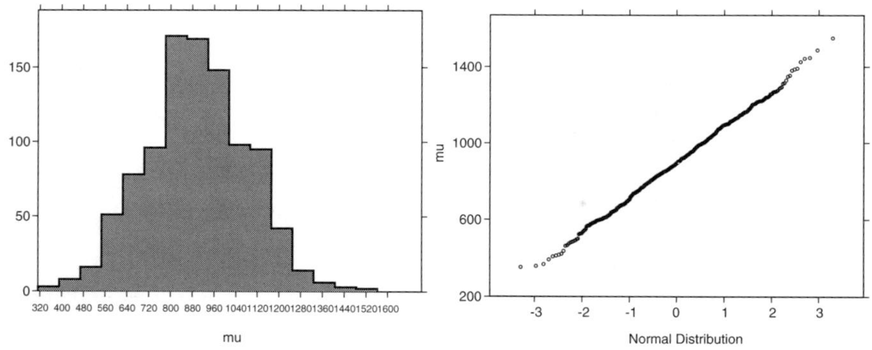

Fig. 11. Histogram and normal probability plot of $\hat{\mu}$ for 1000 simulation runs.

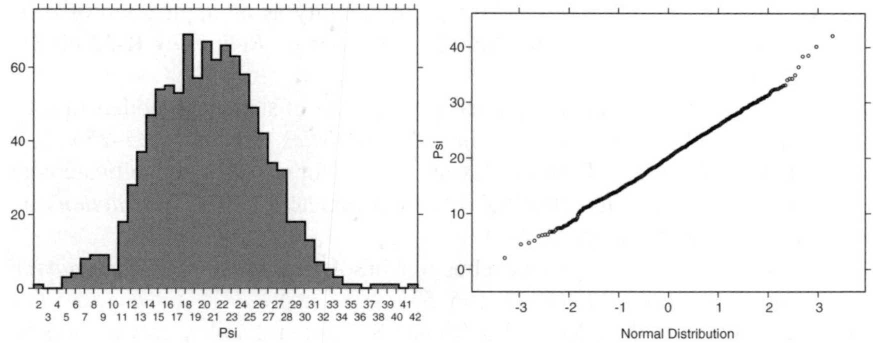

Fig. 12. Histogram and normal probability plot of $\hat{\psi}$ for 1000 simulation runs.

References

1. Z. Jelinski and P. B. Moranda, Software reliability research, *Statistical Computer Performance Evaluation*, ed. W. Freiberger (Academic Press, 1972), pp. 465–484.
2. M. Xie, *Software Reliability Modelling* (World Scientific, 1991).
3. N. D. Singpurwalla and S. P. Wilson, Software reliability modelling, *International Statistical Review* **62** (1994) 289–317.
4. Y. Chen and N. D. Singpurwalla, Unification of software reliability models by self-exciting point process, *Advances in Applied Probability* **29** (1997) 337–352.
5. D. L. Snyder and M. I. Miller, *Random Point Processes*, 2nd edn. (Springer-Verlag, 1991).
6. A. L. Goel and K. Okumoto, Time-dependent error-detection rate model for software reliability and other performance measures, *IEEE Transactions on Reliability* **R-28** (1979) 206–210.
7. J. Jacod, Multivariate point processes: Predictable projection, Radon–Nikodym derivatives, representation of martingales, *Zeitschrift für Wahrscheinlichkeitstheorie* **34** (1975) 225–244.
8. O. Aalen, Inference for counting processes, *Annals of Statistics* **6** (1978) 701–726.
9. P. Brémaud, *Point Processes and Queues, Martingale Dynamics* (Springer-Verlag, 1981).
10. A. F. Karr, *Point Processes and their Statistical Inference* (Marcel-Dekker, Inc., 1986).
11. P. K. Andersen, Ø. Borgan, R. D. Gill and N. Keiding, *Statistical Models Based on Counting Processes* (Springer-Verlag, 1993).

12. G. Koch and P. J. C. Spreij, Software reliability as an application of martingale and filtering theory, *IEEE Transactions on Reliability* **R-32** (1983) 342–345.
13. M. C. van Pul, Asymptotic properties of a class of statistical models in software reliability, *Scandinavian Journal of Statistics* **19** (1992) 235–253.
14. I. Fakhre-Zakeri and E. Slud, Mixture models for reliability of software with imperfect debugging: Identifiability of parameters, *IEEE Transactions on Reliability* **R-44** (1995) 104–113.
15. E. Slud, Testing for imperfect debugging in software reliability, *Scandinavian Journal of Statistics* **24** (1997) 555–572.
16. P. Zeephongsekul, G. Xia and S. Kumar, Software reliability growth models: Primary-failures generate secondary-faults under imperfect debugging, *IEEE Transactions on Reliability* **R-43** (1994) 408–413.
17. P. Zeephongsekul and W. Bodhisuwan, On a generalized dual process software reliability growth model, *International Journal of Reliability, Quality and Safety Engineering* **6** (1999) 19–30.
18. H. Pham, L. Nordman and X. Zhang, A general imperfect-software-debugging model with *S*-shaped fault detection rate, *IEEE Transactions on Reliability* **R-48** (1999) 169–175.
19. X. Teng and H. Pham, A software-reliability growth model for n-version programming systems, *IEEE Transactions on Reliability* **R-51** (2002) 311–321.
20. T. G. Kurtz, Gaussian approximations for Markov chains and countable processes, *Bulletin of the International Statistics Institute* **50** (1983) 361–375.
21. Ø. Borgan, Maximum likelihood estimation in parametric counting process models with applications to censored failure time data, *Scandinavian Journal of Statistics* **1** (1984) 1–16.
22. P. Billingsley, *Statistical Inference for Markov Processes* (The University of Chicago Press, Chicago, 1961).

CHAPTER 14

A Two-Level Continuous Sampling Plan for Software Systems

Seheon Hwang and Hoang Pham

Department of Industrial and Systems Engineering, Rutgers University, 96 Frelinghuysen Rd., Piscataway, New Jersey, 08854-8018, USA

1. Introduction

Software engineering has been developing so fast over the past three decades even though its history is relatively short compared with that of other engineering areas. The high-dependence to computers and the rapid-expansion of software applications lead a new tendency of software development in modern society — less time, less cost, and more reliable.

A software product requires several steps until it is produced. Lee *et al.*[1] described a software life cycle consists in five successive phases; analysis, design, coding, testing and operation phase. Since the operation phase usually begins when the software product is delivered to customers, the preceding phases up to testing phase can be considered as the substantial software development process.

According to the survey investigated by Zhang and Pham,[2] the time allocation for analysis, design, coding, and testing phase showed around 25%, 18%, 36% and 21% of the entire development time, respectively. Testing phase helps a software product be affirmed good quality and reliability by detecting and correcting faults in the program. Software testing is one of the most important processes in software development since it can finally verifies and validates the products before it releases to the customers.[3] In addition to improve the quality of the product by executing the test cases created at design phase, software testing also provides the failure-time data so that it can be used to predict the future failures that might occur during operation phase. The reliability measure estimated by the failure data during system testing is also an important criterion in deciding when to release the software product.[4–10]

In a large software system, insufficient development time and budget would often lead software developers to make a difficult decision on how to allocate the remaining time for testing software products. It is likely to happen, for instance, that the development team has spent most of given time and budget on solving unexpected problems or sometimes, some of team members ceased from the work before completing development. As a result, developers are obliged to test only some portions of the scheduled test cases because of the limitation of time and budget. In such a case, partial testing based on sampling is one of the possible ways to be chosen as an alternative option. However, the partial testing might bring about such an issue that it cannot ensure the quality of the software product comparing with the one conforming to the scheduled testing.

This paper discusses a new testing methodology for software systems based on continuous sampling plan. Even though this testing process performs the testing only for the partially selected test cases, it guarantees the percentage of remaining defects not to exceed a predetermined value as well as reduces testing efforts in terms of testing time and cost. This is, to our knowledge, the first study to incorporate the concepts of continuous sampling method for software development and testing process.

The proposed methodology, called a two-level testing plan, is constructed based on the continuous sampling plan (CSP). This new plan performs sampling testing and 100% testing alternately in accordance with the frequency of software failure like the CSP plans conduct partial and 100% inspections in turns. There are two notable features draw a distinction between our proposed plan and the existing CSP plans. In the two-level testing plan, there are three different characteristics that lead the conversion between 100% testing phase and each level of partial testing phase, whereas there are basically only two outcomes, defective or nondefective throughout all the variations and modifications of CSP plans. In the two-level testing plan, the software failures detected are classified as minor, major and critical according to the severity of impact on its functional feature or difficulties on both detecting and correcting faults. Another feature discerned this plan from the CSP plans is that the decision of transition depends on the type of error when level 2 fractional testing is being conducted. Consequently, this plan makes it possible not only to reduce the number of test cases executed for system testing but also to prevent the quality of final product from exceeding a certain worst level.

The literature reviews on various CSP plans as well as the first CSP plan is presented in Sec. 2. Two specific CSP plans that have two fractional inspection levels are also described in Sec. 2. Detailed procedure and transition policy of our proposed testing plan is discussed in Sec. 3. The model formulation and performance measures of proposed testing plan based on Markov chain approach is derived in Sec. 4. Numerical examples are given to illustrate the testing performance measures in Sec. 5 and some remarks are discussed in Sec. 6.

2. Reviews of Existing Methodologies

The continuous sampling plan designated CSP-1 has been first proposed by Dodge.[11] This plan originally intended for inspection suggested for a product consisting of individual units manufactured continuously. The procedure is as follows: First, all units are inspected one by one. When i, called clearance number, consecutive units of product are found to be free of defects, then a 100% inspection is stopped and then only a fraction (f) of the units is inspected. If a defective unit is found, then 100% inspection is resumed. All defective units found are either reworked or replaced with good ones.

As Dodge pointed out, the objective of continuous sampling plan is "to establish a limiting value of *AOQ* (average outgoing quality) expressed in percent defective which will not be exceeded no matter what quality is submitted to the inspector".[11] The *AOQL* (average outgoing quality limit) defined as the limiting value of *AOQ* may be interpreted as the worst possible long run quality level of the continuous products. The same *AOQL* can be obtained by different combination of i and f and the determination of i and f is usually based on practical consideration in the manufacturing process. Figure 1 depicts the *AOQ* values with respect to p which shows the average outgoing quality has the maximum value, 0.010895 when p is 0.0207 for $i = 100$ and $f = 0.1$.[12] The value of *AOQL* can be determined based on both the clearance number i and the sampling fraction f.

Dodge and Torrey[13] introduced CSP-2 and CSP-3 plans, in which the transition into 100% inspection defers until finding another evidence of poor quality when a defective unit is found. In CSP-2, for instance, sampling inspection cannot be switched to 100% inspection until a second defect occurs in the next k or less sample units even though a defect unit is found during sampling inspection period.

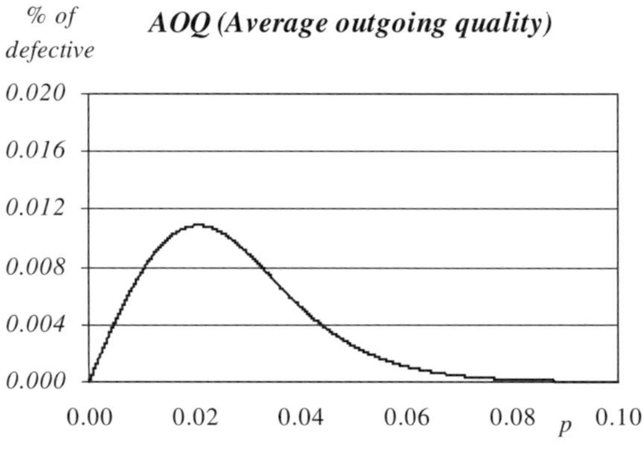

Fig. 1. The value of *AOQ* with respect to p.

Lieberman and Solomon[14] presented multi-level CSP plans that allow for any number of sampling levels, subject to the provision that transition can only occur between adjacent levels. Derman *et al.*[15] proposed tightened multi-level CSP that transition can occur between more than one level (generally r levels) when a defective unit is found in any sampling level. The tightest case in this plan provides more strict policy to assure the quality of production as reverting to 100% inspection immediately whenever a defect is found at any level. These plans can reduce the amount of units inspected compared with other CSP plan when the incoming fraction of defective is relatively small.

Recently, Balamurali and Govindaraju[16] proposed a modified tightened two-level continuous sampling plan based on the existing tightened multi-level continuous sampling plan of Derman *et al.*[15] According to their plan, the transition from one sampling level to another sampling level can occur only by going back to 100% inspection. If the tightened two-level continuous sampling plan of Derman *et al.*[15] is regarded as one of the three tightened multi-level continuous sampling plans with two levels, comparison of it with the modified tightened two-level continuous sampling plan will help readers understand the differences between these two plans. Figure 2 depicts the flow charts of these two continuous sampling plans.[16]

3. New Two-Level Testing Plan

The proposed two-level testing plan consists of three testing phases: a 100% testing, level 1 testing, and level 2 testing, are discussed as follows:

(i) 100% Testing Phase

(a) At the beginning, a test case is randomly selected from the scheduled test cases and tested one at a time. It will continue until i consecutive number of test cases are tested without failure.

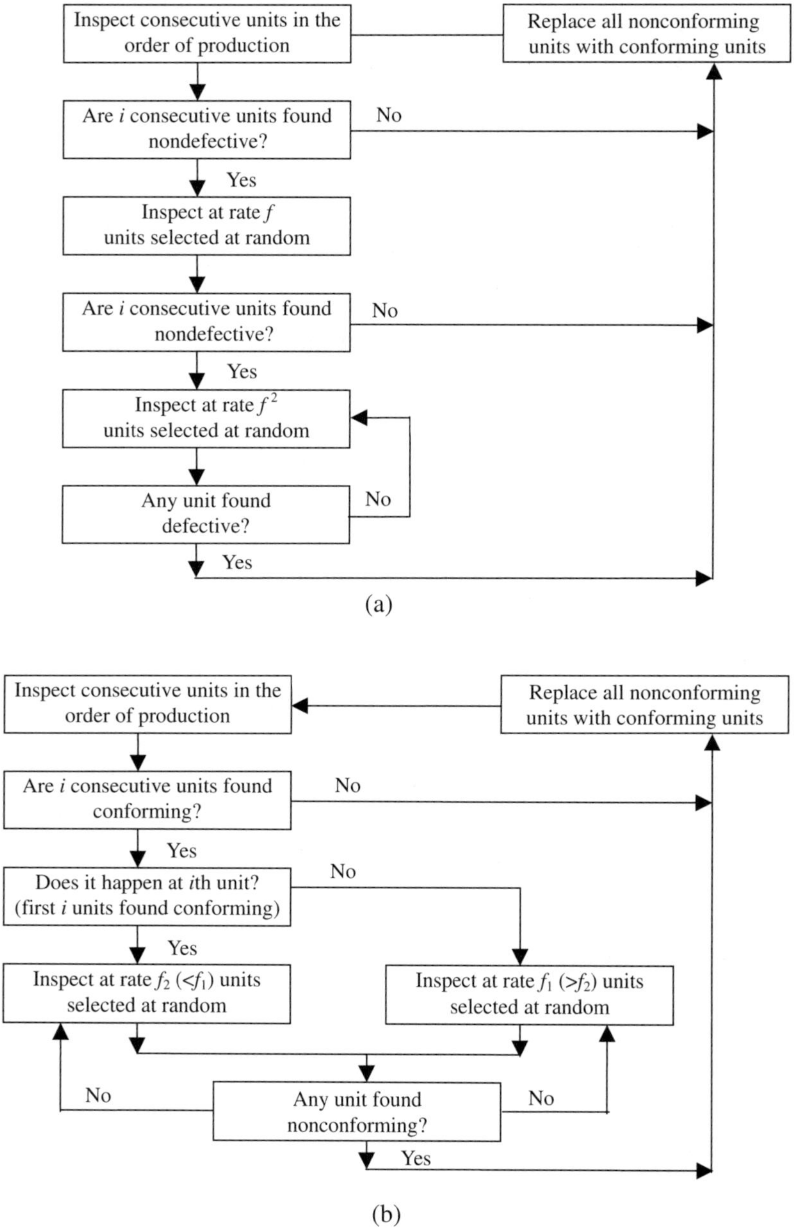

Fig. 2. Flow chart for (a) tightened two-level CSP and (b) modified tightened two-level CSP.

(b) When i consecutive test cases are tested to be free of failures, then 100% testing is stopped and only a fraction of test cases is tested, which means converting to level 1 testing.

(ii) Level 1 Testing Phase (the frequency of testing is f_1)

(a) Select only a fraction f_1 of test cases at random and test it. In practice, a test case is selected at random from $1/f_1$ of test cases and tested. The rests are passed on without replacement. This is defined as level 1 testing phase.
(b) When i consecutive test cases are obtained without failure during level 1 testing phase, then, level 1 testing is stopped and switched to level 2 testing. If a failure occurs before obtaining i consecutive test cases with no failure, then, the 100% testing is resumed regardless of the types of errors.

(iii) Level 2 Testing Phase (the frequency of testing is f_2)

(a) Select only a fraction f_2 ($< f_1$) of test cases at random and test it. Likewise, a test case is selected at random from $1/f_2$ of test cases and tested. The rests are passed on without replacement. This is defined as level 2 testing phase and continued until a failure is found.
(b) If a critical error is found during level 2 testing, the 100% testing is resumed immediately. If the detected error is either minor or major, then, level 2 testing is switched to level 1 testing.

The flow chart, which describes the procedure of two-level testing plan, is shown in Fig. 3. The proposed plan has a specific feature that, as we can see, the type of detected error plays an important role to determine the next testing phase. In the existing CSP plans, the criteria of decision is just "nondefective" or "defective" as Dodge stated "Go–No Go" basis, whereas the decision in the proposed plan is classified into four criteria including three different types of errors

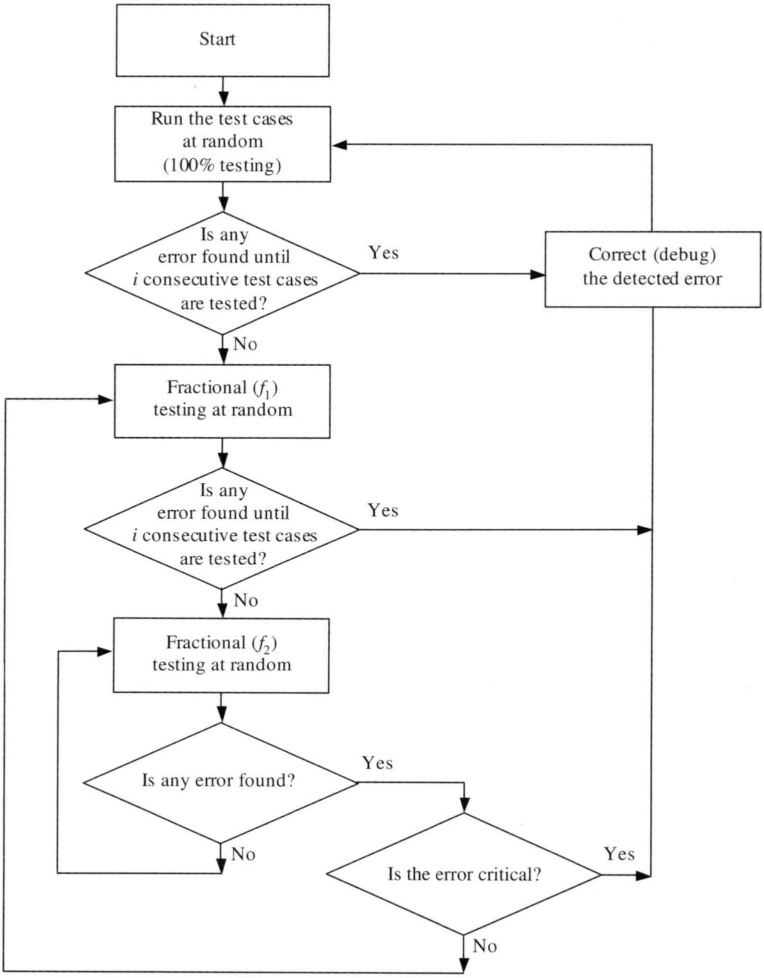

Fig. 3. Flow chart for the two-level fractional testing plan.

in the software program according to the difficulty for correcting them. The types of errors are defined as follows:

- Minor error: suggestion for improvement, cosmetic issue/easy to correct,
- Major error: minor deviation in the functionality/difficult to correct,
- Critical error: critical function is not working/very difficult to correct of error.

Notation

p: Fraction of defects in a program.

q: $1 - p$.

p_1: Fraction of test cases that cause failures due to minor errors or fraction of defects that cause minor problems in a program.

p_2: Fraction of test cases that cause failures due to major errors or fraction of defects that cause major problems in a program.

p_3: Fraction of test cases that cause failures due to critical errors or fraction of defects that cause critical problems in a program

$$p = p_1 + p_2 + p_3.$$

i: Consecutive number of good test cases to be obtained before switching to the next level, integer.

f_1: Fraction of test cases selected in level 1 testing phase.

f_2: Fraction of test cases selected in level 2 testing phase.

FTT: Average fraction of test cases that are finally tested throughout the plan.

FTN: Average fraction of test cases that are finally passed throughout the plan.

PRD: Average percentage of remaining defects after getting through the proposed testing plan.

PRD_{max}: Maximum value of *PRD*.

4. New Methodology

In software engineering, many researchers have been focusing on how to select test set or test cases from the entire input domain in order to build more reliable software product by finding failure region, effectively.[17] In our testing methodology, however, the test cases are assumed to be already selected using a certain method. Therefore, we restrict our consideration how to reduce testing effort effectively by selecting a portion of scheduled test cases. The formulation of the proposed plan is based on the following assumptions:

(a) The scheduled test cases are assumed to be representative of all possible sets of test cases.
(b) The probability of finding an error by running a randomly selected test case is p throughout the entire testing procedure.
(c) The detected error is assumed to be removed immediately. There are no new errors introduced.

The second assumption is formed based on the following fact: The total number of possible test cases in even a simple program is essentially infinite, which implies that the scheduled test cases are selected from infinite number of population. Since the population is considerably large, the probability of finding a test case with failure can be considered as constant p at any time.

In this section, we model the two-level testing plan to determine the quality of the software products using the Markov approach.[16] The transition diagram, shown in Fig. 4, depicts a Markov transition-state of the two-level testing plan. The transition states are mutually exclusive. The definitions of all states are discussed below. This process can be considered as discrete, finite, recurrent, irreducible and aperiodic Markov chain model. The transition probability for each state is also shown in Table 1.

Definition of state

A_j: 100% testing is being performed. The consecutive number of test cases that has been tested with no failure is j.

Ig_j: Level 1 testing is being performed. A test case selected was tested and caused no failure, as the result, j consecutive number of test cases has been tested with no failure so far.

Ie_j: Level 1 testing is being performed. A test case selected was tested and caused a failure and j consecutive number of test cases has been tested with no failure so far.

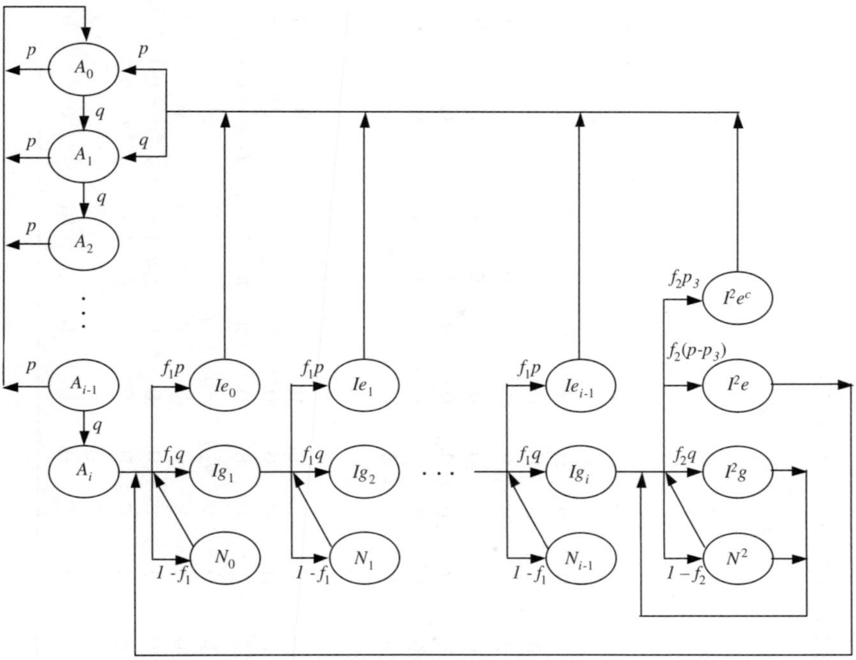

Fig. 4. The transition diagram of the two-level fractional testing plan.

N_j: Level 1 testing is being performed. A test case selected was not tested due to the feature of fractional testing and still j consecutive number of test cases has been tested with no failure.

I^2g: Level 2 testing is being performed. A test case selected was tested and caused no failure.

I^2e: Level 2 testing is being performed. A test case selected was tested and caused a failure due to minor or major error.

I^2e^c: Level 2 testing is being performed. A test case selected was tested and caused a failure due to critical error.

N^2: Level 2 testing is being performed. A test case selected was not tested due to the feature of fractional testing.

Table 1. The transition probability matrix of the two-level testing plan.

	A_0	A_1	A_2	$\cdots$	A_{i-1}	A_i	Ig_1	N_0	Ie_0	Ig_2	N_1	Ie_1	$\cdots$	Ig_{i-1}	N_{i-2}	Ie_{i-2}	Ig_i	N_{i-1}	Ie_{i-1}	I^2g	N^2	I^2e	I^2e^c
A_0	p	q	0	...	0	0	0	0	0	0	0	0	...	0	0	0	0	0	0	0	0	0	0
A_1	p	0	q	...	0	0	0	0	0	0	0	0	...	0	0	0	0	0	0	0	0	0	0
A_2	p	0	0	...	0	0	0	0	0	0	0	0	...	0	0	0	0	0	0	0	0	0	0
...	...	...	...	...	...	...	...	...	...	...	...	...	...	...	...	...	...	...	...	...	...	...	...
A_{i-1}	p	0	0	...	0	q	0	0	0	0	0	0	...	0	0	0	0	0	0	0	0	0	0
A_i	0	0	0	...	0	0	f_1q	$1-f_1$	f_1p	0	0	0	...	0	0	0	0	0	0	0	0	0	0
Ig_1	0	0	0	...	0	0	0	0	0	f_1q	$1-f_1$	f_1p	...	0	0	0	0	0	0	0	0	0	0
N_0	0	0	0	...	0	0	f_1q	$1-f_1$	f_1p	0	0	0	...	0	0	0	0	0	0	0	0	0	0
Ie_0	p	q	0	...	0	0	0	0	0	0	0	0	...	0	0	0	0	0	0	0	0	0	0
Ig_2	0	0	0	...	0	0	0	0	0	0	0	0	...	0	0	0	0	0	0	0	0	0	0
N_1	0	0	0	...	0	0	0	0	0	f_1q	$1-f_1$	f_1p	...	0	0	0	0	0	0	0	0	0	0
Ie_1	p	q	0	...	0	0	0	0	0	0	0	0	...	0	0	0	0	0	0	0	0	0	0
...	...	...	...	...	...	...	...	...	...	...	...	...	...	...	...	...	...	...	...	...	...	...	...
Ig_{i-1}	0	0	0	...	0	0	0	0	0	0	0	0	...	0	0	0	f_1q	$1-f_1$	f_1p	0	0	0	0
N_{i-2}	0	0	0	...	0	0	0	0	0	0	0	0	...	f_1q	$1-f_1$	f_1p	0	0	0	0	0	0	0
Ie_{i-2}	p	q	0	...	0	0	0	0	0	0	0	0	...	0	0	0	0	0	0	0	0	0	0
Ig_i	0	0	0	...	0	0	0	0	0	0	0	0	...	0	0	0	0	0	0	f_2q	$1-f_2$	$f_2(p-p_3)$	f_2p_3
N_{i-1}	0	0	0	...	0	0	0	0	0	0	0	0	...	0	0	0	f_1q	$1-f_1$	f_1p	0	0	0	0
Ie_{i-1}	p	q	0	...	0	0	0	0	0	0	0	0	...	0	0	0	0	0	0	0	0	0	0
I^2g	0	0	0	...	0	0	0	0	0	0	0	0	...	0	0	0	0	0	0	f_2q	$1-f_2$	$f_2(p-p_3)$	f_2p_3
N^2	0	0	0	...	0	0	0	0	0	0	0	0	...	0	0	0	0	0	0	f_2q	$1-f_2$	$f_2(p-p_3)$	f_2p_3
I^2e	0	0	0	...	0	0	f_iq	$1-f_1$	f_1p	0	0	0	...	0	0	0	0	0	0	0	0	0	0
I^2e^c	p	q	0	...	0	0	0	0	0	0	0	0	...	0	0	0	0	0	0	0	0	0	0

The steady-state probability of each state is given as follows:

$$P(A_0) = p\left[P(A_0) + P(A_1) + \sum_{j=2}^{i-1} P(A_j) + P(Ie_0) + \sum_{j=1}^{i-1} P(Ie_j) + P(I^2e^c)\right], \quad (1)$$

$$P(A_1) = q\left[P(A_0) + P(Ie_0) + \sum_{j=1}^{i-1} P(Ie_j) + P(I^2e^c)\right], \quad (2)$$

$$P(A_j) = q[P(A_{j-1})] \quad \text{for } j = 2, 3, 4, \ldots, i\,, \quad (3)$$

$$P(Ig_1) = f_1 q[P(A_i) + P(N_0) + P(I^2e)]\,, \quad (4)$$

$$P(Ig_j) = f_1 q[P(Ig_{j-1}) + P(N_{j-1})] \quad \text{for } j = 2, 3, 4, \ldots, i\,, \quad (5)$$

$$P(N_0) = (1 - f_1)[P(A_i) + P(N_0) + P(I^2e)]\,, \quad (6)$$

$$P(N_j) = (1 - f_1)[P(Ig_j) + P(N_j)] \quad \text{for } j = 1, 2, 3, \ldots, i-1\,, \quad (7)$$

$$P(Ie_0) = f_1 p[P(A_i) + P(N_0) + P(I^2e)]\,, \quad (8)$$

$$P(Ie_j) = f_1 p[P(Ig_j) + P(N_j)] \quad \text{for } j = 1, 2, 3, \ldots, i-1\,, \quad (9)$$

$$P(I^2g) = f_2 q[P(Ig_i) + P(I^2g) + P(N^2)]\,, \quad (10)$$

$$P(N^2) = (1 - f_2)[P(Ig_i) + P(I^2g) + P(N^2)]\,, \quad (11)$$

$$P(I^2e) = f_2(p - p_3)[P(Ig_i) + P(I^2g) + P(N^2)]\,, \quad (12)$$

$$P(I^2e^c) = f_2 p_3[P(Ig_i) + P(I^2g) + P(N^2)]\,. \quad (13)$$

Finally, the sum of the probabilities for all states should be 1, that is,

$$\sum_{j=0}^{i} P(A_j) + \sum_{j=1}^{i} P(Ig_j) + \sum_{j=0}^{i-1} P(Ie_j) + \sum_{j=0}^{i-1} P(N_j) + P(I^2g) + P(N^2) + P(I^2e) + P(I^2e^c) = 1\,. \quad (14)$$

From Eq. (3),

$$P(A_j) = q^{j-1}[P(A_1)] \quad \text{for } j = 2, 3, 4, \ldots, i\,,$$

where (15)

$$P(A_2) = qP(A_1) \quad \text{and} \quad P(A_i) = qP(A_{i-1}) = q^{i-1}P(A_1)\,.$$

The probability summed up from $P(A_2)$ to $P(A_{i-1})$ can be formulated as follows:

$$\begin{aligned}\sum_{j=2}^{i-1} P(A_j) &= P(A_2) + P(A_3) + P(A_4) + \cdots + P(A_{i-1}) \\ &= q \cdot P(A_1) + q^2 P(A_1) + q^3 P(A_1) + \cdots + q^{i-2} P(A_1) \\ &= q \cdot \left(\frac{1-q^{i-2}}{1-q}\right) \cdot P(A_1)\,. \end{aligned} \tag{16}$$

From Eq. (2),

$$P(Ie_0) + \sum_{j=1}^{i-1} P(Ie_j) + P(I^2e^c) = \frac{1}{q}P(A_1) - P(A_0)\,. \tag{17}$$

By substituting Eq. (16) into Eq. (1), $P(A_0)$ can be written in terms of $P(A_1)$

$$\begin{aligned}P(A_0) &= p \cdot \Bigg[P(A_0) + P(A_1) + q \cdot \left(\frac{1-q^{i-2}}{1-q}\right) \cdot P(A_1) \\ &\qquad + \frac{1}{q} \cdot P(A_1) - P(A_0) \Bigg] \\ &= p \cdot \left[P(A_1) + q \cdot \left(\frac{1-q^{i-2}}{1-q}\right) \cdot P(A_1) + \frac{1}{q} \cdot P(A_1) \right] \\ &= \frac{1-q^i}{q} P(A_1)\,. \end{aligned} \tag{18}$$

From Eq. (7), we have

$$P(Ig_j) = \frac{f_1}{1-f_1} P(N_j) \quad \text{for } j = 1, 2, 3, \ldots, i-1\,. \tag{19}$$

By substituting Eq. (19) into Eq. (5), we obtain

$$P(Ig_j) = f_1 q \left[P(Ig_{j-1}) + \frac{1-f_1}{f_1} P(Ig_{j-1}) \right]$$
$$= q \cdot P(Ig_{j-1}) \quad \text{for } j = 2, 3, 4, \ldots, i\,. \tag{20}$$

Therefore,

$$\sum_{j=1}^{i} P(Ig_j) = \frac{1-q^i}{1-q} P(Ig_1)\,. \tag{21}$$

From Eqs. (4), (6) and (19), we obtain

$$\sum_{j=1}^{i-1} P(N_j) = \frac{1-f_1}{f_1} \sum_{j=1}^{i-1} P(Ig_j)$$
$$= \left(\frac{1-f_1}{f_1}\right)\left(\frac{1-q^{i-1}}{1-q}\right) P(Ig_1)\,, \tag{22}$$

$$P(N_0) = \frac{1-f_1}{f_1 q} P(Ig_1)\,. \tag{23}$$

Substituting Eq. (23) into Eq. (21), then

$$P(N_0) = \frac{1}{q} P(N_1)\,. \tag{24}$$

From Eq. (6), we can get the steady-state probability $P(I^2e)$ with respect to $P(Ig_1)$ and $P(A_1)$.

$$P(I^2e) = \frac{f_1}{1-f_1} P(N_0) - P(A_i)$$
$$= \left(\frac{f_1}{1-f_1}\right)\left(\frac{1-f_1}{f_1 q}\right) P(Ig_1) - q^{i-1} P(A_1)$$
$$= \frac{1}{q} P(Ig_1) - q^{i-1} P(A_1)\,. \tag{25}$$

By substituting Eq. (25) into Eq. (8), we have

$$\begin{aligned} P(Ie_0) &= f_1 p\left[q^{i-1}P(A_1) + P(N_0) + \frac{1}{q}P(Ig_1) - q^{i-1}P(A_1)\right] \\ &= f_1 p\left[P(N_0) + \frac{1}{q}P(Ig_1)\right] \\ &= \frac{p}{q}P(Ig_1)\,. \end{aligned} \tag{26}$$

From Eq. (19) and Eq. (9), we can easily obtain

$$P(Ie_j) = \frac{f_1 p}{1-f_1}P(N_j) \quad \text{for } j = 1, 2, 3, 4, \ldots, i\,. \tag{27}$$

Thus, the probability summed up from $P(Ie_1)$ to $P(Ie_{i-1})$ can be formulated as follows:

$$\begin{aligned} \sum_{j=1}^{i-1} P(Ie_j) &= \frac{f_1 p}{1-f_1}\sum_{j=1}^{i-1} P(N_j) \\ &= \left(\frac{f_1 p}{1-f_1}\right)\left(\frac{1-f_1}{f_1}\right)\left(\frac{1-q^{i-1}}{1-q}\right)P(Ig_1) \\ &= (1-q^{i-1})P(Ig_1)\,. \end{aligned} \tag{28}$$

From Eqs. (10)–(13),

$$P(I^2 g) = \frac{P(Ig_1) - q^i P(A_1)}{p - p_3} = \frac{q^i}{p}P(Ig_1)\,, \tag{29}$$

$$\begin{aligned} P(N^2) &= \left(\frac{1-f_2}{f_2(p-p_3)}\right)\left(\frac{(p-p_3)q^i}{pq}\right)P(Ig_1) \\ &= \frac{(1-f_2)q^{i-1}}{f_2 p}P(Ig_1)\,, \end{aligned} \tag{30}$$

$$\begin{aligned} P(I^2 e) &= \frac{1}{q}P(Ig_1) - q^{i-1}\left(\frac{p-(p-p_3)q^i}{pq^i}\right)P(Ig_1) \\ &= \frac{(p-p_3)q^{i-1}}{p}P(Ig_1)\,, \end{aligned} \tag{31}$$

$$P(I^2e^c) = \left(\frac{p_3}{p - p_3}\right) \cdot \frac{(p - p_3)q^i}{pq} P(Ig_1) = \frac{p_3 q^{i-1}}{p} P(Ig_1) \,. \quad (32)$$

By substituting Eqs. (17), (28), (29) and (31) into Eq. (2), we obtain

$$P(A_1) = q \cdot \left[\frac{1 - q^i}{q} P(A_1) + \frac{p}{q} P(Ig_1) + (1 - q^{i-1}) P(Ig_1)\right]$$
$$+ q \cdot \left[\left(\frac{p_3}{p - p_3}\right) \cdot \left(\frac{1}{q} P(Ig_1) - q^{i-1} P(A_1)\right)\right] . \quad (33)$$

Accordingly, we obtain $P(A_1)$ written in terms of $P(Ig_1)$ by solving the above equation.

$$P(A_1) = \frac{p - (p - p_3)q^i}{pq^i} P(Ig_1) \,. \quad (34)$$

Now, we have the probabilities of all states expressed with respect to $P(Ig_1)$. Substituting these all equations into Eq. (14), therefore, we obtain

$$P(Ig_1) = \frac{f_1 f_2 p^2 q^{i+1}}{f_1 f_2 (1 - q^i)(p - (p - p_3)q^i) + (f_2 + (f_1 - f_2)q^i) pq^i} \,. \quad (35)$$

For the simplicity, let D be

$$f_1 f_2 (1 - q^i)(p - (p - p_3)q^i) + (f_2 + (f_1 - f_2)q^i) pq^i \,.$$

Hence, the steady-state probabilities of all the states can be written as follows:

$$P(A_0) = \frac{f_1 f_2 p (1 - q^i)[p - (p - p_3)q^i]}{D} \,, \quad (36)$$

$$P(A_1) = \frac{f_1 f_2 pq (p - (p - p_3)q^i)}{D} \,, \quad (37)$$

$$\sum_{j=2}^{i} P(A_j) = \frac{f_1 f_2 q^2 (1 - q^{i-1})(p - (p - p_3)q^i)}{D} \,, \quad (38)$$

$$P(Ig_1) = \frac{f_1 f_2 p^2 q^{i+1}}{D}, \tag{39}$$

$$\sum_{j=1}^{i} P(Ig_j) = \frac{f_1 f_2 p(1 - q^i) q^{i+1}}{D}, \tag{40}$$

$$P(Ie_0) = \frac{f_1 f_2 p^3 q^i}{D}, \tag{41}$$

$$\sum_{j=1}^{i-1} P(Ie_j) = \frac{f_1 f_2 p^2 (1 - q^{i-1}) q^{i+1}}{D}, \tag{42}$$

$$P(N_0) = \frac{(1 - f_1) f_2 p^2 q^i}{D}, \tag{43}$$

$$\sum_{j=1}^{i-1} P(N_j) = \frac{(1 - f_1) f_2 p(1 - q^{i-1}) q^{i+1}}{D}, \tag{44}$$

$$P(I^2 e) = \frac{f_1 f_2 p(p - p_3) q^{2i}}{D}, \tag{45}$$

$$P(I^2 e^c) = \frac{f_1 f_2 p p_3 q^{2i}}{D}, \tag{46}$$

$$P(I^2 g) = \frac{f_1 f_2 p q^{2i+1}}{D}, \tag{47}$$

$$P(N^2) = \frac{f_1 (1 - f_2) p q^{2i}}{D}. \tag{48}$$

Let $P(100\%)$ be the steady-state probability that testing is conducted in 100% testing phase, then

$$P(100\%) = P(A_0) + P(A_1) + \sum_{j=2}^{i} P(A_j)$$
$$= \frac{f_1 f_2 [p - (p - p_3) q^i](1 - q^i)}{D}. \tag{49}$$

Thus, the steady-state probability that testing is conducted in level 1 fractional testing phase is,

$$\begin{aligned} P(\text{Level 1 fractional testing}) &= \sum_{j=1}^{i} P(Ig_j) + P(Ie_0) + \sum_{j=1}^{i-1} P(Ie_j) \\ &\quad + P(N_0) + \sum_{j=1}^{i-1} P(N_j) \\ &= \frac{f_2 p(1-q^i)q^i}{D}. \end{aligned} \tag{50}$$

Likewise, steady-state probability that testing is conducted in level 2 fractional testing phase is,

$$\begin{aligned} P(\text{Level 2 fractional testing}) &= P(I^2 g) + P(N^2) + P(I^2 e) \\ &\quad + P(I^2 e^c) \\ &= \frac{f_1 p q^{2i}}{D}. \end{aligned} \tag{51}$$

Now, we can formulate the performance measures of this plan in terms of i, f_1, f_2, p and p_3. The average fraction of test cases finally passed on without testing is obtained as follows:

$$\begin{aligned} FTN &= P(N_0) + \sum_{j=1}^{i-1} P(N_j) + P(N^2) \\ &= \frac{(1-f_1)f_2 p^2 q^i}{D} + \frac{(1-f_1)f_2 p(1-q^{i-1})q^{i+1}}{D} \\ &\quad + \frac{f_1(1-f_2)pq^{2i}}{D} \\ &= \frac{pq^i[(1-f_1)f_2 + (f_1-f_2)q^i]}{D} \end{aligned} \tag{52}$$

and, the average fraction of test cases that are finally tested throughout this testing plan is

$$FTT = 1 - FTN = \frac{f_1 f_2[p - (p-p_3)q^i(1-q^i)]}{D}. \tag{53}$$

Hence, *PRD*, the average percentage of remaining defects of the software product after completing this testing plan, can be obtained as follows:

$$\begin{aligned} PRD &= p \cdot (FTN) \\ &= p \cdot \left(\frac{pq^i[(1-f_1)f_2 + (f_1 - f_2)q^i]}{D} \right). \end{aligned} \tag{54}$$

5. Numerical Examples

Two-level testing plan has been originally proposed to be an alternative testing method for a large software product when, because of certain restrictions, scheduled testing cannot be allowed. In other words, a partial testing, choosing some parts of the scheduled test cases and executing them, is assumed to be the only way under the circumstance. Therefore, in order to present the benefits of the proposed plan, we first need to consider another type of partial testing, for instance, a uniform random testing that has the same sampling frequency of test cases as our plan. Here, we assume that all scheduled test cases are equally likely to be selected in the uniform random testing.

Let us suppose that a software program can be divided into five parts in accordance with programmers. For the testing of the functionality of the final product, we can classify the test cases by programmers, either. Based on the survey,[2] each part is assumed to have different fractions of test cases that cause failures during the testing. However, it is not possible for us to be aware of the fraction of test cases causing failures for each part until completing the testing process and estimating it. For the purpose of comparison, however, the fraction of defect for each part is assumed as in Table 2, then, the performance measures of the two-level testing plan are obtained using the formula derived in Sec. 4.

The desired quality level of the final product is assumed never to drop down by 0.99. Among various choices of the combination

Table 2. The performance measures of two-level testing plan when $i = 121$, $f_1 = 0.1$, $f_2 = 0.05$ (assuming $p_3 = 0.1p$).

Fraction of size		p_i	p_3	Two-level testing *FTT*	Two-level testing *PRD*	Partial random testing *FTT*	Partial random testing *PRD*
Part I	0.3	0.0025	0.00025	0.0639	0.0023	0.3115	0.0017
Part II	0.2	0.0050	0.00050	0.0897	0.0046	0.3115	0.0034
Part III	0.1	0.0100	0.00100	0.1864	0.0081	0.3115	0.0069
Part IV	0.2	0.0150	0.00150	0.3403	0.0099	0.3115	0.0103
Part V	0.2	0.0400	0.00400	0.9387	0.0025	0.3115	0.0275
Total	1.0			0.3115	0.0049	0.3115	0.0095

in i, f_1, and f_2 that has the same $PRD_{\max}$ of 0.01, the value of 121, 0.1, and 0.05 are selected as i, f_1, and f_2, respectively. Once these controllable parameters are determined, the performance measures containing *PRD*, *FTT*, and *FTN* are only the function of p and p_3.

Suppose each part of test cases has the fraction of size, such as 0.3, 0.2, 0.1, 0.2, and 0.2, from Part I to Part V, respectively. The percentage of remaining defects of the software product is given in Eq. (55) is given by

$$\begin{aligned} PRD(p)_{\text{two-level}} &= \sum_{j=I}^{V} F_j \cdot PRD(p_j)_j \\ &= 0.3(0.0023) + 0.2(0.0046) + 0.1(0.0081) \\ &\quad + 0.2(0.0099) + 0.2(0.0025) \\ &= 0.00490\,, \end{aligned} \tag{55}$$

where F_j is the fraction of size for part j.

The most important benefit of the two-level testing plan is that the part that has better incoming quality (p) is experienced less amount of testing effort while the part that has worse incoming quality is performed more testing effort. As the results of numerical analysis, *FTT*, the average fraction of test cases conducted by this plan, is calculated as 0.3115, and thus, the average percentage of remaining defect that the software product finally has through this plan, $PRD_{\text{two-level}}$, is found as 0.0049.

As for the choice of partial random testing plan, a random sampling with fixed fraction is considered. For the comparison, the equivalent fraction of the test cases is assumed to be executed as the equivalent two-level testing plan is randomly selected with the same fraction regardless of the of p value. In this case, the fixed fraction, $FTT_{\text{two-level}} = FTT_{\text{partial-random}} = 0.3115$, is applied for each part regardless of its fraction of defect, p. Therefore, the average percentage of remaining defect that the software product finally has by this partial random testing plan is given by

$$PRD(p)_{\text{partial-random}} = \sum_{j=I}^{V} F_j \cdot p_j \cdot (1 - FTT_{\text{partial-random}})\,. \quad (56)$$

Since $FTT_{\text{partial-random}}$ represents the fraction of test cases executed by partial random testing, $(1 - FTT_{\text{partial-random}})$ can be explained as the fraction of test cases never executed, and, therefore, $p_j(1 - FTT_{\text{partial-random}})$ means the fraction of remaining defects of each part of the software product. Accordingly, the numerical result of Eq. (56) is

$$\begin{aligned}
PRD(p)_{\text{partial-random}} &= \sum_{j=I}^{V} F_j \cdot p_j \cdot (1 - FTT_{\text{partial-random}}) \\
&= (0.3)(0.0025)(0.6885) \\
&\quad + (0.2)(0.005)(0.6885) \\
&\quad + (0.1)(0.010)(0.6885) \\
&\quad + (0.2)(0.015)(0.6885) \\
&\quad + (0.2)(0.040)(0.6885) \\
&= 0.00947\,.
\end{aligned}$$

The results show that the application of two-level testing plan can reduce the remaining defect in the software product from 0.00947 to 0.00490, which brings 93% of improvement in software quality, compared to the partial random testing that has the same testing efforts.

6. Concluding Remarks

In this paper, a new partial testing methodology, called two-level testing plan, for large software systems is proposed based on continuous sampling plan. The performance measures of this plan are derived using Markov approach. Two-level testing plan has been basically proposed to be an alternative testing method for a large software product when, because of certain restrictions, scheduled testing cannot be allowed. The proposed testing plan can reduce the testing effort and, furthermore, it shows better effectiveness in improving the quality of the final software product comparing to a partial random testing method that has the equivalent testing efforts.

The most important benefit of the proposed testing plan is that when a software program is tested the part that has the smaller p is experienced less testing effort while the part that has the larger p is performed more testing effort. Through the numerical example, we found that our plan can reduce the remaining defect, consequently, 93% of improvement in software quality compared to the partial random testing with the same testing efforts.

References

1. M. Lee, H. Pham and X. Zhang, A methodology of priority settings and its application on software development, *European Journal of Operational Research* **118** (1999) 375–389.
2. X. Zhang and H. Pham, An analysis of factors affecting software reliability, *The Journal of Systems and Software* **50** (2000) 43–56.
3. H. Pham, *Software Reliability* (Springer, Berlin, 2000).

4. W. Ehrlich, B. Prasanna, J. Stampfel and J. Wu, Determining the cost of a stop-testing decision, *IEEE Software* **10** (1993) 33–42.
5. S. R. Dalal and A. A. McIntosh, When to stop testing for large software systems with changing code, *IEEE Transactions on Software Engineering* **20** (1994) 318–323.
6. H. Pham and X. Zhang, Software release policies with gain in reliability justifying the costs, *Annals of Software Engineering* **8** (1999) 147–166.
7. H. Pham and X. Zhang, NHPP software reliability and cost models with testing coverage, *European Journal of Operational Research* **145** (2003) 443–454.
8. X. Zhang and H. Pham, A software cost model with warranty and risk costs, *IEEE Transactions on Computers* **48** (1999) 71–75.
9. X. Zhang, X. Teng and H. Pham, Considering fault removal efficiency in software reliability assessment, *IEEE Transactions on Systems, Man, and Cybernetics — Part A* **33** (2003) 114–120.
10. H. Pham and H. Wang, A quasi renewal process for software reliability and testing costs, *IEEE Transactions on Systems, Man, and Cybernetics — Part A* **31** (2001) 623–631.
11. H. F. Dodge, A sampling inspection plan for continuous production, *Annals of Mathematical Statistics* **14** (1943) 264–279.
12. D. C. Montgomery, *Introduction to Statistical Quality Control*, 3rd edn. (John Wiley & Sons, 1996).
13. H. F. Dodge and M. N. Torrey, Additional continuous sampling inspection plans, *Industrial Quality Control* **7** (1951) 7–12.
14. G. J. Lieberman and H. Solomon, Multi-level continuous sampling plans, *Annals of Mathematical Statistics* **26** (1955) 686–704.
15. C. Derman, S. Littauer and H. Solomon, Tightened multi-level continuous sampling plans, *Annals of Mathematical Statistics* **28** (1957) 395–404.
16. S. Balamurali and K. Govindaraju, Modified tightened two-level continuous sampling plans, *Journal of Applied Statistics* **27** (2000) 397–409.
17. P. Frankl, R. Hamlet, B. Littlewood and L. Strigini, Evaluating testing methods by delivered reliability, *IEEE Transactions on Software Engineering* **24** (1998) 58–65.

CHAPTER 15

Software Reliability Analysis and Optimal Release Problem Based on a Flexible Stochastic Differential Equation Model in Distributed Development Environment

Masaya Uchida[*,‡], Yoshinobu Tamura[†,§] and Shigeru Yamada[*,¶]

[*]*Department of Social Systems Engineering,*
Faculty of Engineering, Tottori University,
Tottori-shi, 680-8552 Japan

[†]*Department of Information Systems,*
Faculty of Environmental and Information Studies,
Tottori University of Environmental Studies,
Tottori-shi, 689-1111 Japan

[‡]*98t7006@sse.tottori-u.ac.jp*
[§]*tamura@kankyo-u.ac.jp*
[¶]*yamada@sse.tottori-u.ac.jp*

1. Introduction

At present, a software development environment has been changing into a distributed one because of the spread of Internet and the software development using the object-oriented programming languages. It is known that software systems are difficult to be developed under distributed development environment since the architecture of such systems can have different development styles, i.e., software composition under the conventional host-concentrated one is substantially different from software composition under distributed one.[1] As mentioned above, the effective testing method for distributed development environment has never presented.

Basically, software reliability can be evaluated by the number of detected faults or the software failure-occurrence time in the testing phase which is the last phase of the development process, and it can be also estimated in the operational phase. Many software reliability growth models (SRGMs) based on a nonhomogeneous Poisson process (NHPP) have been proposed by many researchers.[2] NHPP models have treated the software fault-detection process in the testing phase as the discrete state space. However, if the size of the software system is large, the number of faults detected during the testing phase becomes large, and the change of the number of faults which are detected and removed through each debugging becomes sufficiently small compared with the initial fault content at the beginning of the testing. Therefore, in such a case, we can use a stochastic process model with continuous state space in order to describe the stochastic behavior of the fault-detection process.[3–5]

In this chapter, we propose a flexible stochastic differential equation (SDE) model from an inflection S-shaped SRGM based on an NHPP. And we derive several software reliability assessment measures based on our new model. Furthermore, we analyze actual software fault data to show numerical examples of software reliability measurement for application of our model, and perform sensitivity analysis for various software reliability assessment measures by using actual fault-detection count data.

Also, it is an important concern for the software management that we decide the optimum delivery time to user. Such a decision problem is called an optimal software release problem. Many optimal software release problems for the conventional software development environment have been presented. On the other hand, the effective optimal software release problems under distributed development environment has never presented.[6] Thus, we discuss the optimal software release problem based on a flexible SDE model based on the reusable rate in the system testing phase of the distributed development environment.

2. Model Description

Let $M(t)$ be the number of faults remaining in the software system at the testing time t $(t \geq 0)$. Suppose that $M(t)$ takes on continuous real values. Since latent faults in the software system are detected and eliminated during the testing phase, $M(t)$ gradually decreases as the testing procedures go on. Thus, under common assumptions for software reliability growth modeling, we consider the following linear differential equation:

$$\frac{dM(t)}{dt} = -b(t)M(t)\,, \tag{1}$$

where $b(t)$ is a fault-detection rate per unit time per fault at testing time t and is a non-negative function.

In this chapter, we suppose that $b(t)$ in Eq. (1) has the irregular fluctuation, that is, we extend Eq. (1) to the following stochastic differential equation:

$$\frac{dM(t)}{dt} = -\{b(t) + \sigma\gamma(t)\}M(t)\,, \tag{2}$$

where σ is a positive constant representing a magnitude of the irregular fluctuation and $\gamma(t)$ a standardized Gaussian white noise.

Therefore, we obtain the solution of Eq. (2) under the initial condition $M(0) = m_0$ as follows:

$$M(t) = m_0 \cdot \exp\left[-\int_0^t b(t)dt - \sigma W(t)\right]\,, \tag{3}$$

where $W(\cdot)$ is a one-dimensional Wiener process which is formally defined as an integration of the white noise $\gamma(t)$ with respect to time t. The Wiener process is a Gaussian process and it has the following properties:

$$\Pr[W(0) = 0] = 1\,, \tag{4}$$

$$\mathrm{E}[W(t)] = 0\,, \tag{5}$$

$$\mathrm{E}[W(t)W(t')] = \mathrm{Min}[t, t']\,. \tag{6}$$

Next, we apply the inflection S-shaped SRGM to assessing software reliability in the module testing-phase. Generally, the mean value function of the inflection S-shaped SRGM, which represents the expected cumulative number of faults in the time-interval $(0, t]$, is given by the following equation:

$$H(t) = \frac{a(1 - e^{-bt})}{1 + c \cdot e^{-bt}}\,, \tag{7}$$

where $a(> 0)$ is the expected number of initial inherent faults, $b(> 0)$ the software failure rate per inherent fault, $c(\geq 0)$ the prespecified inflection parameter.

We assume that the flexible NHPP model for distributed development environment is based on the following assumptions[7]:

(a) A software system consists of $(n + m)$ software components.
(b) A software failure-occurrence phenomenon is described by an NHPP.
(c) Software faults detected during the testing-phase are corrected certainly and completely, i.e., no new faults are introduced into the software system during the debugging.

Thus, we consider the following structure of the mean value function because an NHPP model is characterized by its mean value function:

$$H_{dde}(t) = a\left[\sum_{i=1}^{n+m} \frac{p_i(1 - e^{-b_i t})}{1 + c_i \cdot e^{-b_i t}}\right] \times \left(a > 0,\; b_i > 0,\; c_i \geq 0,\; p_i > 0,\; \sum_{i}^{n+m} p_i = 1\right), \tag{8}$$

where a is the expected number of initial inherent faults, b_i ($i = 1, 2, \ldots, n + m$) the software failure rate per inherent fault for the ith software component, and p_i ($i = 1, 2, \ldots, n + m$) the weight parameters which mean the proportion of the total testing load for the software components. Moreover, c_i is represented $c_i = (1-r_i)/r_i$, and r_i the inflection rate for the ith software component.

Moreover, the fault-detection rate per remaining fault derived from Eq. (8) is given by:

$$b_{dde}(t) \equiv \frac{\frac{dH_{dde}(t)}{dt}}{a - H_{dde}(t)} = \frac{\sum_{i=1}^{n+m} \frac{p_i b_i e^{-b_i t}(1 + c_i)}{(1 + c_i e^{-b_i t})^2}}{\sum_{i=1}^{n+m} \frac{p_i e^{-b_i t}(1 + c_i)}{1 + c_i e^{-b_i t}}} \,. \tag{9}$$

By applying Eq. (9) to $b(t)$ in Eq. (3), we can obtain the following solution process:

$$M_s(t) = m_0 \left[\sum_{i=1}^{n+m} \frac{p_i e^{-b_i t}(1 + c_i)}{1 + c_i e^{-b_i t}} \right] e^{-\sigma W(t)} \,. \tag{10}$$

Using solution process $M_s(t)$ in Eq. (10), we can derive several software reliability measures.

3. Software Reliability Assessment Measures

Information on the current number of remaining/detected faults in the system is important to estimate the situation of the progress on the software testing procedures. Since it is a random variable in our model, its expected value and variance can be useful measures. We can calculate them from Eq. (10) as follows:

$$\mathrm{E}[M_s(t)] = m_0 \left[\sum_{i=1}^{n+m} \frac{p_i e^{-b_i t}(1+c_i)}{1+c_i e^{-b_i t}} \right] e^{\sigma^2 t/2}, \tag{11}$$

$$\begin{aligned} \mathrm{E}[N_s(t)] &\equiv \mathrm{E}[m_0 - M_s(t)] \\ &= m_0 \left\{ 1 - \left[\sum_{i=1}^{n+m} \frac{p_i e^{-b_i t}(1+c_i)}{1+c_i e^{-b_i t}} \right] e^{\sigma^2 t/2} \right\}, \end{aligned} \tag{12}$$

$$\begin{aligned} \mathrm{Var}[M_s(t)] &= \mathrm{Var}[N_s(t)] \\ &= m_0^2 \left[\sum_{i=1}^{n+m} \frac{p_i e^{-b_i t}(1+c_i)}{1+c_i e^{-b_i t}} \right]^2 \\ &\quad \cdot e^{\sigma^2 t}(e^{\sigma^2 t} - 1), \end{aligned} \tag{13}$$

where $\mathrm{E}[N_s(t)]$ is the expected number of faults detected up to testing time t.

The instantaneous mean time between software failures (denoted by $MTBF_I$) is useful to measure the property of the frequency of software failure-occurrence. $MTBF_I$ is given by the following equation as an approximation.

$$\begin{aligned} MTBF_I(t) &\equiv \frac{1}{\frac{d}{dt}\mathrm{E}[N_s(t)]} \\ &= \frac{1}{m_0\left(b(t) - \frac{1}{2}\sigma^2\right)\left[\sum_{i=1}^{n+m} \frac{p_i e^{-b_i t}(1+c_i)}{1+c_i e^{-b_i t}} \right] e^{\sigma^2 t/2}}. \end{aligned} \tag{14}$$

We have the following the cumulative mean time between software failures (denoted by $MTBF_C$).

$$\begin{aligned} MTBF_C(t) &\equiv \frac{t}{\mathrm{E}[N_s(t)]} \\ &= \frac{t}{m_0 \left\{ 1 - \left[\sum_{i=1}^{n+m} \frac{p_i e^{-b_i t}(1+c_i)}{1+c_i e^{-b_i t}} \right] \cdot e^{\sigma^2 t/2} \right\}}. \end{aligned} \tag{15}$$

4. Method of Maximum-Likelihood

In this chapter, the estimation method of unknown parameters m_0, b_i, b_{n+j}, and σ $(i = 1, 2, \ldots, n;\ j = 1, 2, \ldots, m)$ in Eq. (12) is presented. Let us denote the joint probability distribution function of the process $N_s(t)$ as:

$$\begin{aligned} &P(t_1, y_1; t_2, y_2; \ldots; t_K, y_K) \\ &\quad \equiv \Pr[N_s(t_1) \leq y_1, \ldots, N_s(t_K) \leq y_K | N_s(0) = 0, M_s(0) = m_0]\,, \end{aligned} \tag{16}$$

where $N_s(t)$ is the cumulative number of faults detected up to the testing time t $(t \geq 0)$, and denote its density as:

$$\begin{aligned} &p(t_1, y_1; t_2, y_2; \ldots; t_K, y_K) \\ &\quad \equiv \frac{\partial^K P(t_1, y_1; t_2, y_2; \ldots; t_K, y_K)}{\partial y_1 \partial y_2 \cdots \partial y_K}\,. \end{aligned} \tag{17}$$

Since $N_s(t)$ takes on continuous values, we construct the likelihood function l for the observed data $(t_k, y_k)(k = 1, 2, \ldots, K)$ as follows:

$$l = p(t_1, y_1; t_2, y_2; \ldots; t_K, y_K)\,. \tag{18}$$

For convenience in mathematical manipulations, we use the following logarithmic likelihood function:

$$L = \log l\,. \tag{19}$$

The maximum-likelihood estimates m_0^*, b_i^*, b_{n+j}^*, and σ^* are the values making L in Eq. (19) maximize. These can be obtained as the solutions of the following simultaneous likelihood equations:

$$\begin{aligned} &\frac{\partial L}{\partial m_0} = 0\,, \quad \frac{\partial L}{\partial b_i} = 0\,, \quad \frac{\partial L}{\partial b_{n+j}} = 0\,, \\ &\frac{\partial L}{\partial \sigma} = 0 \quad (i = 1, 2, \ldots, n;\ j = 1, 2, \ldots, m)\,. \end{aligned} \tag{20}$$

5. Numerical Examples

5.1. *Results of estimation of model parameters*

A set of fault-detection count data used in this section is obtained from the actual software development project which was developed the software system and consists of nine software components. Also, we adopt the case of ($p_i = 0.3$, $p_j = 0.7$) which maximizes the value of logarithmic likelihood function. The following model parameters have been estimated by solving the likelihood equations.

$$\hat{m}_0 = 45.57\,, \quad \hat{b}_i = 0.2252\,, \quad \hat{b}_j = 0.09354\,,$$
$$\hat{\sigma} = 0.04780\,, \quad r_i = 0.85\,, \quad r_j = 0.15\,,$$

where we consider that r_i represents approximately the reusable rate of software components.

The estimated expected number of remaining faults in Eq. (11), $\hat{\mathrm{E}}[M_s(t)]$, are plotted in Fig. 1. Figure 2 shows the estimated variance of the number of remaining faults in Eq. (13), $\widehat{\mathrm{Var}}[M_s(t)]$. Moreover, the estimated $MTBF_I$ in Eq. (14), $\widehat{MTBF}_I(t)$, and the estimated $MTBF_C$ in Eq. (15), $\widehat{MTBF}_C(t)$, are plotted in Figs. 3 and 4, respectively.

5.2. *Sensitivity analysis in terms of weight parameters*

From the results of the former section, we have verified that our SDE model can be applied to evaluate quantitatively software quality in the system testing-phase of distributed development environment. The estimated expected number of detected faults, $\mathrm{E}[N_s(t)]'s$, with changing parameter p_i to every 0.2 is shown in Fig. 5. Also, the estimated $\mathrm{Var}[N_s(t)]'s$ and the estimated $MTBF_I(t)'s$ with changing parameter p_i to every 0.2 is shown in Figs. 6 and 7, respectively.

From the above results, this model can widely describe both the exponential growth curve and the S-shaped one. Therefore, if we estimate the values of the parameters p_i $(i = 1, 2, \ldots, n + m)$

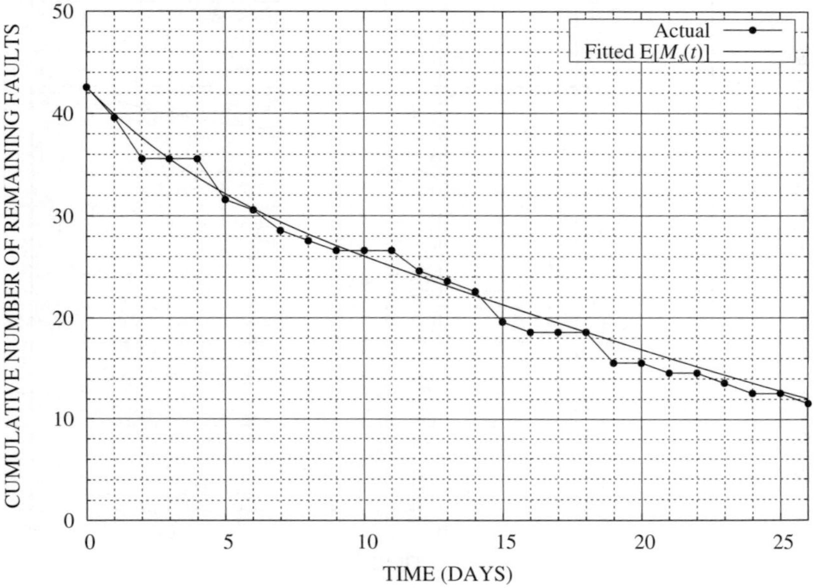

Fig. 1. The estimated expected number of remaining faults, $\hat{\mathrm{E}}[M_s(t)]$.

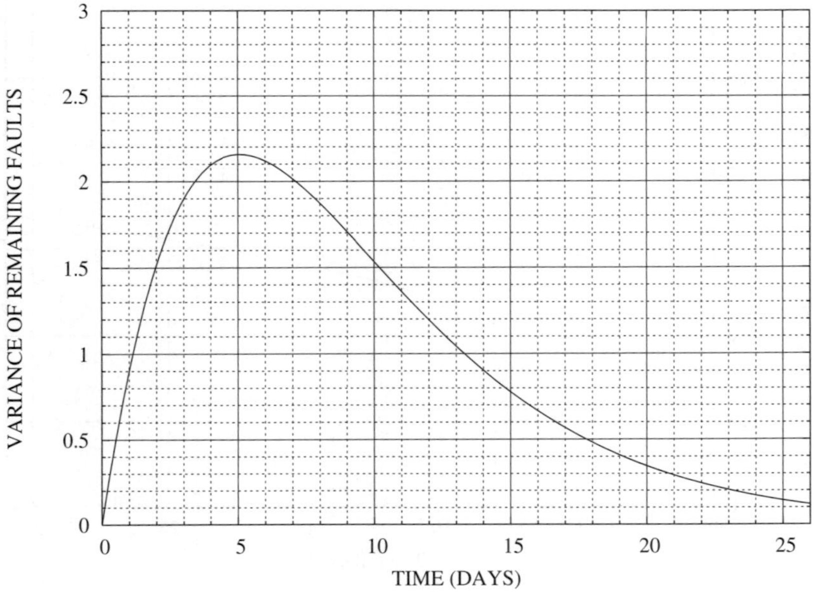

Fig. 2. The estimated variance of the number of remaining faults, $\widehat{\mathrm{Var}}[M_s(t)]$.

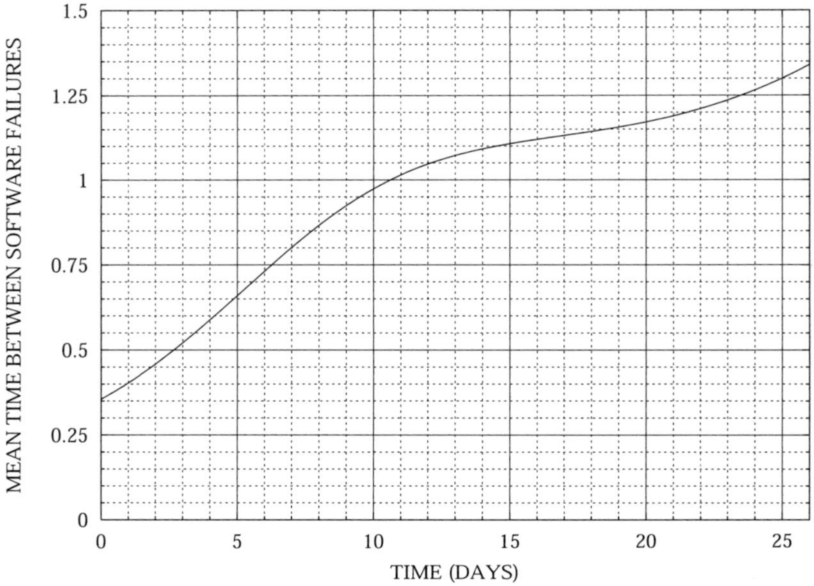

Fig. 3. The estimated $MTBF_I$, $\widehat{MTBF}_I(t)$.

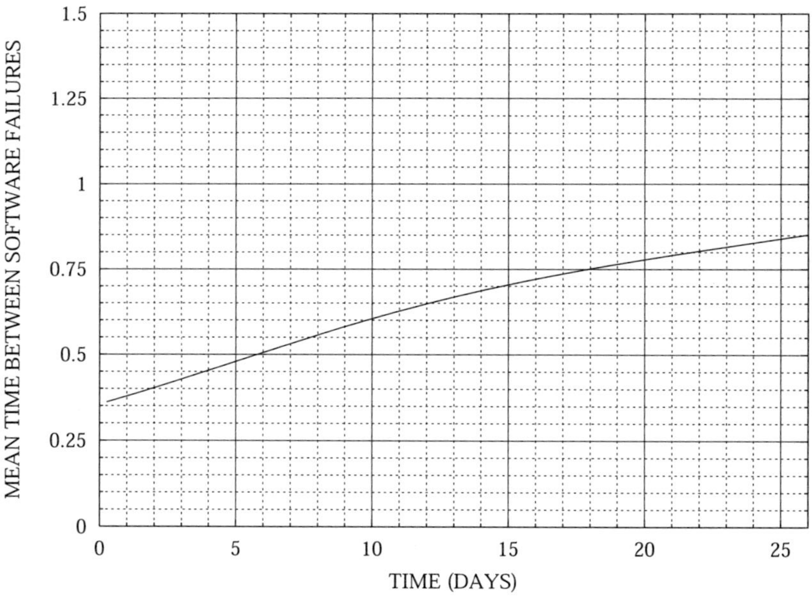

Fig. 4. The estimated $MTBF_C$, $\widehat{MTBF}_C(t)$.

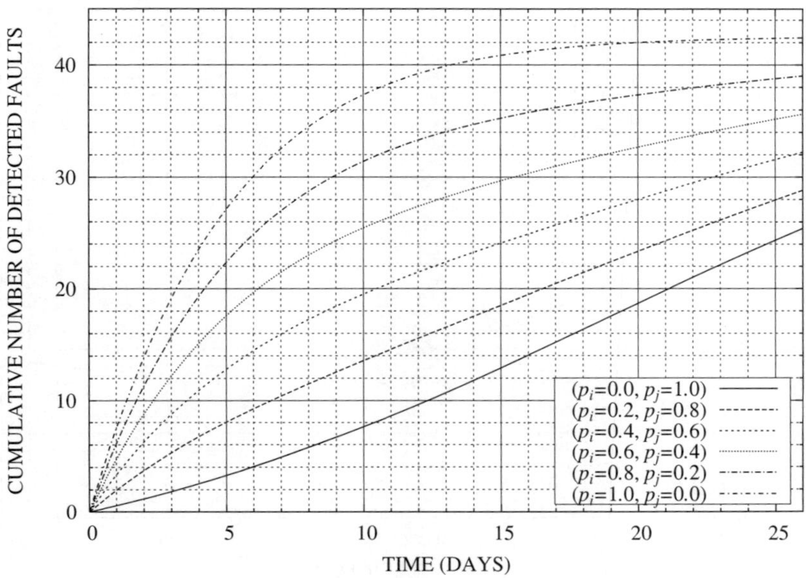

Fig. 5. Dependence of parameter p_i in the estimated expected number of detected faults, $\hat{\mathrm{E}}[N_s(t)]'s$.

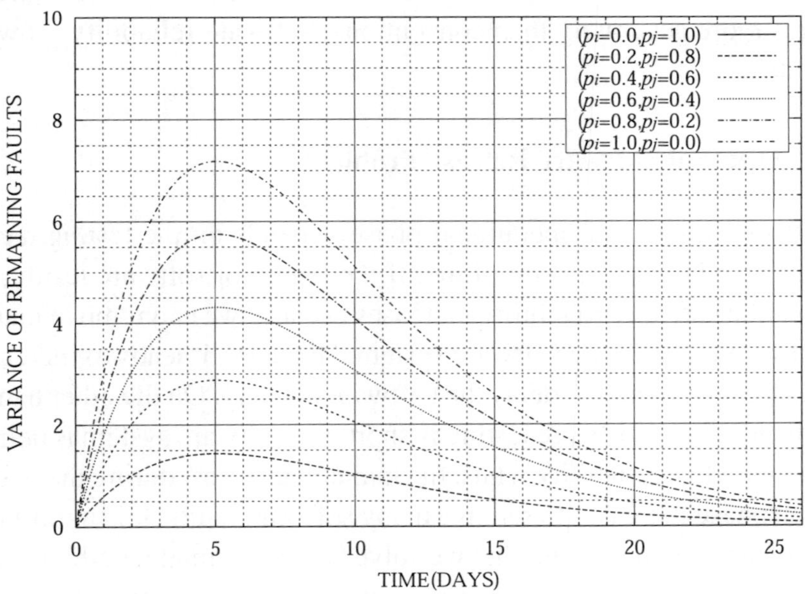

Fig. 6. Dependence of parameter p_i in the estimated variance of the number of remaining faults, $\widehat{\mathrm{Var}}[N_s(t)]'s$.

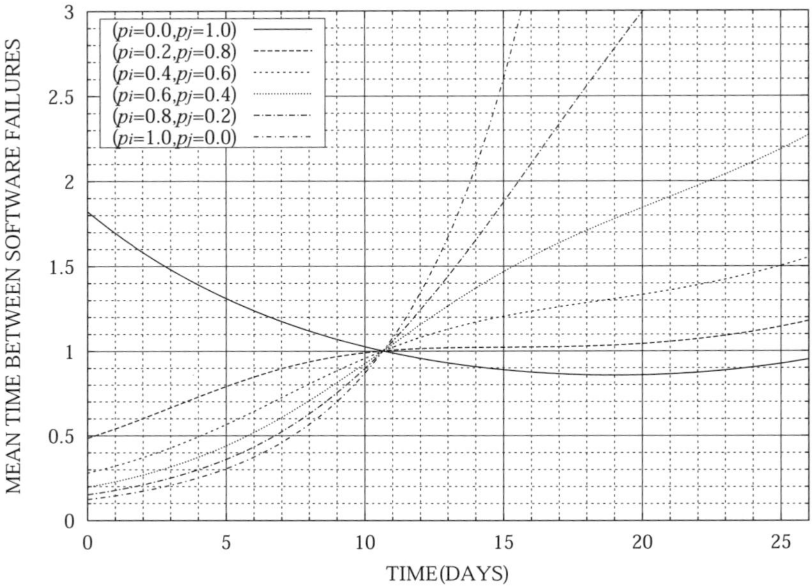

Fig. 7. Dependence of parameter p_i in the estimated $MTBF_I$, $\widehat{MTBF}_I(t)'s$.

reasonably, we can give the software reliability assessment measures more accurately than conventional software reliability growth models.

6. Optimal Software Release Problems

Software quality of operational phase depends on the testing techniques and the total testing-time. If the length of software testing is long, the software reliability increases because we can remove many software faults in the software system. However, it leads to increase the testing cost and to delay the software delivery. On the other hand, if the length of software testing is short, the software system is delivered with low reliability. Thus, the maintenance cost during the operation phase increases to remove many software faults. It is important in software management that we solve for the optimal length of software testing to shift from the testing-phase to the operation phase, which is called an optimum release time. Such a decision problem is called an optimal software release problem. In this section, we discuss optimal software release problems about optimum total testing times minimizing the expected total software maintenance cost.

6.1. *Formulation of expected software cost*

The following notations are defined:

$c_{1,i}$: Fixing cost per fault during the module testing-phase of component i ($c_{1,i} > 0$).
$c_{2,i}$: Testing cost per unit time in the module testing-phase of component i ($c_{2,i} > 0$).
c_{1c}: Fixing cost per fault during the system testing-phase ($c_{1c} > 0$).
c_{2c}: Testing cost per unit time in the system testing-phase ($c_{2c} > 0$).
c_{3c}: Maintenance cost per fault during the operational phase ($c_{3c} > 0, c_{3c} > c_{1i}, c_{3c} > c_{1c}$).

Moreover, we assume that if the end time of module testing-phase exceeds over the beginning time of system testing-phase, the penalty cost is imposed. We define the penalty cost function as follows:

$$G_i(t_i) = \begin{cases} c_{3i}\left[e^{k_i(t_i - t_{di})} - 1\right] & (t_i > t_{di}), \\ 0 & (t_i \leq t_{di}), \end{cases} \tag{21}$$

where t_i is measured from the beginning time of module testing-phase to the end time of module testing-phase, t_{di} the delivery time of software component i ($i = 1, 2, \ldots, n + m$) for system testing-phase, and $c_{3i}(> 0)$ and $k_i(> 0)$ represent constant parameters. Thus, we formulate the expected total software cost in the module testing-phase of each component as follows:

$$C_i(t_i) = c_{1,i}H_i(t_i) + c_{2,i}t_i + G_i(t_i) \quad (i = 1, 2, \ldots, n + m), \tag{22}$$

where $H(t)$ is the mean value function in Eq. (7).

Next, the expected software cost in the system testing-phase is given by:

$$C_c(t_c) = c_{1c}\mathrm{E}[N_s(t_c)] + c_{2c}t_c \,. \tag{23}$$

Also, the expected software maintenance cost after the operational phase is given by:

$$C_d(t_c) = c_{3c}\mathrm{E}[M_s(t_c)] \,. \tag{24}$$

Then, from Eqs. (22)–(24), the expected total software cost is given as follows:

$$C(t_1, \ldots, t_{n+m}, t_c) = \sum_{i=1}^{n+m} C_i(t_i) + C_c(t_c) + C_d(t_c) \,. \tag{25}$$

Therefore, software release time t_c which minimize Eq. (25) is the optimum release time t_c^*.

Next, we discuss a numerical example on the expected total software cost, and we assume the values of parameters in the cost factor as follows:

$$\begin{aligned}
&c_{11} = 1\,, \quad c_{12} = 1\,, \quad c_{13} = 1\,, \quad c_{14} = 1\,, \quad c_{15} = 1\,,\\
&c_{16} = 1\,, \quad c_{17} = 2\,, \quad c_{18} = 1\,, \quad c_{19} = 2\,,\\
&c_{21} = 2\,, \quad c_{22} = 2\,, \quad c_{23} = 2\,, \quad c_{24} = 2\,, \quad c_{25} = 2\,,\\
&c_{26} = 2\,, \quad c_{27} = 4\,, \quad c_{28} = 2\,, \quad c_{29} = 4\,,\\
&c_{1c} = 10\,, \quad c_{2c} = 20\,, \quad c_{3c} = 50\,.
\end{aligned}$$

The module testing-periods of each software component are given by:

$$\begin{aligned}
&t_{d1} = 22\,, \quad t_{d2} = 5\,, \quad t_{d3} = 32\,, \quad t_{d4} = 20\,, \quad t_{d5} = 24\,,\\
&t_{d6} = 4\,, \quad t_{d7} = 33\,, \quad t_{d8} = 12\,, \quad t_{d9} = 42\,.
\end{aligned}$$

In this chapter, we discuss a case that the module testing-phase of seventh and ninth software components are delayed, assume the delay interval times of seventh and ninth software components and

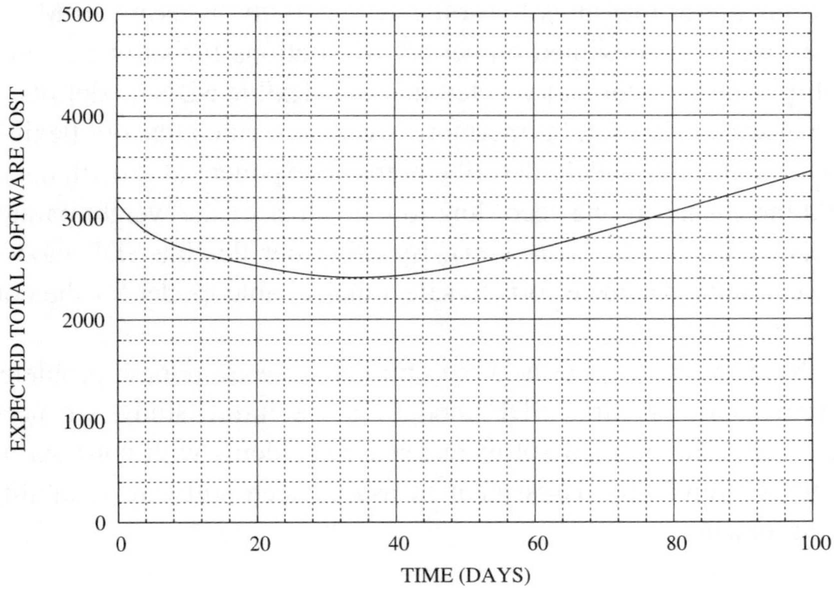

Fig. 8. The estimated expected total software cost.

the constant parameters of the penalty cost as follows:

$$(t_7 - t_{d7}) = (38 - 33) = 5\,, \quad c_{37} = 0.5\,, \quad k_7 = 1\,,$$
$$(t_9 - t_{d9}) = (47 - 42) = 5\,, \quad c_{39} = 0.5\,, \quad k_9 = 1\,.$$

Thus, the estimated expected total software cost is shown in Fig. 8. The minimized expected total software cost is 2412.81 from Fig. 8, and then optimum release time t_c^* is 32.7348.

7. Concluding Remarks

In this chapter, we have treated the event of fault-detection in distributed development environment as continuous state space. Especially, we have proposed a flexible SDE model describing a fault-detection process during the system-testing phase of the distributed development environment by applying a mathematical technique of stochastic differential equations. Next, we have derived several useful measures for software reliability assessment. Moreover, we have presented numerical illustrations for software reliability measurement and verified that our flexible SDE model fits to the observed data set. Furthermore, we have verified that our flexible SDE model can widely describe both the exponential growth curve and the S-shaped one according to the values of the weight parameters p_i $(i = 1, 2, \ldots, n + m)$. By using our flexible SDE model, we can reduce some efforts to select the suitable model for the collected data sets.

Next, we have discussed the optimal software release problems based on our flexible SDE model. In the future study, we need to discuss the optimal software release problems with considering software reliability assessment measures such software reliability requirement.

Acknowledgment

This work was supported in part by, the Grant-in-Aid for the Scientific Research(C)(2) from the Ministry of Education, Culture, Sports, Science and Technology of Japan under Grant No. 15510129.

References

1. N. Nagano and T. Miyachi (eds.), *Distributed Software Development* (Kyoritsu Shuppan, Tokyo, 1996) (in Japanese).
2. S. Yamada, *Software Reliability Models: Fundamentals and Applications* (JUSE Press, Tokyo, 1994) (in Japanese).
3. L. Arnold, *Stochastic Differential Equations: Theory and Applications* (John Wiley & Sons, New York, 1974).
4. S. Yamada, M. Kimura, H. Tanaka and S. Osaki, Software reliability measurement and assessment with stochastic differential equations, *IEICE Transactions on Fundamentals* **E77-A** (1994) 109–116.
5. Y. Tamura, M. Kimura and S. Yamada, A software reliability growth model based on stochastic differential equations for distributed development environment, *Proc. the 32nd ISCIE Int. Symp. Stochastic Systems Theory and Its Applications* (2000), pp. 155–160.

6. S. Yamada and S. Osaki, Optimal software release policies with simultaneous cost and reliability requirements, *European J. Operational Research* **31** (1987) 46–51.
7. S. Yamada, Y. Tamura and M. Kimura, A software reliability growth model for distributed development environment, *Electronics and Communications in Japan*, Part 3 **83** (2000) 1–8.

CHAPTER 16

An Extended Delayed *S*-Shaped Software Reliability Growth Model Based on Infinite Server Queuing Theory

Shinji Inoue* and Shigeru Yamada†

Department of Social Systems Engineering, Faculty of Engineering, Tottori University, 4-101 Minami, Koyama-cho, Tottori-shi, Tottori 680-8552, Japan

**ino@sse.tottori-u.ac.jp*

†yamada@sse.tottori-u.ac.jp

1. Introduction

Quantitative assessment of software reliability in the testing phase is important to provide a software system keeping high degree of reliability for the user because the testing phase is the final stage of software development process. Up to now, as a mathematical model to assess software reliability, several SRGMs have been utilized for assessing the degree of the achievement of software quality, deciding the time to release for operational use, and evaluating the maintenance cost for faults undetected during the testing phase. Most of SRGMs have been modeled by any stochastic process to describe the software fault-detection phenomenon or the software failure-occurrence phenomenon, especially, it is known that an NHPP model can describe the software reliability growth process easily by supposing mean value function of an NHPP intuitively. Accordingly, the NHPP models have been utilized in many software houses and computer manufacturers. On the other hand, for most of the NHPP models, it has been pointed out that it is difficult to understand physical interpretation for fault-detection phenomenon by many researchers. As one of the method for solving the problem, general methods for SRGMs such as generalized order statistics models[1] have been proposed. In recent years, Dohi *et al.*[2] proposed a general approach to existing SRGMs by regarding the software failure-occurrence phenomenon as an infinite server queue.

The delayed S-shaped SRGM[3,4] is one of the SRGMs which can analyze the physical interpretation for the fault-detection phenomenon. This SRGM has been developed by supposing that the fault-detection phenomenon is consisted of successive software failure-detection and fault-isolation processes. In the actual testing phase, we can consider that the time for analyzing or isolating the causes of software failures do not always take a constant values.

In this chapter, we first discuss the delayed S-shaped SRGM which is a basic concept for our extended delayed S-shaped SRGM. Secondly, before developing our extended delayed S-shaped SRGM, we discuss a concept of conditional distribution of arrival times, which is utilized for developing our model. Thirdly, based on the concepts of the delayed S-shaped SRGM and conditional distribution of arrival times, we develop an infinite server queuing model considering the time distribution of the fault-isolation process for software reliability assessment. Finally, we also mention that this model is a general approach for several SRGMs described on NHPPs, and show numerical examples for our model by using actual fault count data.

2. Delayed S-Shaped SRGM

In this section, we discuss an SRGM based on an NHPP, and a concept of the delayed S-shaped SRGM,[3,4] which is the basic concept for developing our infinite server queuing model.

Let $\{Z(t), t \geq 0\}$ be the counting process representing the cumulative number of faults detected up to time $t(t \geq 0)$. Supposing that $Z(t)$ obeys an NHPP, we can formulate the fault-detection phenomenon as follows:

$$\left.\begin{aligned} \Pr\{Z(t) = n\} &= \frac{\{H(t)\}^n}{n!} \exp\{-H(t)\} \qquad (n = 0, 1, 2, \ldots) \\ H(t) &= \int_0^t h(t)dt \end{aligned}\right\}, \tag{1}$$

where $H(t)$ is mean value function which indicates the expectation of $Z(t)$, i.e., the expected cumulative number of faults detected up to time t, and $h(t)$ called an intensity function which indicates the instantaneous fault-detection rate at time t. Equation (1) implies that the software reliability growth process in the testing phase is characterized by mean value function $H(t)$ or intensity function $h(t)$.

Generally, the cause analysis to detect software faults occurring software failures is practiced in the testing phase. Accordingly, the delayed S-shaped SRGM has been developed by supposing that the fault-detection process is consisted of successive software failure-detection and fault-isolation processes, that is, this SRGM regards analyzing the software failure-occurrence phenomenon and isolating the faults causing software failures as the fault-detection.

The delayed S-shaped SRGM is derived by the following procedure: First, in the software failure-detection process, letting $m(t)$ be the expected cumulative number of software failures detected up to time t, we can obtain the following differential equation by assumptions of the delayed S-shaped SRGM:

$$\frac{dm(t)}{dt} = b_1[a - m(t)], \tag{2}$$

where a indicates the expected initial fault content in the software system, and $b_1(>0)$ the failure-occurrence rate. Next, in the fault-isolation process, letting $M(t)$ be the expected cumulative number of faults isolated (or detected) up to time t, we can also obtain the following differential equation by assumptions of the delayed S-shaped

SRGM:

$$\frac{dM(t)}{dt} = b_2[m(t) - M(t)], \tag{3}$$

where $b_2(>0)$ represents the fault-detection rate. Supposing that $b = b_1 = b_2$ approximately, $M(t)$ can be derived from Eqs. (2) and (3) as follows:

$$M(t) = a[1 - (1 + bt)\exp\{-bt\}]. \tag{4}$$

The mean value funtion $M(t)$ in Eq. (4) is called a *delayed S-shaped software reliability growth model*.[3,4] Figure 1 shows a concept of the delayed S-shaped SRGM. Additionally, letting $M_G(t)$ be the expected cumulative number of faults detected up to time t in case of $b_1 \neq b_2$, mean value function of the NHPP model called a *generalized delayed S-shaped software reliability growth model*[5] is derived by solving Eqs. (2) and (3) with respect to $M(t)$ as follows:

$$M_G(t) = a\left[1 - \frac{1}{1-v}(\exp\{-b_1 vt\} - v\exp\{-b_1 t\})\right], \tag{5}$$

where $v = b_2/b_1$ which represents a relative measure between the frequency of the fault-occurrence and the isolation progress rate.

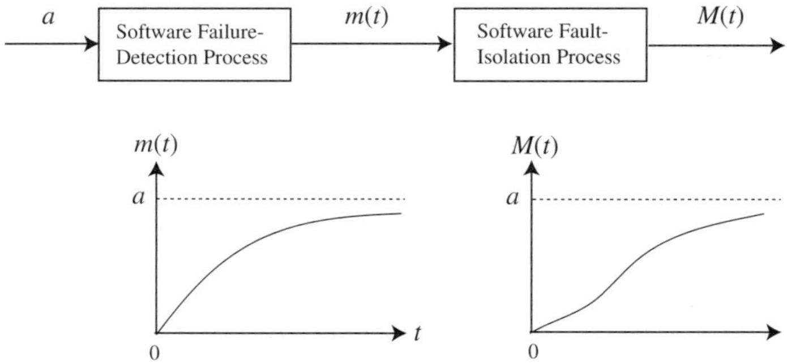

Fig. 1. The delayed S-shaped software reliability growth modeling.

3. Infinite Server Queuing Modeling

Two events of software failure-occurrence phenomenon and fault-detection process have different meaning each other. Thus, the faults are not always detected even if the software failures are occurred. The time spent by analyzing the causes of each software failure is randomly behaved by difference of the difficulty of isolating and detecting each fault.

In this section, we first introduce a concept of conditional distribution of arrival times which is need for developing an infinite server queuing model. After that, utilizing the concept, we develop an infinite server queuing model[6–8] to treat above situation comprehensively. We also propose a SRGM considering the time distribution of fault-isolation process.

3.1. *Conditional arrival times distribution*

Before developing an infinite server queuing model, we need to discuss a concept of the conditional distribution of arrival times. In this section, we discuss the conditional arrival times distribution in case that the events occur in accordance with an NHPP formulated as Eq. (1).

Let $S_1, S_2, \ldots, S_n$ be the n arrival times of a counting process $\{Z(t),\ t \geq 0\}$ which obeys an NHPP with mean value function $H(t)$ and its intensity function $h(t)$ in Eq. (1). Now we consider the conditional distribution of the first arrival time S_1 given that there was an event in the time-interval $[0, t]$. For $s < t$, the conditional distribution is derived as:

$$\begin{aligned} \Pr\{S_1 \leq s_1 \mid Z(t) = 1\} &= \frac{H(s_1)}{H(t)} \\ &= \int_0^{s_1} \frac{h(x)}{H(t)} dx \,. \end{aligned} \tag{6}$$

Similarly, we can derive the joint conditional distribution of S_1 and S_2 as follows:

$$\begin{aligned}
&\Pr\{S_1 \le s_1, S_2 \le s_2 \mid Z(t) = 2\} \\
&= 2!\frac{H(s_1)[H(s_2) - H(s_1)]}{[H(t)]^2} \\
&= 2!\int_{s_1}^{s_2}\int_0^{s_1} \frac{\prod_{i=1}^{2} h(x_i)}{[H(t)]^2} dx_1\, dx_2\,, \qquad (7)
\end{aligned}$$

where $s_1 < s_2 \le t$. Generalizing these fact, if we condition that $Z(t) = n$, the joint conditional distribution of n arrival times is given by:

$$\begin{aligned}
&\Pr\{S_1 \le s_1, S_2 \le s_2, \ldots, S_n \le s_n \mid Z(t) = n\} \\
&= n!\int_{s_0}^{s_1}\int_{s_1}^{s_2} \cdots \int_{s_{n-1}}^{s_n} \frac{\prod_{i=1}^{n} h(x_i)}{[H(t)]^n} dx_1\, dx_2 \cdots dx_n\,. \qquad (8)
\end{aligned}$$

Therefore, given that $Z(t) = n$, the joint conditional density of n arrival times is derived as follows:

$$f(t_1,\ t_2, \ldots,\ t_n \mid Z(t) = n) = n!\frac{\prod_{i=1}^{n} h(t_i)}{[H(t)]^n}\,. \qquad (9)$$

Equation (9) implies that unordered random variables of n arrival times $S_1, S_2, \ldots, S_n$ are independent and identically distributed with the density

$$f(x) = \begin{cases} \dfrac{h(x)}{H(t)} & (0 \le x \le t)\,, \\ 0 & (\text{otherwise})\,, \end{cases} \qquad (10)$$

if we condition that $Z(t) = n$.[7] Of course, if $Z(t)$ obeys a *homogeneous Poisson process* (abbreviated as HPP) which is a special case for an NHPP, the n arrival times given $Z(t) = n$ are independent and identically distributed uniformly on the interval $[0, t]$.

Additionally, we also introduce a useful conditional probability relative to the conditional arrival times distribution discussed above.

If $s < t$ and $0 \leq m \leq n$, then

$$\Pr\{Z(s) = m \mid Z(t) = n\} = \binom{n}{m} \left(\frac{H(s)}{H(t)}\right)^m \left(1 - \frac{H(s)}{H(t)}\right)^{n-m}. \tag{11}$$

Equation (11) implies that m events occured by time $s(< t)$ are independent and have a probability $H(s)/H(t)$ respectively, given that $Z(t) = n$, that is, the conditional distribution of $Z(s)$ given that $Z(t) = n$ obeys a binomial distribution with parameters $(n, \ H(s)/H(t))$.

These conditional distributions discussed above are directly applied to several probability models such as inventory and queuing models. Then, using these properties, we discuss an infinite server queuing modeling for software reliability assessment in the next subsection.

3.2. *Infinite server queuing modeling*

We develop an infinite server queuing model based on the following assumptions:

(A-1) The expected cumulative number of software failures are observed according to an NHPP with mean value function $\Lambda(t)$ and intensity function $\lambda(t)$.

(A-2) The observed software failure is directly analyzed in the fault-isolation process when the software failure is observed. After the software failure analysis, the software fault is detected.

(A-3) The fault-isolation times are assumed to be independent with a common distribution $F(t)$.

Let a counting process $\{X(t), t \geq 0\}$ be the random variable indicating the cumulative number of software failures observed up to time t, and also a counting process $\{N(t), \ t \geq 0\}$ be the one indicating the cumulative number of faults detected up to time t. If the test was begun at $t = 0$, the distribution function of $N(t)$ is given by:

$$\Pr\{N(t) = n\} = \sum_{j=0}^{\infty} \Pr\{N(t) = n \mid X(t) = j\} \Pr\{X(t) = j\}$$
$$= \sum_{j=0}^{\infty} \Pr\{N(t) = n \mid X(t) = j\} \frac{[\Lambda(t)]^j}{j!} e^{-\Lambda(t)} . \quad (12)$$

For j software failures observed up to time t, the probability that n faults are detected via the fault-isolation process is given as:

$$\Pr\{N(t) = n \mid X(t) = j\} = \binom{j}{n} \{p(t)\}^n \{1 - p(t)\}^{j-n} , \quad (13)$$

where $p(t)$ means the probability that an arbitrary one faults is detected by time t, and also $p(t)$ is given by:

$$p(t) = \int_0^t F(t-x) \frac{d\Lambda(x)}{\Lambda(t)} , \quad (14)$$

from the concept of the conditional arrival times distribution discussed in Sec. 3.1. Thus, substituting Eqs. (13) and (14) into Eq. (12), we obtain the distribution function of the cumulative number of faults detected up to time t as:

$$\Pr\{N(t) = n\} = \frac{\left[\int_0^t F(t-x)d\Lambda(x)\right]^n}{n!} \times \exp\left[-\int_0^t F(t-x)d\Lambda(x)\right] . \quad (15)$$

Equation (15) is equivalent to the NHPP in Eq. (1) with mean value function $\int_0^t F(t-x)d\Lambda(x)$, that is, $N(t)$ has the NHPP with mean value function $\int_0^t F(t-x)d\Lambda(x)$. Figure 2 shows the concept of the infinite server queuing model.

3.3. *Relationship to existing SRGMs*

We have developed the infinite server queuing model by incorporating the time distribution of the fault-isolation process in the preceding

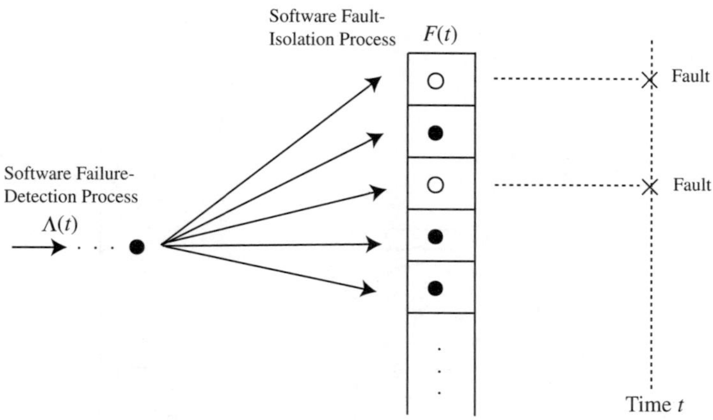

Fig. 2. An infinite server queuing model with the time distribution of fault-isolation process.

subsection. Using Eq. (15), we can characterize the time-dependent behavior of the fault-detection phenomenon by $\Lambda(t)$ and $F(t)$ which indicate the expected cumulative number of failures observed up to time t and the time distribution function of the fault-isolation process, respectively. Thus, we can easily reflect the physical phenomenon in the fault-detection phenomenon on the SRGM.

Also, Eq. (15) can be considered as an general description for several NHPP models. Specifically, for example, Eq. (15) is equivalent to the generalized delayed S-shaped SRGM in Eq. (5) essentially if $\Lambda(t)$ and $F(t)$ in Eq. (15) are supposed as follows:

$$\Lambda(t) = a(1 - e^{-bt}), \quad F(t) = 1 - e^{-\alpha t}, \tag{16}$$

where a represents the expected initial fault content in the software system, b the failure-occurrence rate, $\alpha(>0)$ the reciprocal of the expectation of the exponential distribution, i.e., the expectation of the Poisson distribution. Furthermore, if we suppose that $\eta = b = \alpha$ in Eq. (16), Eq. (15) is equivalent to the delayed S-shaped SRGM in Eq. (4). Table 1 summarizes the relationships between the infinite

Table 1. The infinite server queuing models versus existing NHPP models.

$\Lambda(t)$	$F(t)$	$M_S(t)$	Ref.
$a(1-\exp[-bt])$	$\mathbf{1}(t)$	$a(1-\exp[-bt])$	10, 11
$a(1-\exp[-bt])$	$T \sim EXP(\alpha)$	$a\left[1-\frac{1}{b-\alpha}(b\exp\{-\alpha t\}-\alpha\exp\{-bt\})\right]$	5, 12
$a(1-\exp[-\eta t])$	$T \sim EXP(\eta)$	$a[1-(1+\eta t)\exp\{-\eta t\}]$	3, 4
μt	$T \sim WEI(\alpha, m)$	$\mu\left[t-\frac{1}{m\alpha}\left\{m\Gamma_1\left(1+\frac{1}{m}\right)-\Gamma_2\left(\frac{1}{m},(\alpha t)^m\right)\right\}\right]$	
$a[1-(1+bt) \times\exp\{-bt\}]$	$T \sim EXP(\alpha)$	$a\left[1-\left(1-bt+\frac{b^2(bt-\alpha t+1)}{(b-\alpha)^2}\right)\exp\{-bt\} -\frac{b^2}{(b-\alpha)^2}\exp\{-\alpha t\}\right]$	
$a(1-r^t)$	$T \sim EXP(\alpha)$	$a\left[(1-r^t)+\frac{(r^t-\exp\{-\alpha t\})\log r}{\alpha+\log r}\right]$	

$(\alpha,\ \eta > 0,\ 0 < r < 1)$
$\mathbf{1}(t)$: Unit function.
Γ_1 : Gamma function, Γ_2 : Incomplete gamma function.
EXP : Exponential distribution.
WEI : Weibull distribution.

server queuing model and existing NHPP models where $M_S(t) = \int_0^t F(t-x)d\Lambda(x)$.

4. Numerical Examples

In this section, we show several numerical examples for software reliability assessment by using actual observed data. First, we suppose that $\Lambda(t)$ and $F(t)$ are given as:

$$\Lambda(t) = a(1 - r^t) \quad (0 < r < 1)\,, \tag{17}$$

and

$$F(t) = 1 - \exp[-\alpha t]\,, \tag{18}$$

respectively. We employ the maximum-likelihood estimation (abbreviated as MLE) method to estimate the model parameters. Supposing that we observed K data pairs $(t_k, y_k)(k = 0, 1, 2, \ldots, K)$ in respect of the total faults y_k detected during constant time-interval $(0, t_k]$, we can derive the following logarithmic likelihood function from the properties of the NHPP:

$$\begin{aligned} \ln L = & \sum_{k=1}^{K} (y_k - y_{k-1}) \cdot \ln[M_S(t_k) - M_S(t_{k-1})] - M_S(t_K) \\ & - \sum_{k=1}^{K} \ln[y_k - y_{k-1}]\,. \end{aligned} \tag{19}$$

Furthermore, we can derive the following simultaneous equations by partially differentiating above logarithmic likelihood funtion with respect to the parameters a, r, and α:

$$\frac{\partial \ln L}{\partial a} = \frac{\partial \ln L}{\partial r} = \frac{\partial \ln L}{\partial \alpha} = 0\,. \tag{20}$$

Accordingly, by solving numerically the above equations, we can estimate $\hat{a}$, $\hat{r}$, and $\hat{\alpha}$, which are the estimates of a, r, and α, respectively.

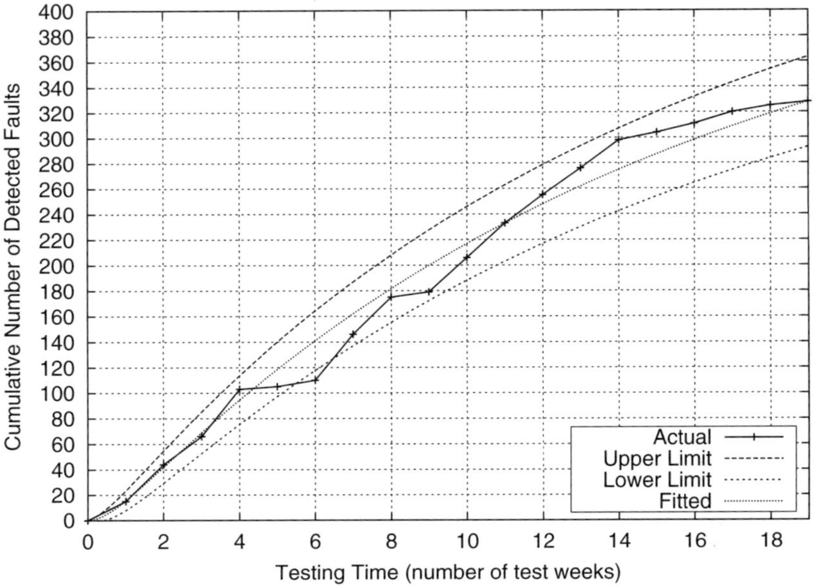

Fig. 3. The estimated mean value function $\widehat{M_S}(t)$.

We use a PL/I programming to test the data consisting of 19 data pairs $(t_k, y_k)(k = 1, 2, \ldots, 19; t_{19} = 19, y_{19} = 328)$.[9] Figure 3 shows the estimated mean value function $\widehat{M_S}(t)$ and the 95% confidence limits of it where the estimated parameters of $\widehat{M_S}(t)$ are $\hat{a} = 459.08$, $\hat{r} = 0.1916$, and $\hat{\alpha} = 0.0682$. The $100\gamma\%$ confidence limits is derived as:

$$\widehat{M_S}(t) \pm K_\gamma \sqrt{\widehat{M_S}(t)}\,, \tag{21}$$

where K_γ indicates the $100(1 + \gamma)/2$ percent point of the standard normal distribution. We also apply the Kolmogorov–Smirnov (abbreviated as K–S) goodness-of-fit test[11,13] to evaluate whether $\widehat{M_S}(t)$ fits statistically to the observed data. This statistical testing is considered to be efficient even if the data set observed is small.[11] We verified that $\widehat{M_S}(t)$ fits to the observed data with the 5% level of significance by the K–S goodness-of-fit testing.

4.1. *Software reliability function*

Given that the testing or the operation has been going up to time t, the probability that a software failure does not occur in the time-interval $[t, t + x)(x \geq 0,\ t \geq 0)$ is derived as:

$$R_S(x \mid t) = \exp[-\{M_S(t + x) - M_S(t)\}] \tag{22}$$

from Eq. (1). $R_S(x \mid t)$ in Eq. (22) is called a *software reliability function*. We can estimate the software reliability by using this equation. Figure 4 shows the software reliability with respect to $t = 19$(weeks) which is the termination time of the testing. Assuming that the software users operate the software system under the same environment as the testing, we can estimate the software reliability $\widehat{R_S}(0.1 \mid 19)$ to be about 0.410.

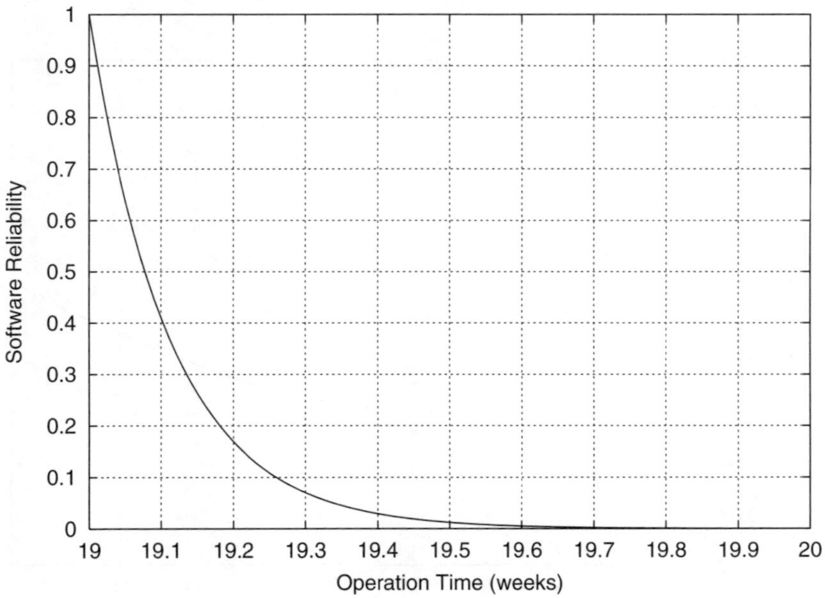

Fig. 4. The estimated software reliability $\widehat{R_S}(x \mid 19)$.

4.2. *Instantaneous MTBF*

We also estimate an instantaneous MTBF (mean time between software failures or fault-detections) which has been used as one of the substitution of measures of MTBF. An instantaneous MTBF can be obtained as:

$$MTBF_I(t) = \frac{1}{h_S(t)}, \tag{23}$$

where $h_S(t)$ indicates the intensity function. Figure 5 shows the time-dependent behavior of the instantaneous MTBF in Eq. (23). Using Eq. (23), we can estimate the mean time between failures $\mathrm{MTBF}_I(19)$ to be about 0.112 (weeks).

5. Concluding Remarks

In this chapter, we have discussed an infinite server queuing model considering the time distribution of the fault-isolation process based on the concept of the delayed S-shaped SRGM. Generally, it is

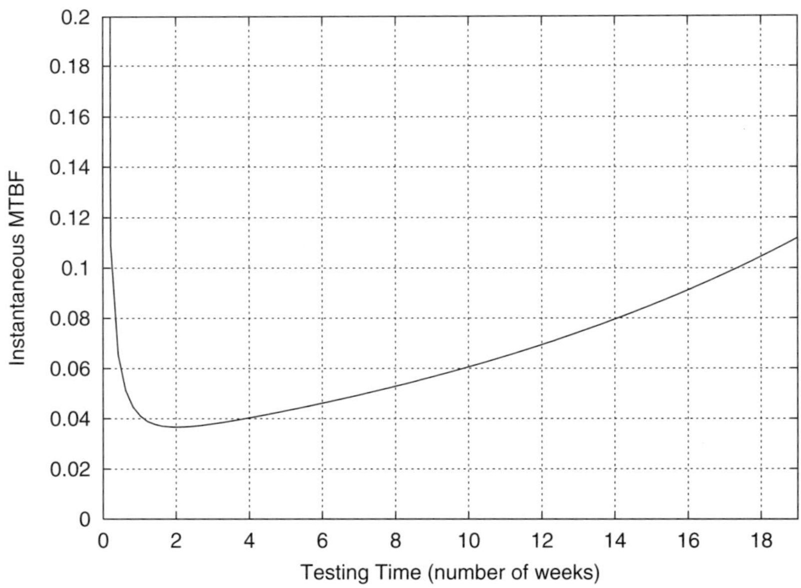

Fig. 5. The estimated instantaneous MTBF, $\widehat{MTBF_I}(t)$.

considered that the time spent by analyzing the causes of each software failure is randomly behaved by difference of the difficulty of isolating and detecting each fault. Accordingly, this chapter has treated with the random behavior of isolation times for each software fault by developing an infinite server queuing model. Additionally, this chapter has shown that this model can easily express the physical description for the fault-detection phenomenon, and can describe several NHPP models as the special cases. Finally, having assumed that $\Lambda(t) = a(1 - r^t)$ and $F(t) = 1 - \exp[-\alpha t]$, we have shown the goodness-of-fit evaluation to the actual data and serveral numerical examples by using fault count data observed in an actual testing phase.

By using the SRGM proposed in this chapter, the time-dependent behavior of the fault-detection phenomenon is characterized by $\Lambda(t)$ and $F(t)$ which indicate mean value function of the software failure-occurrence phenomenon and the distribution function of isolation time, respectively. But, there are several problems for application this SRGM to the actual testing phase. We have shown numerical examples in Sec. 4 by assuming that $\Lambda(t) = a(1 - r^t)$ and $F(t) = 1 - \exp[-\alpha t]$ intuitively. However, we have to decide how to assume $\Lambda(t)$ and $F(t)$ in actual testing phase, which is an important issue for actual software development managers as future studies.

Acknowledgment

This work was supported in part by the Research Grant from the Telecommunications Advancement Foundation, and the Grant-in-Aid for Scientific Research (C)(2) from the Ministry of Education, Culture, Sports, Science and Technology of Japan under Grant No. 15510129.

References

1. N. Langberg and N. D. Singpurwalla, Unification of some software reliability models, *SIAM J. Scientist. Comput.* **6** (1985) 781–790.

2. T. Dohi, T. Matsuoka and S. Osaki, An infinite server queuing model for assessment of the software reliability, *Trans. IEICE* **J83-A** (2000) 536–544 (in Japanese).
3. S. Yamada, M. Ohba and S. Osaki, *S*-shaped reliability growth modeling for software error detection, *IEEE Trans. Reliability* **R-32** (1983) 475–478.
4. S. Yamada and S. Osaki, Software reliability growth modeling: Models and applications, *IEEE Trans. Software Engineering* **SE-11** (1985) 1431–1437.
5. S. Yamada, T. Hangai and S. Osaki, A generalized delayed *S*-shaped software reliability growth model and evaluation of its goodness-of-fit, *Trans. IEICE* **J76-D-I** (1993) 613–620 (in Japanese).
6. S. Osaki, *Applied Stochastic System Modeling* (Springer-Verlag, Berlin, Heidelberg, 1992).
7. S. M. Ross, *Applied Probability Models with Optimization Applications* (Dover Publications, New York, 1992).
8. S. M. Ross, *Introduction to Probability Models* (Academic Press, San Diego, California, 1993).
9. M. Ohba, Software reliability analysis models, *IBM J. Research and Development* **28** (1984) 428–443.
10. S. Yamada and H. Ohtera, *Software Reliability: Theory and Application* (Soft Research Center, Tokyo, 1990) (in Japanese).
11. S. Yamada, *Software Reliability Models: Fundamentals and Applications* (JUSE Press, Tokyo, 1994) (in Japanese).
12. S. Yamada, H. Ohtera and M. Ohba, Testing-domain dependent software reliability models, *Computers and Mathematics with Applications* **24** (1992) 679–686.
13. H. Pham, *Software Reliability* (Springer-Verlag, Singapore, 2000).

CHAPTER 17

Disappointment Probability Based on the Number of Debuggings for Operational Software Availability Measurement

Yutaka Saitoh

Graduate School of Engineering,
Tottori University,
Tottori-shi, 680-8552 Japan
99t7024@sse.tottori-u.ac.jp

Koichi Tokuno* and Shigeru Yamada†

Department of Social Systems Engineering,
Faculty of Engineering, Tottori University,
Tottori-shi, 680-8552 Japan
**toku@sse.tottori-u.ac.jp*
†yamada@sse.tottori-u.ac.jp

1. Introduction

Software quality/performance evaluation from user's viewpoint grows in importance. One of the user-oriented software quality characteristics is **software availability**; this is defined as the characteristic that the software systems are available whenever the users want to use them. Studies on software availability measurement and assessment have been conducted for a decade.[1,2]

Existing software availability models have often paid attention to the stochastic behaviors of only software systems themselves. Several software availability measures representing the probability that the system is operating at a given time point have been derived from the models considering several operational environments. However, from the viewpoint of users, these measures are not always appropriate. It will be enough for users if the system is available only when usage demands occur. In other words, the users do not care about the state of the system when the users do not want to use it. For example, in the software system of the mobile communication system, the whole system is required nonstop operation, but from the viewpoint of users, each of them uses the system intermittently. Then the system failure time for users can be defined as the time to a software failure during a usage period, or to occurrence of a usage demand during a system inoperable period, whichever occurs first. Gaver[3] has called such a time the disappointment time and derived the Laplace–Stieltjes transform of the distribution of this time. Osaki[4] has discussed the disappointment time of a two-unit standby redundant system when it is used intermittently. Furthermore, Tokuno and Yamada[5] have proposed the new measure for software availability measurement; this is called the **disappointment probability** and defined as the probability that the user is made to interrupt the usage of the system due to software failure-occurrences.

In this chapter, based on the model proposed in Ref. 5, we discuss the disappointment probability for operational use. In the operation phase, the restoration of the system includes various works: the data recovery, the re-installation of the system, the debugging activities in order to remove the software faults having caused the system down, and soon. However, debugging activities are not always performed for all of software failures. It is often that the restoration actions with debugging prolong the down time and this much affects the users. In such case, the system is restored with emergency countermeasures such as only the data recovery and the re-installation of the system, and no debugging. This is a different policy from the testing phase. We consider two kinds of restoration actions; one involves debugging and the other does not involve debugging.[6]

Then we derive the disappointment probabilities as the functions of time and the number of debuggings.[7] The stochastic behavior of the system and the user is described by a Markov process.[8] The software reliability growth process, the upward tendency in difficulty of fault removal, and the imperfect debugging environment are also incorporated into the model.[9] Several numerical examples of these measures are presented.

2. Model Description

The following assumptions are made for software availability modeling:

(A1) The software system is unavailable and starts to be restored as soon as a software failure occurs, and the system is unavailable until the restoration action is complete.

(A2) The system is not in use at time point zero. The time to occurrence of a usage demand, X, and the usage time, Y, follow the exponential distributions with means $1/\theta$ and $1/\eta$, respectively.

(A3) When a software failure occurs, the restoration action with the debugging activity is performed with the probability $p(0 < p < 1)$, on the other hand, one without the debugging activity is performed with the probability $q(= 1 - p)$.

(A4) The restoration action with the debugging activity is performed perfectly with the probability $a(0 < a < 1)$ and imperfectly with the probability $b(= 1 - a)$. One fault is removed from the system when the debugging activity is perfect, and then the software reliability growth and the rise of difficulty in debugging occur.

(A5) When n faults have been corrected, the next software failure time-interval, Z_n, and the restoration time with the debugging activity, T_n, the exponential distributions with means $1/\lambda_n$ and $1/\mu_n$, respectively. λ_n and μ_n are nonincreasing functions of n.

(A6) The restoration time without debugging activity, T, follows the exponential distribution with mean $1/\gamma$.

(A7) The probability that two or more software failures occur simultaneously is negligible.

(A8) The usage demands occurring when the system is restored are canceled.

Consider the stochastic process $\{X(t), t \geq 0\}$ representing the state of the system at the time point t. The state space of the process $\{X(t), t \geq 0\}$ is defined as follows:

$\mathbf{W} = \{W_n; n = 0, 1, 2, \ldots\}$: the system is available but not used,

$\mathbf{U} = \{U_n; n = 0, 1, 2, \ldots\}$: the system is available and used,

$\mathbf{R^1} = \{R_n^1; n = 0, 1, 2, \ldots\}$: the system is restored with the debugging activity,

$\mathbf{R^2} = \{R_n^2; n = 0, 1, 2, \ldots\}$: the system is restored without the debugging activity,

and it is denoted that $\mathbf{R} = \{\mathbf{R^1}, \mathbf{R^2}\}$.

From assumption (A3), a software failure occurs,

$$X(t) = \begin{cases} R_n^1 & \text{(with probability } p) \,, \\ R_n^2 & \text{(with probability } q) \,. \end{cases} \tag{1}$$

Furthermore, from assumption (A4), when the restoration action with debugging has been complete in $\{X(t) = R_n^1\}$,

$$X(t) = \begin{cases} W_n & \text{(with probability } b) \,, \\ W_{n+1} & \text{(with probability } a) \,. \end{cases} \tag{2}$$

Figure 1 illustrates the sample state transition diagram of $X(t)$.

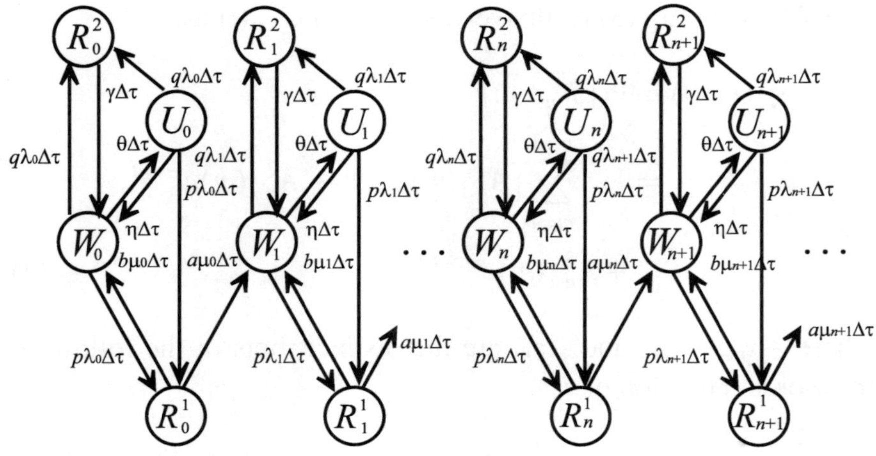

Fig. 1. A sample state transition diagram of $X(t)$.

3. Derivation of Software Availability Measures

3.1. *Distribution of transition time between up states*

Let $S_{i,n}$ $(i \leq n)$ be the random variable representing the transition time from state W_i to state W_n, and $G_{i,n}(t)$ be the distribution function of $S_{i,n}$, respectively. Then, we obtain the following renewal equation:

$$\left.\begin{aligned} G_{i,n}(t) = {} & Q_{W_i,U_i} * Q_{U_i,W_i} * G_{i,n}(t) \\ & + Q_{W_i,U_i} * Q_{U_i,R_i^1} * Q_{R_i^1,W_i} * G_{i,n}(t) \\ & + Q_{W_i,U_i} * Q_{U_i,R_i^2} * Q_{R_i^2,W_i} * G_{i,n}(t) \\ & + Q_{W_i,R_i^1} * Q_{R_i^1,W_i} * G_{i,n}(t) \\ & + Q_{W_i,R_i^2} * Q_{R_i^2,W_i} * G_{i,n}(t) \\ & + Q_{W_i,R_i^1} * Q_{R_i^1,W_{i+1}} * G_{i+1,n}(t) \\ & + Q_{W_i,U_i} * Q_{U_i,R_i^1} * Q_{R_i^1,W_{i+1}} * G_{i+1,n}(t) \\ & (i = 0, 1, 2, \ldots, n-1) \end{aligned}\right\}, \tag{3}$$

where $*$ denotes the Stieltjes convolution and $Q_{A,B}(t)(A, B \in \{\mathbf{W}, \mathbf{U}, \mathbf{R^1}, \mathbf{R^2}\})$ denotes the one-step transition probability from state A to state B. Furthermore, $G_{n,n}(t) \equiv \mathbf{1}(t)$ (the step function; $n = 0, 1, 2, \ldots$).

By solving Eq. (3) recursively, we obtain $G_{i,n}(t)$ as:

$$\begin{aligned} G_{i,n}(t) &\equiv \Pr\{S_{i,n} \leq t\} \\ &= 1 - \sum_{m=i}^{n-1} \big[A^1_{n,i}(m)e^{-x_m t} + A^2_{n,i}(m)e^{-y_m t} \\ &\quad + A^3_{n,i}(m)e^{-z_m t}\big], \end{aligned} \tag{4}$$

where $-x_m$, $-y_m$, and $-z_m$ are the distinct roots of the following third-order equation of s:

$$\begin{aligned} &s^3 + (\lambda_m + \mu_m + \gamma)s^2 + [p\lambda_m\gamma + \mu_m\gamma + (1 - pb)\lambda_m\mu_m]s \\ &\quad + ap\lambda_m\mu_m\gamma = 0, \end{aligned} \tag{5}$$

and constant coefficients $A^1_{n,i}(m)$, $A^2_{n,i}(m)$, and $A^3_{n,i}(m)$ are given by:

$$\left.\begin{aligned} A^1_{n,i}(m) &= \frac{\prod_{j=i}^{n-1} ap\lambda_j\mu_j(\gamma - x_m)}{x_m \prod_{\substack{j=i \\ j\neq m}}^{n-1}(x_j - x_m)\prod_{j=i}^{n-1}(y_j - x_m)(z_j - x_m)} \\ A^2_{n,i}(m) &= \frac{\prod_{j=i}^{n-1} ap\lambda_j\mu_j(\gamma - y_m)}{x_m \prod_{\substack{j=i \\ j\neq m}}^{n-1}(x_j - x_m)\prod_{j=i}^{n-1}(y_j - x_m)(z_j - x_m)} \\ A^3_{n,i}(m) &= \frac{\prod_{j=i}^{n-1} ap\lambda_j\mu_j(\gamma - z_m)}{x_m \prod_{\substack{j=i \\ j\neq m}}^{n-1}(x_j - x_m)\prod_{j=i}^{n-1}(y_j - x_m)(z_j - x_m)} \\ &(m = i, i+1, \ldots, n-1) \end{aligned}\right\}, \tag{6}$$

respectively. It is noted that

$$\sum_{m=i}^{n-1} \big[A^1_{n,i}(m) + A^2_{n,i}(m) + A^3_{n,i}(m)\big] = 1, \tag{7}$$

and that Eq. (4) has no bearing on parameters θ and η associated with the usage characteristic. Furthermore, the expectation and the variance of $S_{i,n}$ are given by:

$$\mathrm{E}[S_{i,n}] = \sum_{m=i}^{n-1} \left(\frac{1}{x_m} + \frac{1}{y_m} + \frac{1}{z_m} - \frac{1}{\gamma} \right), \tag{8}$$

$$\mathrm{Var}[S_{i,n}] = \sum_{m=i}^{n-1} \left(\frac{1}{{x_m}^2} + \frac{1}{{y_m}^2} + \frac{1}{{z_m}^2} - \frac{1}{\gamma^2} \right), \tag{9}$$

respectively.

3.2. *State occupancy probability*

Let $P_{A,B}(t)$ be the state occupancy probability that the system is in state B at the time point t on the condition that the system was in state A at time point zero, i.e.,

$$P_{A,B}(t) \equiv \Pr\{X(t) = B | X(0) = A\} \quad (A, B \in \{\mathbf{W}, \mathbf{U}, \mathbf{R^1}, \mathbf{R^2}\}). \tag{10}$$

We obtain the following renewal equation of $P_{W_i,W_n}(t)$:

$$\left.\begin{aligned} P_{W_i,W_n}(t) &= G_{i,n} * P_{W_n,W_n}(t) \\ P_{W_n,W_n}(t) &= e^{-(\lambda_n+\theta)t} \\ &\quad + Q_{W_n,R_n^1} * Q_{R_n^1,W_n} * P_{W_n,W_n}(t) \\ &\quad + Q_{W_n,R_n^2} * Q_{R_n^2,W_n} * P_{W_n,W_n}(t) \\ &\quad + Q_{W_n,U_n} * Q_{U_n,W_n} * P_{W_n,W_n}(t) \\ &\quad + Q_{W_n,U_n} * Q_{U_n,R_n^1} * Q_{R_n^1,W_n} * P_{W_n,W_n}(t) \\ &\quad + Q_{W_n,U_n} * Q_{U_n,R_n^2} * Q_{R_n^2,W_n} * P_{W_n,W_n}(t) \\ &\quad (i = 0, 1, 2, \ldots, n-1) \end{aligned}\right\}. \tag{11}$$

By solving Eq. (11), we obtain $P_{W_i,W_n}(t)$ as:

$$\begin{aligned} P_{W_i,W_n}(t) &\equiv \Pr\{X(t) = W_n | X(0) = W_i\} \\ &= B_{n,i} e^{-(\lambda_n+\theta+\eta)t} + \sum_{m=i}^{n} \left[B_{n,i}^1(m) e^{-x_m t} + B_{n,i}^2(m) e^{-y_m t} \right. \\ &\quad \left. + B_{n,i}^3(m) e^{-z_m t} \right], \end{aligned} \tag{12}$$

where constant coefficients $B_{n,i}$, $B^1_{n,i}(m)$, $B^2_{n,i}(m)$, and $B^3_{n,i}(m)$ are given by:

$$\left.\begin{aligned}
B_{n,i} &= \frac{-\theta(\mu_n-\lambda_n-\theta-\eta)(\gamma-\lambda_n-\theta-\eta)^{n-i+1}\prod_{j=i}^{n-1}ap\lambda_j\mu_j}{\prod_{j=i}^{n}(x_j-\lambda_n-\theta-\eta)(y_j-\lambda_n-\theta-\eta)(z_j-\lambda_n-\theta-\eta)}\\
B^1_{n,i}(m) &= \frac{(\lambda_n+\eta-x_m)(\mu_n-x_m)(\gamma-x_m)^{n-i+1}\prod_{j=i}^{n-1}ap\lambda_j\mu_j}{(\lambda_n+\theta+\eta-x_m)\prod_{j=i}^{n}(y_j-x_m)(z_j-x_m)\prod_{\substack{j=i\\ j\neq m}}^{n}(x_j-x_m)}\\
B^2_{n,i}(m) &= \frac{(\lambda_n+\eta-y_m)(\mu_n-y_m)(\gamma-y_m)^{n-i+1}\prod_{j=i}^{n-1}ap\lambda_j\mu_j}{(\lambda_n+\theta+\eta-y_m)\prod_{j=i}^{n}(x_j-y_m)(y_j-x_m)\prod_{\substack{j=i\\ j\neq m}}^{n}(y_j-y_m)}\\
B^3_{n,i}(m) &= \frac{(\lambda_n+\eta-z_m)(\mu_n-z_m)(\gamma-z_m)^{n-i+1}\prod_{j=i}^{n-1}ap\lambda_j\mu_j}{(\lambda_n+\theta+\eta-z_m)\prod_{j=i}^{n}(x_j-z_m)(y_j-z_m)\prod_{\substack{j=i\\ j\neq m}}^{n}(z_j-z_m)}\\
&(m=i,i+1,\ldots,n)
\end{aligned}\right\}, \tag{13}$$

respectively. It is noted that

$$B_{n,i}+\sum_{m=i}^{n}\left[B^1_{n,i}(m)+B^2_{n,i}(m)+B^3_{n,i}(m)\right]=0 \quad (n>i)\,, \tag{14}$$

$$B_{i,i}+B^1_{i,i}(0)+B^2_{i,i}(0)+B^3_{i,i}(0)=1 \quad (n=i)\,. \tag{15}$$

Similarly, we obtain the following renewal equations of $P_{W_i,R^1_n}(t)$ and $P_{W_i,R^2_n}(t)$:

$$\left.\begin{aligned}
P_{W_i,R^1_n}(t) &= G_{i,n}*H_{W_i,R^1_n}*P_{R^1_n,R^1_n}(t)\\
P_{R^1_n,R^1_n}(t) &= e^{-\mu_n t}+Q_{R^1_n,W_n}*H_{W_n,R^1_n}*P_{R^1_n,R^1_n}(t)\\
H_{W_n,R^1_n}(t) &= Q_{W_n,R^1_n}(t)+Q_{W_n,U_n}*Q_{U_n,R^1_n}(t)\\
&\quad+Q_{W_n,U_n}*Q_{U_n,W_n}*H_{W_n,R^1_n}(t)\\
&\quad+Q_{W_n,U_n}*Q_{U_n,R^2_n}*Q_{R^2_n,W_n}*H_{W_n,R^1_n}(t)\\
&\quad+Q_{W_n,R^2_n}*Q_{R^2_n,W_n}*H_{W_n,R^1_n}(t)\\
&\quad(i=0,1,2,\ldots,n-1)
\end{aligned}\right\}, \tag{16}$$

$$\left.\begin{aligned}
&P_{W_i,R_n^2}(t) = G_{i,n} * H_{W_i,R_n^2} * P_{R_n^2,R_n^2}(t)\\
&P_{R_n^2,R_n^2}(t) = e^{-\gamma t} + Q_{R_n^2,W_n} * H_{W_n,R_n^1} * P_{R_n^2,R_n^2}(t)\\
&H_{W_n,R_n^2}(t) = Q_{W_n,R_n^2}(t) + Q_{W_n,U_n} * Q_{U_n,R_n^2}(t)\\
&\qquad + Q_{W_n,U_n} * Q_{U_n,W_n} * H_{W_n,R_n^2}(t)\\
&\qquad + Q_{W_n,U_n} * Q_{U_n,R_n^1} * Q_{R_n^1,W_n} * H_{W_n,R_n^2}(t)\\
&\qquad + Q_{W_n,R_n^1} * Q_{R_n^1,W_n} * H_{W_n,R_n^2}(t)\\
&\qquad (i = 0, 1, 2, \ldots, n-1)
\end{aligned}\right\}, \tag{17}$$

respectively.

By solving Eq. (16), we obtain $P_{W_i,R_n^1}(t)$ as:

$$\begin{aligned}
P_{W_i,R_n^1}(t) &\equiv \Pr\{X(t) = R_n^1|X(0) = W_i\}\\
&= \frac{g_{i,n+1}(t)}{a\mu_n},
\end{aligned} \tag{18}$$

where $g_{i,n}(t) \equiv \mathrm{d}G_{i,n}(t)/\mathrm{d}t$ denotes the probability density function of $S_{i,n}$. Furthermore solving Eq. (17), we obtain $P_{W_i,R_n^2}(t)$ as:

$$\begin{aligned}
P_{W_i,R_n^2}(t) &\equiv \Pr\{X(t) = R_n^2|X(0) = W_i\}\\
&= \sum_{m=i}^{n-1}\left[C_{i,n}^1(m)e^{-x_m t} + C_{i,n}^2(m)e^{-y_m t} + C_{i,n}^3(m)e^{-z_m t}\right],
\end{aligned} \tag{19}$$

where constant coefficients $C_{i,n}^1(m)$, $C_{i,n}^2(m)$, and $C_{i,n}^3(m)$ are given by:

$$\left.\begin{aligned}
&C_{i,n}^1(m) = \frac{q\lambda_n(\mu_n - x_m)\prod_{j=i}^{n-1} ap\lambda_j\mu_j(\gamma - x_m)}{\prod_{\substack{j=i\\ j\neq m}}^{n}(x_j - x_m)\prod_{j=i}^{n}(y_j - x_m)(z_j - x_m)}\\
&C_{i,n}^2(m) = \frac{q\lambda_n(\mu_n - y_m)\prod_{j=i}^{n-1} ap\lambda_j\mu_j(\gamma - y_m)}{\prod_{\substack{j=i\\ j\neq m}}^{n}(x_j - x_m)\prod_{j=i}^{n}(y_j - x_m)(z_j - x_m)}\\
&C_{i,n}^3(m) = \frac{q\lambda_n(\mu_n - z_m)\prod_{j=i}^{n-1} ap\lambda_j\mu_j(\gamma - z_m)}{\prod_{\substack{j=i\\ j\neq m}}^{n}(z_j - z_m)\prod_{j=i}^{n}(x_j - z_m)(y_j - z_m)}\\
&(m = i, i+1, \ldots, n-1)
\end{aligned}\right\}. \tag{20}$$

Let $\{Y(t), t \geq 0\}$ be the counting process representing the cumulative number of faults corrected up to the time interval $(0, t]$. Suppose that the system was in state W_i at time point $t = 0$. Then we have the conditional probability of $Y(t)$ as:

$$\Pr\{Y(t) = n - i | X(0) = W_i\} = G_{i,n}(t) - G_{i,n+1}(t) \quad (i \leq n)\,. \tag{21}$$

Furthermore, we have the following relationship:

$$\begin{aligned} &\{Y(t) = n - i | X(0) = W_i\} \\ &\quad \Leftrightarrow \{X(t) = W_n | X(0) = W_i\} \cup \left\{X(t) = R_n^1 | X(0) = W_i\right\} \\ &\quad \cup \left\{X(t) = R_n^2 | X(0) = W_i\right\} \cup \{X(t) = U_n | X(0) = W_i\}\,. \end{aligned} \tag{22}$$

Therefore, we have $P_{W_i,U_n}(t)$ as:

$$\begin{aligned} P_{W_i,U_n}(t) &\equiv \Pr\{X(t) = U_n | X(0) = W_i\} \\ &= G_{i,n}(t) - G_{i,n+1}(t) - P_{W_i,W_n}(t) - P_{W_i,R_n^1}(t) \\ &\quad - P_{W_i,R_n^2}(t)\,. \end{aligned} \tag{23}$$

3.3. *Disappointment probability*

Hereafter, we have the discussion under the condition that the system was in state W_i at time point zero. It is denoted that

$$\Pr\{X(t) \in \mathbf{W} | X(0) = W_i\} = \sum_{n=i}^{\infty} P_{W_i,W_n}(t)\,, \tag{24}$$

$$\Pr\{X(t) \in \mathbf{R} | X(0) = W_i\} = \sum_{n=i}^{\infty} \left[P_{W_i,R_n^1}(t) + P_{W_i,R_n^2}(t)\right], \tag{25}$$

$$\Pr\{X(t) \in \mathbf{U} | X(0) = W_i\} = \sum_{n=i}^{\infty} P_{W_i,U_n}(t)\,, \tag{26}$$

respectively.

When n faults have been corrected, the probability that a software failure occurs when the system is used is given by:

$$\Pr\{Z_n < Y\} = \frac{\lambda_n}{\eta + \lambda_n}. \tag{27}$$

On the other hand, the probabilities that a usage demand occurs when the system is restored with and without debugging are given by:

$$\Pr\{X < T_n\} = \frac{\theta}{\theta + \mu_n}, \tag{28}$$

$$\Pr\{X < T\} = \frac{\theta}{\theta + \gamma}, \tag{29}$$

respectively.

Let Z_t be the random variable representing the software failure-occurrence time measured from the arbitrary time point t. Then **the disappointment probability in use** is defined as the conditional probability that a software failure occurs during a usage period, provided the system is used at the time point t (see Fig. 2), and given by:

$$H_u(t, i) = \frac{\Pr\{Z_t < Y, X(t) \in \mathbf{U} | X(0) = W_i\}}{\Pr\{X(t) \in \mathbf{U} | X(0) = W_i\}}. \tag{30}$$

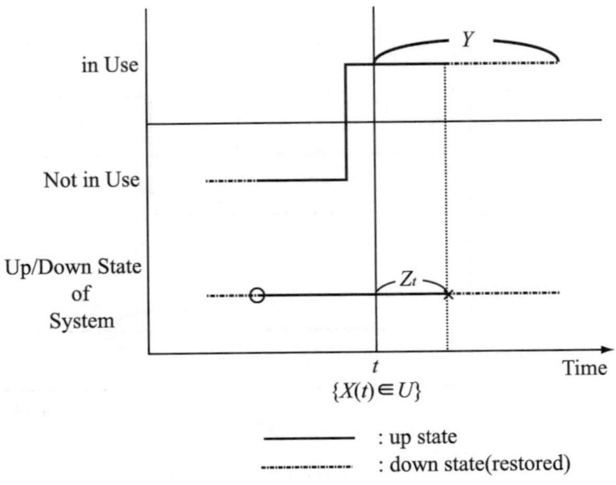

Fig. 2. An example of a system failure in use.

On the other hand, let T_t be the random variable representing the restoration time measured from the arbitrary time point t. Then **the disappointment probability due to demand rejection** is defined as the probability that a usage demand is canceled when the restoration action is performed at the time point t (see Fig. 3), and given by:

$$H_{dr}(t,i) = \sum_{n=i}^{\infty} \big[\Pr\{X < T_n\} \cdot \Pr\{X(t) = R_n^1 | X(0) = W_i\} + \Pr\{X < T\} \cdot \Pr\{X(t) = R_n^2 | X(0) = W_i\} \big]. \tag{31}$$

Furthermore, **the disappointment probability under restoration** is defined as the conditional probability that a usage demand occurs before a restoration action is complete, provided the restoration action is performed at the time point t, and given by:

$$H_r(t,i) = \frac{H_{dr}(t,i)}{\Pr\{X(t) \in \mathbf{R} | X(0) = W_i\}}. \tag{32}$$

In the above discussion, we have assumed that the system was in state W_i at time point zero. However, we should note that the cumulative number of faults corrected at the completion of the lth debugging activity, C_l, is not explicitly observed since the imperfect

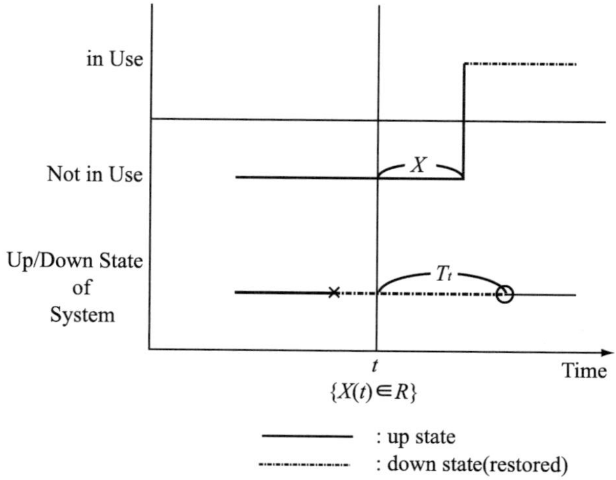

Fig. 3. An example of a system under restoration.

debugging environment is assumed throughout this chapter. However, C_l follows the binomial distribution with the following probability mass function:

$$\Pr\{C_l = i\} = \binom{l}{i} a^i b^{l-i} \quad (i = 0, 1, 2, \ldots, l), \tag{33}$$

where $\binom{l}{i} \equiv l!/[i!(l-i)!]$ denotes the binomial coefficient.

Accordingly, the disappointment probability in use after the completion of the lth debugging, $H_u(t; l)$, is given by:

$$\begin{aligned} H_u(t; l) &\equiv \sum_{i=0}^{l} \Pr\{C_l = i\} H_u(t, i) \\ &= \frac{\sum_{i=0}^{l} \binom{l}{i} a^i b^{l-i} \sum_{n=i}^{\infty} \frac{\lambda_n P_{W_i, U_n}(t)}{\eta + \lambda_n}}{\sum_{n=i}^{\infty} P_{W_i, U_n}(t)}. \end{aligned} \tag{34}$$

Similarly, the disappointment probability due to demand rejection, $H_{dr}(t; l)$, and the disappointment probability under restoration, $H_r(t; l)$, after the completion of the lth debugging are given by:

$$\begin{aligned} H_{dr}(t; l) &\equiv \sum_{i=0}^{l} \Pr\{C_l = i\} H_{dr}(t, i) \\ &= \sum_{i=0}^{l} \binom{l}{i} a^i b^{l-i} \sum_{n=i}^{\infty} \left(\frac{\theta P_{W_i, R_n^1}(t)}{\theta + \mu_n} + \frac{\theta P_{W_i, R_n^2}(t)}{\theta + \gamma} \right), \end{aligned} \tag{35}$$

$$\begin{aligned} H_r(t; l) &\equiv \sum_{i=0}^{l} \Pr\{C_l = i\} H_r(t, i) \\ &= \frac{\sum_{i=0}^{l} \binom{l}{i} a^i b^{l-i} \sum_{n=i}^{\infty} \left(\frac{\theta P_{W_i, R_n^1}(t)}{\theta + \mu_n} + \frac{\theta P_{W_i, R_n^2}(t)}{\theta + \gamma} \right)}{\sum_{n=i}^{\infty} \left[P_{W_i, R_n^1}(t) + P_{W_i, R_n^2}(t) \right]}, \end{aligned} \tag{36}$$

respectively.

4. Numerical Examples

We show several numerical examples of the software availability analysis, where we apply the model of Moranda[10] to λ_n and μ_n, i.e., $\lambda_n \equiv Dk^n$ $(D > 0, 0 < k < 1)$ and $\mu_n \equiv Er^n$ $(E > 0, 0 < r \leq 1)$, respectively.

Figure 4 shows the dependence of the disappointment probability in use, $H_u(t; l)$ in Eq. (34) on the number of the debuggings, l. This figure tells us that the disappointment probability in use decreases with time and increase in the number of debuggings, that is, the probability that the user can finish a process or a job (i.e., the user is not disappointed) before the system is down increases.

Figure 5 shows the dependence of $H_u(t; 5)$ on the parameter η associated with the usage time. This figure indicates that $H_u(t; 5)$ increases with the decreasing η. The smaller η means that the usage time tends to be longer. Accordingly, the probability that the user is disappointed is larger when the user tends to use the system for the longer period.

Figures 6 and 7 show the dependences of the disappointment probabilities due to demand rejection, $H_{dr}(t; 0)$ and $H_{dr}(t; 5)$ in Eq. (35) on the parameter p associated with the restoration scenario,

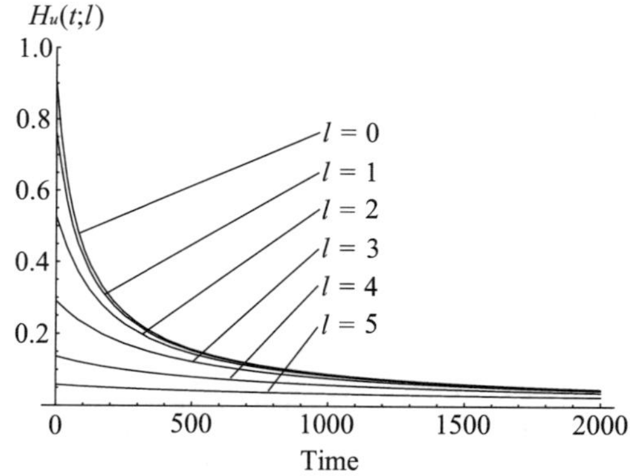

Fig. 4. Dependence of $H_u(t; l)$ on l ($a = 0.9,\ p = 0.9,\ D = 0.1,\ k = 0.3,\ E = 0.5,\ r = 0.9,\ \theta = 0.9,\ \eta = 0.01,\ \gamma = 1.0$).

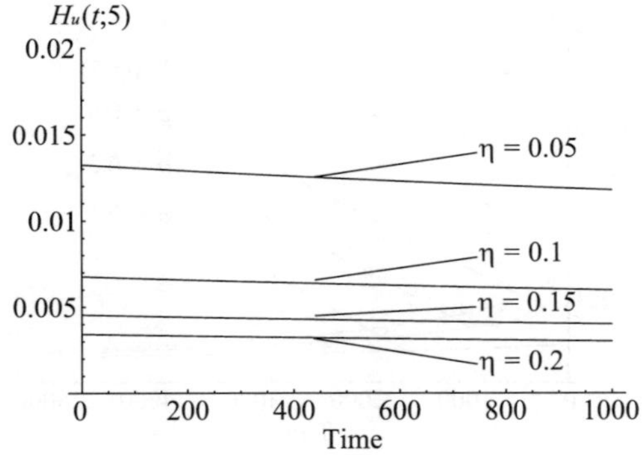

Fig. 5. Dependence of $H_u(t;5)$ on η ($a = 0.9,\ p = 0.9,\ D = 0.1,\ k = 0.3,\ E = 0.5,\ r = 0.9,\ \theta = 0.9,\ \gamma = 1.0$).

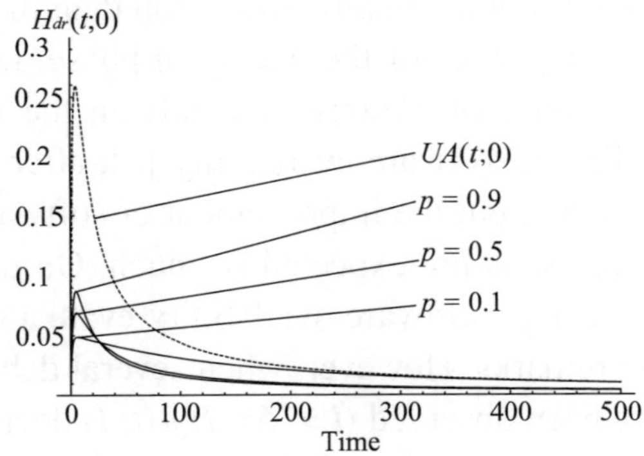

Fig. 6. Dependence of $H_{dr}(t;0)$ on p ($a = 0.9, D = 0.1,\ k = 0.3,\ E = 0.5,\ r = 0.9,\ \theta = 0.3,\ \eta = 0.01,\ \gamma = 1.0$).

respectively, where

$$UA(t;l) \equiv \sum_{i=0}^{l} \binom{l}{i} a^i b^{l-i} \sum_{n=i}^{\infty} \left[P_{W_i,R_n^1}(t) + P_{W_i,R_n^2}(t) \right]$$

is called **the software unavailability**; this represents the probability that the system is down at the time point t when the lth debugging has been complete at time point zero. These figures indicate that $H_{dr}(t;l)$ decrease with time and the increasing number of debuggings and we can see the disappointment probability due to demand

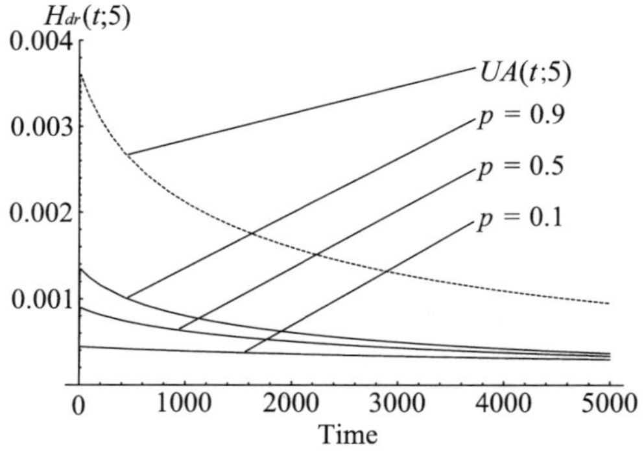

Fig. 7. Dependence of $H_{dr}(t;5)$ on p ($a = 0.9$, $D = 0.1$, $k = 0.3$, $E = 0.5$, $r = 0.9$, $\theta = 0.3$, $\eta = 0.01$, $\gamma = 1.0$).

rejection gives more optimistically evaluation than the conventional measure. In the early stage of the operation phase, i.e., in the case of $l = 0$, the behavior of $H_{dr}(t;l)$ depends on the value of p. In the case of smaller p, software availability just after the beginning of operation is higher, but the improvement of software availability in the operational use is not expected so much. On the other hand, in the case of larger p, software availability evaluation is opposite to the preceding remarks. However, when several debugging activities have already been observed ($l = 5$), $H_{dr}(t;l)$ decreases with the decreasing p. This reasoning is as follows: the increase in the number of debuggings means that the possibility that software reliability has already been improved is higher. Therefore, software availability is higher when the restoration actions without debugging are performed since these shorten the down time.

Figure 8 shows the dependence of $H_{dr}(t;5)$ on the parameter θ associated with the frequency of the usage demand. This figure indicates that $H_{dr}(t;5)$ decreases with the decreasing θ. The larger θ means that the frequency of a usage demand is higher.

Figure 9 shows the dependence of the disappointment probability under restoration, $H_r(t;l)$ in Eq. (36) on l. This figure tells us that $H_r(t;l)$ increases with time and the increasing number of debuggings, i.e., the probability that a usage demand occurs before a restoration

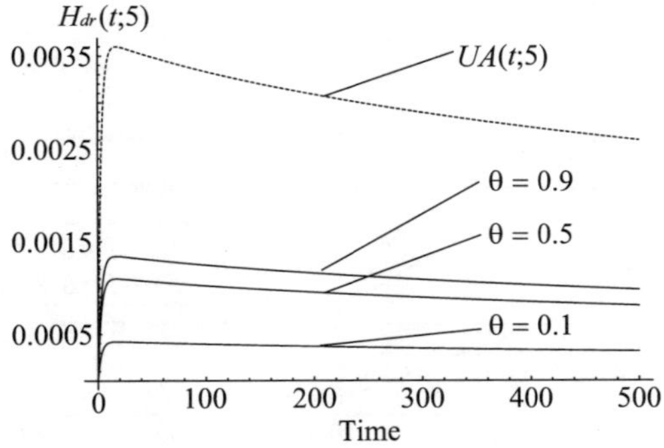

Fig. 8. Dependence of $H_{dr}(t; 5)$ on θ ($a = 0.9$, $p = 0.9$, $D = 0.1$, $k = 0.3$, $E = 0.5$, $r = 0.9$, $\eta = 0.01$, $\gamma = 1.0$).

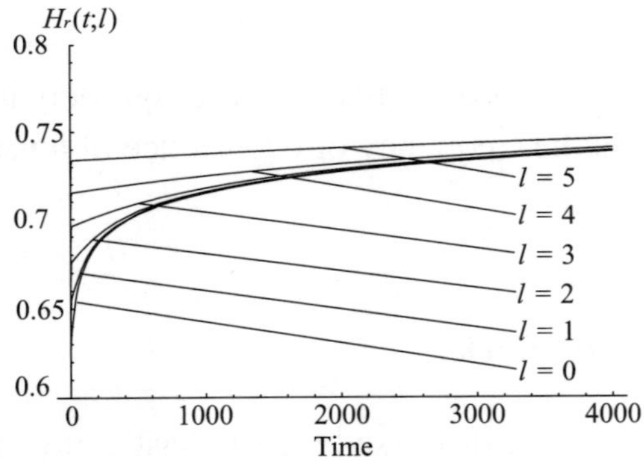

Fig. 9. Dependence of $H_r(t; l)$ on l ($a = 0.9$, $p = 0.9$, $D = 0.1$, $k = 0.3$, $E = 0.5$, $r = 0.9$, $\theta = 0.3$, $\eta = 0.01$, $\gamma = 1.0$).

action is complete increases. This result is opposite to Figs. 6, 7 and 8. The reasoning is due to the consideration of the upward tendency in difficulty of debugging, i.e., we assume that μ_n is a decreasing function of n. This assumption means that the restoration time tends to be longer with the progress of debugging. Comparing Figs. 6, 7 and 8 with Fig. 9, we can see that software availability evaluation changes, depending on whether or not we get the information that the system is under restoration.

Figure 10 shows the dependence of $H_r(t; 0)$ and $H_r(t; 5)$ on θ. As shown in this figure, we can see that $H_r(t; l)$ increases with the

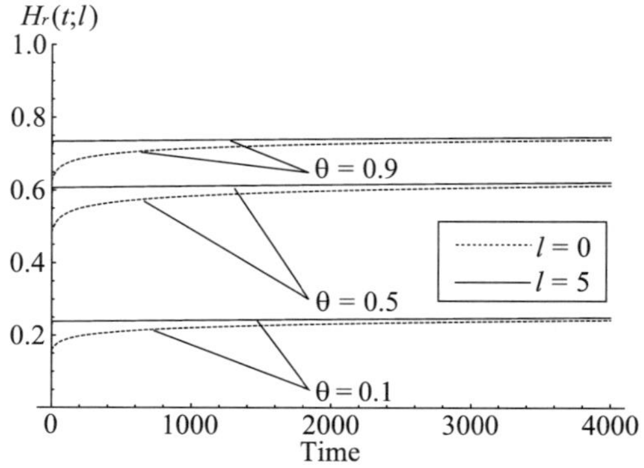

Fig. 10. Dependence of $H_r(t;5)$ on θ ($a = 0.9, D = 0.1, k = 0.3, E = 0.5, r = 0.9, \eta = 0.01, \gamma = 1.0$).

increasing θ, i.e., the probability that a usage demand is canceled increases more as the frequency of occurrence of a usage demand is larger.

5. Concluding Remarks

In this chapter, we have discussed the stochastic modeling for operational software availability measurement, considering that the system is used intermittently. From the model, the following software availability measures based on the number of debugging activities from user's viewpoint have been derived: the disappointment probabilities in use, due to demand rejection, and under restoration. Considering the software reliability growth process, the upward tendency in difficulty of debugging, and the imperfect debugging environment, we have described the time-dependent behaviors of the user and the system with a Markov process.

We have assumed that the time to usage demand and the usage time follow the exponential distributions. However, the actual users are diversified and their characteristics are more complicated. We need to reflect the actual usage characteristics to the model.

References

1. M. R. Lyu (ed.), *Handbook of Software Reliability Engineering* (IEEE Computer Society Press, Los Alamitus, CA, 1996).
2. K. Tokuno and S. Yamada, Software availability theory and its applications, *Handbook of Reliability Engineering*, ed. H. Pham (Springer-Verlag, Berlin, 2003), Chapter 13, pp. 235–244.
3. D. P. Gaver, Jr., A probability problem arising in reliability and traffic studies, *Operations Research* **12** (1964) 534–542.
4. S. Osaki, Reliability analysis of a system when it is used intemittently, *Trans. IECE* **54-C** (1971) 83–89 (in Japanese).
5. K. Tokuno and S. Yamada, Markovian modeling for software availability analysis under intermittent use, *Int. J. Reliability, Quality and Safety Engineering* **8** (2001) 249–258.
6. K. Tokuno and S. Yamada, Operational software availability measurment with two kinds of restoration actions, *J. Quality in Maintenance Engineering* **4** (1998) 273–283.
7. K. Tokuno and S. Yamada, Markovian software availability measurement based on the number of restoration actions, *IEICE Trans. Fundamentals* **E83-A** (2000) 835–841.
8. S. M. Ross, *Applied Probability Models with Optimization Applications* (Dover Publication, New York, 1992).
9. S. Yamada, Software reliability models, *Stochastic Models in Reliability and Maintenance*, ed. S. Osaki (Springer-Verlag, Berling, 2002), Chapter 10, pp. 253–280.
10. P. B. Moranda, Event-altered rate models for general reliability analysis, *IEEE Trans. Reliability* **R-28** (1979) 376–381.

CHAPTER 18

Optimal Random and Periodic Inspection Policies

Toshitsugu Sugiura, Satoshi Mizutani and Toshio Nakagawa
Department of Industrial Engineering,
Aichi Institute of Technology, 1247 Yachigusa, Yagusa-cho,
Toyota 470-0392, Japan

1. Introduction

Most systems in offices and industries are successively executing the work of jobs and the process of computers. For such systems, it would be impossible or impractical to maintain them in a strictly periodic fashion. For example, when a job has a variable working cycle and processing time, it would be better to make some maintenances after it has completed its work and process. Barlow and Proschan[1] considered the random age replacement policy and obtained analytically its reliability quantities using a renewal theory. Pinedo[2] summarized the various schedules of jobs which have random processing times.

This chapter proposes the random inspection policy in which a system is checked at the same random times as its working times. Many papers of inspection models have been already published and were surveyed in extensively Barlow and Proschan,[1] Kaio and Osaki,[3] Valdes-Flores and Feldman,[4] Hariga and Al-Fawzan,[5] and Nakagawa.[6] However, there is no paper in the literature treating with a random inspection model.

At first, we obtain the total expected cost of a system with random checking times until the detection of system failure. However, it would be necessary to check a working system at periodic times in the case where its processing time becomes large. Next, we consider the extended inspection model where a system is checked at both random times and periodic times. Then, the total expected cost is derived, and optimal inspection policies which minimize it are analytically discussed. Finally, numerical examples are given and some useful discussions about results are made. Further, the inspection model with random and successive checking times is introduced.

2. Random Inspection

Suppose that a system works for an infinite time span and is checked at successive times Y_j $(j = 1, 2, \ldots)$, where $Y_0 \equiv 0$ and Y_j $(j = 1, 2, \ldots)$ are independently, identically distributed random variables, and also, are independent of its failure time. It is assumed that each Y_j has a general distribution $G(x)$ with finite mean $1/\mu \equiv \int_0^\infty [1 - G(x)]dx < \infty$, i.e., $\{Y_j\}_{j=1}^\infty$ form a renewal process, and so that, the distribution of $Y_1 + Y_2 + \cdots + Y_j$ is represented by the jth fold convolution $G^{(j)}$ of G with itself, where $G^{(j)}(x) \equiv \int_0^x G^{(j-1)}(x - y)dG(y)$ $(j = 1, 2, \ldots)$ and $G^{(0)}(x) \equiv 1$ for $x \geq 0$.

Further, a system has a failure time distribution $F(t)$ with finite mean $1/\lambda \equiv \int_0^\infty [1 - F(t)]dt < \infty$, and its failure is detected only by some check. It is assumed that the failure rate of a system is not changed by any check, and all times needed for checks are negligible. Then, the mean time to the detection of system failure is

$$\sum_{j=0}^{\infty} \int_0^\infty \left[\int_0^t dG^{(j)}(x) \int_{t-x}^\infty (y + x)dG(y) \right] dF(t)$$
$$= \int_0^\infty dM(x) \int_0^\infty (y + x)[F(x + y) - F(x)]dG(y) , \quad (1)$$

where $M(x) \equiv \sum_{j=0}^{\infty} G^{(j)}(x)$ represents the expected number of checks in $[0, x]$. It is noted that

$$\sum_{j=0}^{\infty} \int_0^{\infty} \left[\int_0^t dG^{(j)}(x) \int_{t-x}^{\infty} dG(y) \right] dF(t)$$
$$= \int_0^{\infty} \left[\int_0^t \bar{G}(t-x) dM(x) \right] dF(t) = 1 \,.$$

where, in general, $\bar{\Phi} \equiv 1 - \Phi$.

Let us introduce the following costs: c_i is the cost for one check and c_d is the cost per unit of time for the time elapsed between a failure and its detection at the next check. Then, the total expected cost until failure detection is

$$C = \sum_{j=0}^{\infty} \int_0^{\infty} \left\{ \int_0^t dG^{(j)}(x) \int_{t-x}^{\infty} [c_i(j+1) \right.$$
$$\left. + c_d(y + x - t)] dG(y) \right\} dF(t)$$
$$= c_i \sum_{j=0}^{\infty} (j+1) \int_0^{\infty} [G^{(j)}(t) - G^{(j+1)}(t)] dF(t)$$
$$+ c_d \int_0^{\infty} dM(x) \int_0^{\infty} [F(x+y) - F(x)] \bar{G}(y) dy \,. \qquad (2)$$

We consider the following three particular cases:

(i) *Periodic inspection*
Suppose that $G(x) \equiv 0$ for $x < T$; 1 for $x \geq T$, i.e., $1/\mu = T$, and $G^{(j)}(x) \equiv 0$ for $x < jT$; 1 for $x \geq jT$ $(j = 1, 2, \ldots)$. Then, the total expected cost given in Eq. (2) can be rewritten as:

$$C(T) = (c_i + c_d T) \sum_{j=0}^{\infty} \bar{F}(jT) - \frac{c_d}{\lambda} \,. \qquad (3)$$

This corresponds to the expected cost of the standard periodic inspection policy.[6]

(ii) *Random inspection*
When $G(x) = 1 - e^{-\mu x}$, the total expected cost is, from Eq. (2),

$$C(\mu) = c_i\left(\frac{\mu}{\lambda} + 1\right) + \frac{c_d}{\mu}. \tag{4}$$

Thus, the optimal mean checking time $1/\mu^*$ which minimizes Eq. (4) is given by:

$$\frac{1}{\mu^*} = \sqrt{\frac{c_i}{\lambda c_d}}. \tag{5}$$

(iii) *Exponential failure time*
When $F(t) = 1 - e^{-\lambda t}$, the total expected cost is

$$C = \left(c_i + \frac{c_d}{\mu}\right)\frac{1}{1 - G^*(\lambda)} - \frac{c_d}{\lambda}, \tag{6}$$

where $G^*(\lambda) \equiv \int_0^\infty e^{-\lambda t} dG(t)$ which is the Laplace Stieltjes transform of G.

3. Random and Periodic Inspections

A system is checked at successive times Y_j $(j = 1, 2, \ldots)$ and also at periodic times kT $(k = 1, 2, \ldots)$ for a specified $T > 0$. The system failure is detected by either random or periodic inspection, whichever occurs first.

The probability that the failure is detected by periodic check is

$$\sum_{k=0}^{\infty} \int_{kT}^{(k+1)T} dF(t) \sum_{j=0}^{\infty} \int_0^t \bar{G}[(k+1)T - x] dG^{(j)}(x), \tag{7}$$

and the probability that it is detected by random check is

$$\sum_{k=0}^{\infty} \int_{kT}^{(k+1)T} dF(t) \sum_{j=0}^{\infty} \int_0^t \{G[(k+1)T - x] - G(t - x)\} dG^{(j)}(x). \tag{8}$$

It is evident that Eq. (7) + Eq. (8) = 1.

Let c_{i1} be the cost for periodic check and c_{i2} be the cost for random check. Then, the total expected cost until failure detection is

$$\begin{aligned}
C(T) = &\sum_{k=0}^{\infty} \int_{kT}^{(k+1)T} dF(t) \sum_{j=0}^{\infty} \{(k+1)c_{i1} + jc_{i2} \\
&+ c_d[(k+1)T - t]\} \int_0^t \bar{G}[(k+1)T - x] dG^{(j)}(x) \\
&+ \sum_{k=0}^{\infty} \int_{kT}^{(k+1)T} dF(t) \sum_{j=0}^{\infty} \int_0^t \left\{ \int_{t-x}^{(k+1)T-x} \right. \\
&\times \left. [kc_{i1} + (j+1)c_{i2} + c_d(x+y-t)] dG(y) \right\} dG^{(j)}(x) \\
= \; & c_{i1} \sum_{k=0}^{\infty} \bar{F}(kT) + c_{i2} \sum_{j=0}^{\infty} j \int_0^{\infty} [G^{(j)}(t) - G^{(j+1)}(t)] dF(t) \\
&- (c_{i1} - c_{i2}) \sum_{k=0}^{\infty} \int_{kT}^{(k+1)T} dF(t) \int_0^t \{G[(k+1)T - x] \\
&- G(t-x)\} dM(x) \\
&+ c_d \sum_{k=0}^{\infty} \int_{kT}^{(k+1)T} dF(t) \int_0^t \left[\int_{t-x}^{(k+1)T-x} \bar{G}(y) dy \right] dM(x) .
\end{aligned} \tag{9}$$

We consider the following two particular cases:

(i) *Random inspection*

If $T \to \infty$, i.e., a system is checked by random inspection, then the total expected cost is

$$\begin{aligned}
\lim_{T\to\infty} C(T) = \; & c_{i2} \sum_{j=0}^{\infty} (j+1) \int_0^{\infty} [G^{(j)}(t) - G^{(j+1)}(t)] dF(t) \\
&+ c_d \int_0^{\infty} dM(x) \int_0^{\infty} [F(x+y) - F(x)] \bar{G}(y) dy ,
\end{aligned} \tag{10}$$

which agrees with Eq. (2) when $c_{i2} = c_i$.

(ii) *Periodic and random inspections*
When $G(x) = 1 - e^{-\mu x}$, the total expected cost $C(T)$ in Eq. (9) is rewritten as:

$$C(T) = c_{i1} \sum_{k=0}^{\infty} \bar{F}(kT) + c_{i2}\frac{\mu}{\lambda} - \left(c_{i1} - c_{i2} - \frac{c_d}{\mu}\right) \times \sum_{k=0}^{\infty} \int_{kT}^{(k+1)T} \left[1 - e^{-\mu((k+1)T-t)}\right] dF(t) \,. \tag{11}$$

We find an optimal checking time T^* which minimizes $C(T)$ in Eq. (11). Differentiating $C(T)$ with respect to T and setting it equal to zero, we have

$$\frac{\sum_{k=0}^{\infty}(k+1)\int_{kT}^{(k+1)T} \mu e^{-\mu((k+1)T-t)}\, dF(t)}{\sum_{k=0}^{\infty} kf(kT)} - (1 - e^{-\mu T}) = \frac{c_{i1}}{c_{i2} + \frac{c_d}{\mu} - c_{i1}}, \tag{12}$$

for $c_{i2} + c_d/\mu > c_{i1}$, where f is a density of F. This is a necessary condition that an optimal T^* minimizes $C(T)$ in Eq. (11).

In particular, when $F(t) = 1 - e^{-\lambda t}$ for $\lambda < \mu$, the expected cost $C(T)$ in Eq. (11) becomes

$$C(T) = \frac{c_{i1}}{1 - e^{-\lambda T}} + c_{i2}\frac{\mu}{\lambda} - \left(c_{i1} - c_{i2} - \frac{c_d}{\mu}\right)\left[1 - \frac{\lambda}{\mu - \lambda}\frac{e^{-\lambda T} - e^{-\mu T}}{1 - e^{-\lambda T}}\right]. \tag{13}$$

It is evident that

$$C(0) \equiv \lim_{T\to 0} C(T) = \infty \,, \qquad C(\infty) \equiv \lim_{T\to\infty} C(T) = c_{i2}\left(\frac{\mu}{\lambda} + 1\right) + \frac{c_d}{\mu}. \tag{14}$$

Equation (12) is simplified as:

$$\frac{\mu}{\mu - \lambda}(1 - e^{-(\mu-\lambda)T}) - (1 - e^{-\mu T}) = \frac{c_{i1}}{c_{i2} + \frac{c_d}{\mu} - c_{i1}}. \tag{15}$$

It can be easily seen that the left-hand side of Eq. (15) is strictly increasing from 0 to $\lambda/(\mu - \lambda)$.

Therefore, if $\lambda/(\mu - \lambda) > c_{i1}/(c_{i2} + c_d/\mu - c_{i1})$, i.e., $c_{i2} + c_d/\mu > (\mu/\lambda)c_{i1}$, then there exists a finite and unique $T^*(0 < T^* < \infty)$ which satisfies Eq. (15), and it minimizes $C(T)$ in Eq. (13). It is noted that the physical meaning of the condition $c_{i2} + c_d/\mu > [(1/\lambda)/(1/\mu)]c_{i1}$ is that the total of the checking cost and the downtime cost of the mean interval between random checks is greater than the periodic cost for the expected number of random checks until system failure. Conversely, if $c_{i2} + c_d/\mu \leq (\mu/\lambda)c_{i1}$ then we need to make no periodic inspection at all.

Further, using the approximation of $e^{-at} \approx 1 - at + (at)^2/2$ for small $a > 0$, we have, from Eq. (15),

$$\tilde{T} = \sqrt{\frac{2}{\lambda\mu} \cdot \frac{c_{i1}}{c_{i2} + \frac{c_d}{\mu} - c_{i1}}}, \tag{16}$$

which gives the approximate time of T^*.

4. Numerical Examples

Suppose that the failure time has a Weibull distribution and the random inspection is exponential, i.e., $F(t) = 1 - e^{-\lambda t^m}$ and $G(x) = 1 - e^{-\mu x}$. Then, an optimal checking time T^* satisfies, from Eq. (12),

$$\frac{\sum_{k=0}^{\infty}(k+1)\int_{kT}^{(k+1)T} \mu e^{-\mu((k+1)T-t)}\lambda m t^{m-1} e^{-\lambda t^m}\,dt}{\sum_{k=0}^{\infty} k\lambda m(kT)^{m-1}e^{-\lambda(kT)^m}}$$

$$-(1 - e^{-\mu T}) = \frac{\frac{c_{i1}}{c_d}}{\frac{c_{i2}}{c_d} + \frac{1}{\mu} - \frac{c_{i1}}{c_d}}. \tag{17}$$

In particular, when $m = 1$, i.e., the failure time is exponential, Eq. (17) is simplified as:

$$\frac{\mu}{\mu-\lambda}(1-e^{-(\mu-\lambda)T})-(1-e^{-\mu T})=\frac{\frac{c_{i1}}{c_d}}{\frac{c_{i2}}{c_d}+\frac{1}{\mu}-\frac{c_{i1}}{c_d}}, \tag{18}$$

for $\mu > \lambda$. Further, when $1/\mu$ tends to infinity, Eq. (17) reduces to

$$\frac{\sum_{k=0}^{\infty} e^{-\lambda(kT)^m}}{\sum_{k=0}^{\infty} k\lambda m(kT)^{m-1}e^{-\lambda(kT)^m}}-T=\frac{c_{i1}}{c_d}. \tag{19}$$

Table 1 gives the optimal checking times T^* for $m = 1, 2, 3$ and $1/\mu = 1, 5, 10, 20, 50, \infty$, and approximate times $\tilde{T}$ in Eq. (16) when $1/\lambda = 100$, $c_{i1}/c_d = 2$ and $c_{i2}/c_d = 1$. This indicates that the optimal times are decreasing with parameters $1/\mu$ and m. However, if the mean time $1/\mu$ exceeds some level, they do not vary remarkably for given m. Thus, it would be useful to check a system at least at the smallest time T^* for large $1/\mu$, which is given by Eq. (19). Approximate times $\tilde{T}$ give a good approximation for large $1/\mu$ when $m = 1$.

Table 1. Optimal checking times T^* when $1/\lambda = 100$ and $c_{i1}/c_d = 2$, $c_{i2}/c_d = 1$.

$\frac{1}{\mu}$	$\tilde{T}$	T^*		
		$m = 1$	$m = 2$	$m = 3$
1	∞	∞	∞	∞
5	22.361	∞	12.264	6.187
10	21.082	∞	8.081	5.969
20	20.520	32.240	6.819	5.861
50	20.203	22.568	6.266	5.794
∞	20.000	19.355	5.954	5.748

Table 2. Values of $\hat{T} = 1/\hat{\mu}$ in Eq. (20).

	$\frac{1}{\hat{\mu}}$	
$m = 1$	$m = 2$	$m = 3$
26.889	11.712	6.687

Further, it is noticed from Table 1 that values of T^* are larger than $1/\mu$ when $1/\mu < 1/\hat{\mu}$ for some $\hat{\mu}$, and *vice versa*. Hence, there would exist numerically a unique $\hat{T}$ which satisfies $T = 1/\mu$ in Eq. (17), and it is given by a solution of the following equation:

$$\left\{\left[\left(\frac{c_{i2}}{c_d} - \frac{c_{i1}}{c_d}\right)\frac{1}{T} + 1\right] \times \frac{\sum_{k=0}^{\infty}(k+1)\int_{kT}^{(k+1)T} e^{-[(k+1)-t/T]}\lambda m t^{m-1} e^{-\lambda t^m}\,dt}{\sum_{k=0}^{\infty} k\lambda m (kT)^{m-1} e^{-\lambda (kT)^m}} - (1 - e^{-1})\right\} = \frac{c_{i1}}{c_d}. \tag{20}$$

We show values of $\hat{T} = 1/\hat{\mu}$ for $m = 1, 2, 3$ in Table 2. If the mean working time $1/\mu$ is previously estimated and is smaller than $1/\hat{\mu}$, then we may check a system at a larger interval than $1/\hat{\mu}$, and *vice versa*.

5. Conclusions

We have considered the random inspection policy and discussed the optimal checking time which minimizes the expected cost. If a working system is checked at successive times T_k $(k = 1, 2, \ldots)$ where $T_0 \equiv 0$ and at random times, the expected cost in Eq. (9) can be easily rewritten as:

$$
\begin{aligned}
&C(T_1, T_2, \ldots) \\
&= c_{i1} \sum_{k=0}^{\infty} \bar{F}(T_k) + c_{i2} \sum_{j=0}^{\infty} j \int_0^{\infty} [G^{(j)}(t) - G^{(j+1)}(t)] dF(t) \\
&\quad - (c_{i1} - c_{i2}) \sum_{k=0}^{\infty} \int_{T_k}^{T_{k+1}} dF(t) \int_0^t [G(T_{k+1} - x) \\
&\quad - G(t-x)] dM(x) + c_d \sum_{k=0}^{\infty} \int_{T_k}^{T_{k+1}} dF(t) \\
&\quad \times \int_0^t \left[\int_{t-x}^{T_{k+1}-x} \bar{G}(y) dy \right] dM(x) . \qquad (21)
\end{aligned}
$$

In particular, when $G(x) = 1 - e^{-\mu x}$,

$$
\begin{aligned}
C(T_1, T_2, \ldots) &= c_{i1} \sum_{k=0}^{\infty} \bar{F}(T_k) + c_{i2} \frac{\mu}{\lambda} - \left(c_{i1} - c_{i2} - \frac{c_d}{\mu} \right) \\
&\quad \times \sum_{k=0}^{\infty} \int_{T_k}^{T_{k+1}} [1 - e^{-\mu(T_{k+1}-t)}] dF(t) . \qquad (22)
\end{aligned}
$$

Further, we may consider the processing time as shock times and maintenance times of other systems. In these cases, a working system is checked at random times when shocks occur and other systems are replaced or are preventively maintained. Further studies of random maintenances should be made for other reliability models.

References

1. R. E. Barlow and F. Proschan, *Mathematical Theory of Reliability* (John Wiley & Sons, New York, 1965).
2. M. Pinedo, *Scheduling Theory, Algorithms, and Systems* (Prentice Hall, New Jersey, 2002).
3. N. Kaio and S. Osaki, Comparison of inspection policies, *J. of Operations Research Soc.* **40** (1989) 499–503.

4. C. Valdes-Flores and R. M. Feldman, A survey of preventive maintenance models for stochastically deteriorating single-system systems, *Naval Research Logistics Quarterly* **36** (1989) 419–446.
5. M. Hariga and M. A. Al-Fawzan, Discounted models for the single machine inspection problem, *Maintenance, Modeling and Optimization*, eds. M. Ben-Daya, S. O. Duffuaa and A. Raouf (Kluwer Academic Publisher, Massachusetts, 2000), pp. 215–243.
6. T. Nakagawa, Maintenance and optimum policy, *Reliability Engineering*, ed. Hoang Pham (Springer, 2003), pp. 367–395.

CHAPTER 19

Screening Scheme for High Performance Products

Wee-Tat Cheong* and Loon-Ching Tang†

Department of Industrial and Systems Engineering,
National University of Singapore,
Engineering Drive 2, Singapore 117576
**engp1013@nus.edu.sg*
†isetlc@nus.edu.sg

1. Introduction

For many mission-critical products, building in redundancies has become a standard practice in ensuring the product quality during the design phase. For some products, redundancies are also introduced to cater for process variation and to maintain high process yield. The concept of redundancy in reliability engineering can be found in most of the reliability engineering books such as those by Elsayed,[1] and Tobias and Trindade.[2] Built-in redundancies not only improve process yield and product reliability but also their overall performance. Thus, products having this feature with low defects per million opportunities (dpmo) quality level can be termed as *high performance products*. This is because their intended functions will not be compromised even if there exists nonconformities within each item; as long as the number of such nonconformities is below a critical threshold.

For example, a simple telecommunications component such as copper transmission line or optical cable, the occurrence of failure in transmitting signal is extremely low, as there are numerous small wires in pair or quad within the core of the cable (see Ref. 3). Minor breakages within a pair or quad would definitely not affect the effectiveness of the current or signal transmission of the cable. Consequently, these products are still conforming when the number of nonconformities within an item is below the critical threshold. Another example of high performance product is computer hard disk. The occurrence of nonconformities is sporadic and rare (see Ref. 4), and a reasonable amount of faulty bits or bad tracks can be marked resulting in usable drives. This is because although the marked ones will not be used again in data storage, the amount of "lost" capacity will be replaced by the spare bits allocated in the disk. Thus, as long as the occurrence of faulty bits is not too frequent, the performance and the total capacity of the disk drive will not be affected.

In order to ensure the number of nonconformities does not exceed the threshold, some screening tests are usually put in place to eliminate products that are out-of-specifications and/or failure-prone. The screening test can be applied at some critical stages of production, or after some stress-testings so as to ensure high quality level and field reliability of the product. The objective is to realize the economic benefits of not having "dead-on-delivery", lower warranty claims and field repairs, and the profits of repeat business from satisfied customers. Here, a decision rule for the screening test is introduced to dispose of nonconforming or potentially nonconforming products and failure-prone products. It may also be used as a process control rule for monitoring the process if the screening test can be done quickly.

In the following, we present a model for defects occurrence for high performance products. Then the reliability screening scheme and its associated decision rules are presented. A numerical example will be given as illustration.

2. A Model for Occurrence of Defects

Here, the production outputs for high performance product is modeled by two subpopulations, one major population with proportion of ω, which is defect-free, and the other population, with proportion of $1 - \omega$ is not defect-free (NDF). If k units of measurement of an item are tested, there would be k opportunities of nonconforming in the test. Such test is carried out to examine the occurrence rate of the nonconformities within a product, the probability of obtaining x nonconformities in each product is thus given by:

$$P(X = x) \begin{cases} \omega + (1 - \omega)(1 - p)^k & x = 0\,, \\ (1 - \omega)\dbinom{k}{x} p^x (1 - p)^{k-x} & x > 0\,. \end{cases} \tag{1}$$

This modified binomial distribution shown in Eq. (1) is referred to as one of the ***Zero-Modified Distributions*** and named as *binomial-with-added-zeros* distribution by Johnson *et al.*[5] The mean and variance for the model (see Ref. 5) are given by:

$$\mu = (1 - \omega)kp\,, \tag{2}$$

$$\sigma^2 = (1 - \omega)kp\{1 - p + \omega kp\}\,. \tag{3}$$

An example of such testing is the read-write error testing of the computer hard disk drives (HDD). The opportunities of nonconforming, k, for such test is interpreted as the total number of bits tested during the test. The parameter, p, is the fraction of error bits within each drive and is expected to be very small.

In the context of reliability screening, this model can be interpreted as having $(1-\omega)$ weak subpopulation which will precipitate an expected fraction of nonconformities within each product after some stress screenings. For example, if a time censored test is planned and products are screened at the end of the test. The fraction of nonconformities, p under exponential assumption is given by:

$$p = 1 - e^{-\lambda At}\,, \tag{4}$$

where λ is the average failure rate (AFR) of each defect opportunity, and A is the acceleration factor of the stress-test. Other models such as Weibull and lognormal can also be used if it is deemed more appropriate (see Ref. 2).

The planning of this type of stress-test will be dealt with in future research. Here, we focus on decision rule and the model.

3. Screening Scheme

From Eq. (1), it is clear that the two critical aspects that need to be monitored are the proportion of the NDF populations as well as the fraction of nonconformities within each item; the respective parameters are ω and p. The frequency of observing the NDF ones is normally not frequent as the overall quality of the product should always be well-maintained at a substantially high level. Thus, the proportion of the NDF population, $1 - \omega$ is expected to be small and usually ranging between 1% to 10%. With the appropriate rational subgroup size and inspection scheme, this minor population can be well-monitored using Shewhart p or np chart (the p is referred to $1 - \omega$ in this case), as discussed in most of the statistical process control books such as Montgomery[6] and Wheeler.[7] On the other hand, among the NDF ones, the fraction of nonconformities within each item, p, should be as small as possible, so that the performance of the NDF ones conforms to the requirement. This is the parameter of interest in this chapter. The screening scheme introduced here is different from the existing process monitoring schemes for high yield processes, such as the Cumulative Counts of Conforming (CCC) chart discussed by Goh and Xie,[8] which considers only conforming and nonconforming items. Here, we consider cases where the classification of nonconforming products are done based on observing the number of nonconformities/errors occurs within a product. Moreover, NDF products are generally more failure-prone especially when the number of nonconformities is approaching the threshold.

Besides for reliability screening purpose, the proposed scheme can be used for discriminating the failure-prone ones among the NDF populations.

3.1. *The decision rules*

The screening scheme presenting here is mainly focused on the fraction of nonconformities, which will affect the reliability and performance of the NDF population if the value of p is larger than expected. Suppose that at the end of the production, in order to ensure the performance of the product conforms to the requirements, the reliability screening is carried out. For illustration, the example of the read–write error testing of the HDD is used here.

After taking into the consideration of the testing cost and cycle time constraint, the number of bits used, k, in the test is normally set by the product designer. If there are read errors (nonconformities) found in the test and the number of errors (nonconformities) found exceeds a critical value x_α, the HDD fails the test and labeled as failure-prone. The rate of observing one nonconformity of the failure-prone drive is considered much higher than the specification. The critical threshold, x_α is determined by obtaining the exact probability limits, which will be discussed in the following. When there are nonconformities found in the product and the number of nonconformities are less than x_α, with a confidence level of $1 - \alpha$, the product will not be categorized as failure-prone.

Failure analysis (FA) should be carried out on each of the failure-prone product to identify the root cause of the nonconformities for continuous improvement; this will provide the start of a closed loop FA and corrective action program for all nonconformities found in the test. If no problem is found (NPF) during FA, for products with high processing cost, it is recommended that a re-test be carried out. If the NPF item passes the test, then it could resume to the production and shipped. This would reduce the wastage of scraping a conforming item. From the production point of view, the rate of NPF product

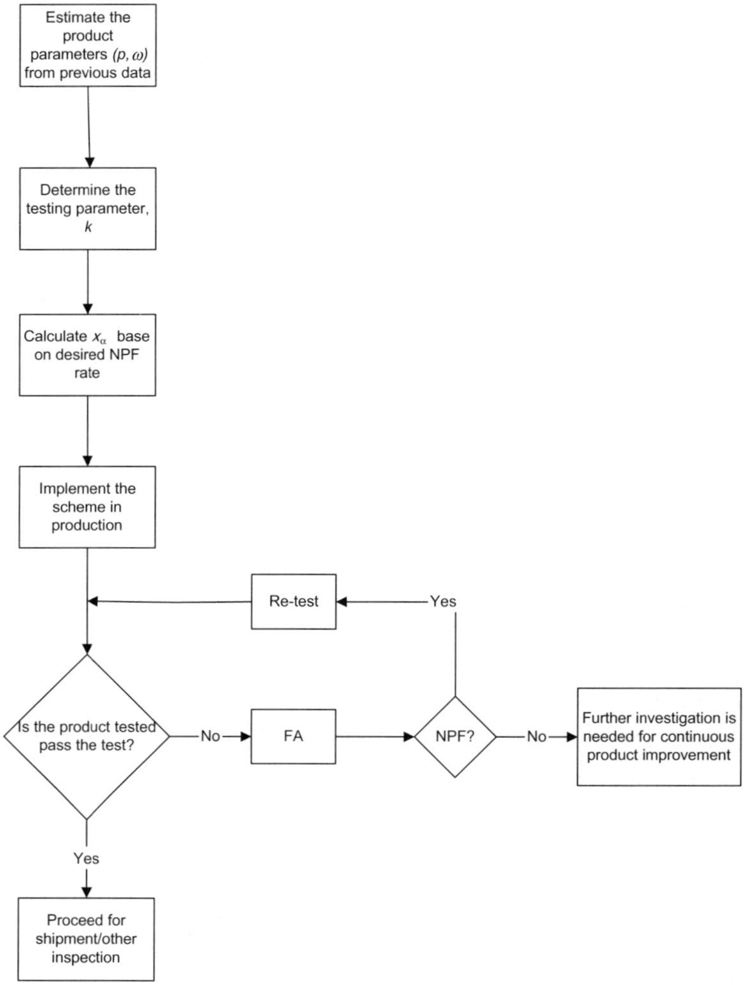

Fig. 1. The decision rules of the screening scheme.

should be as low as possible. Figure 1 presents a simple decision making procedure for the screening scheme.

3.2. *The critical value, x_α*

The critical x_α can be defined as the maximum number of nonconformities allowable during the test. If the number of nonconformities exceeds x_α, it is very likely the fraction of nonconformities of the product is higher than the specifications, i.e., the reliability of the product could not meet the requirements. After deciding the value of k used in the test, the critical value x_α can then be obtained by using the exact probability limits. Let α be the type I error for the screening test,

$$P(X \geq x_\alpha) = \alpha\,, \tag{5}$$

the critical value x_α, can thus be obtained by solving the equation as closely as possible

$$\begin{aligned} P(X \geq x_\alpha) &= 1 - P(X < x_\alpha) \\ &= 1 - \sum_{x=0}^{x_\alpha - 1} \binom{k}{x} p^x (1-p)^{k-x} \\ &= \alpha\,. \end{aligned} \tag{6}$$

The reciprocal of α is the NPF rate, which means that if $\alpha = 0.001$, there will only be one NPF product in 1000 failed products in this screening scheme, on average.

Due to discontinuity in discrete data, for a specific α value the x_α value is the largest integer value so that the exact α value is less than the desired level. Figure 2 shows some x_α values with different combinations of p and k with the desired α value closed to 0.001. From the graph, it is clear that the x_α value increases as p increases for the same α and k. Figure 3 is the x_α values with different combinations of α and k for $p = 10$ ppm.

As for the case of the HDDs, k is usually in the order of 100 millions (100^6) bits and above; and the fraction of nonconformities, p is in the order of parts-per-million (ppm) or even smaller.

4. Numerical Example

Here, we present a numerical example to illustrate the usage of the proposed screening scheme. Consider a screening test of HDD production, using the opportunities of nonconforming $k = 10^9$ and desired α is preferred to be close to 0.005, having the fraction of

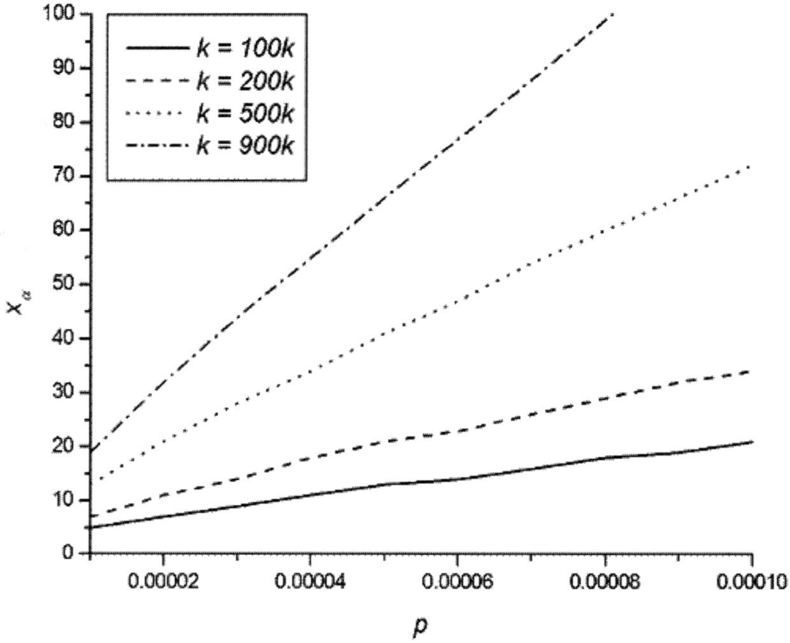

Fig. 2. x_α values for different combinations of p and k with $\alpha \approx 0.001$.

error bits within each drive is $p = 0.01$ ppm. The value of p here is very low because in the case of HDD, which is a highly reliable data storage device, the fraction of error bits found at the end of the production is very low as most of the error ones have already been picked up during the numerous online testings. The suitable critical value x_α is 19, which provides the exact α value of 0.00345, is the closest to the desired α (α for $x = 18$ is 0.00719 whereas for $x = 20$ is 0.00159).

Thus, a product fails the test if the number of nonconformities found in the test is more than 19. The NPF rate for this test is

$$\begin{aligned}\text{NPF rate} &= \frac{1}{\alpha} \\ &= \frac{1}{0.00345} \\ &\approx 290\,, \end{aligned} \tag{7}$$

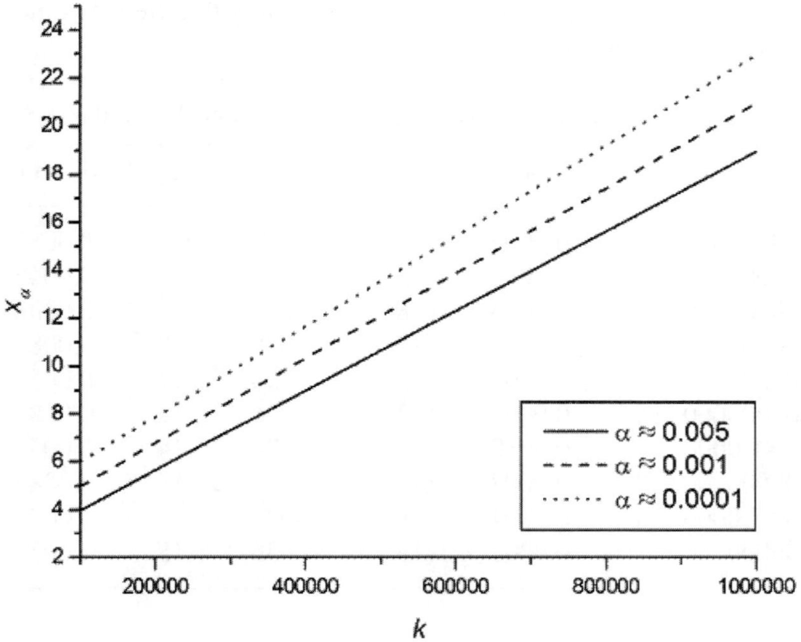

Fig. 3. x_α values for different combinations of α and k with $p = 10$ ppm.

which means that the chance of getting a NPF drive is once every 290 fail drives. Since HDD is usually produced in large volume, α is typically very small.

Table 1 shows some of the exact α values for $p = 0.01$ ppm with different values of defect opportunities, k and 3 different desired levels of α. As discussed before, due to the discontinuity behavior of the discrete data, some of the x_α values are the same for different desired α level. Figure 4 is the α curves with different values of k with $p = 0.01$ ppm. From the curves, it is clear that the α value decreases as x_α increases for the same values of p and k.

The operating-characteristic (OC) curve of the test is calculated from Eq. (5). The OC curve is plotted in Fig. 5. From the graph, it is clear that the test can detect the increase in p effectively, i.e., the probability of getting a failure-prone product increased when p increased from the intended value (0.01 ppm).

Table 1. The exact α values for $p = 0.01$ ppm with different combinations of k and desired α.

	Desired $\alpha = 0.001$		Desired $\alpha = 0.005$		Desired $\alpha = 0.01$	
k	x_α	Exact α	x_α	Exact α	x_α	Exact α
100000000	5	0.0006	4	0.0037	4	0.0037
200000000	8	0.0002	6	0.0045	6	0.0045
300000000	10	0.0003	8	0.0038	8	0.0038
400000000	11	0.0009	10	0.0028	9	0.0081
500000000	13	0.0007	12	0.0020	11	0.0055
600000000	15	0.0005	13	0.0036	12	0.0088
700000000	16	0.0010	15	0.0024	14	0.0057
800000000	18	0.0007	16	0.0037	15	0.0082
900000000	20	0.0004	18	0.0024	17	0.0053
1000000000	21	0.0007	19	0.0035	18	0.0072

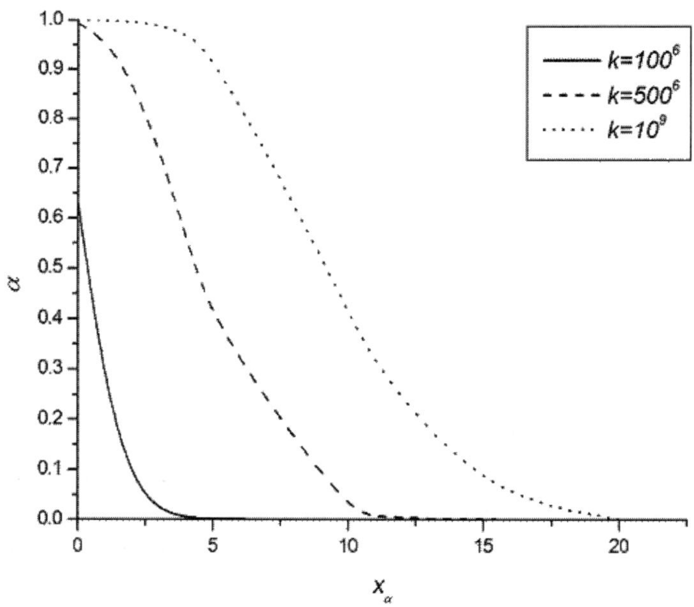

Fig. 4. α values for different values of x_α with $k = 100^6, 500^6$, and 10^9; $p = 0.01$ ppm.

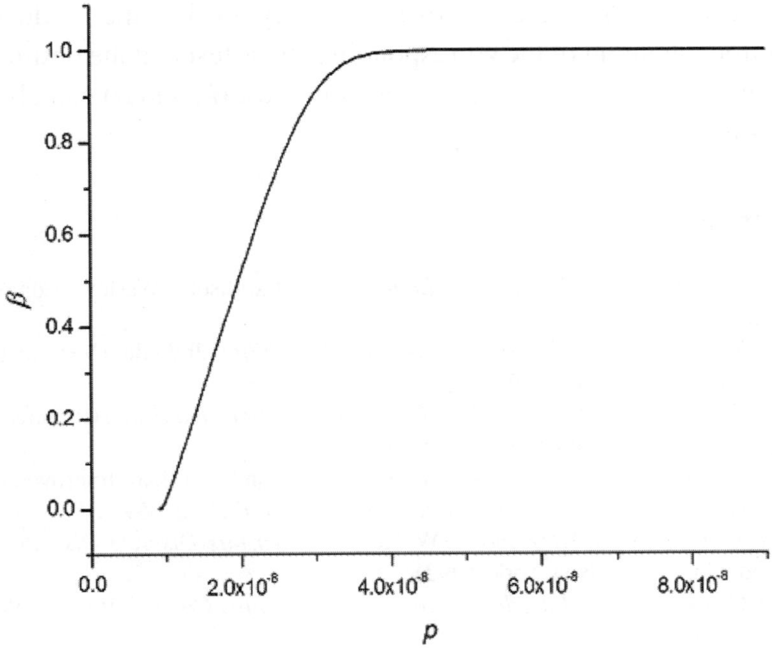

Fig. 5. The OC curve for the screening test with $p = 0.01$ ppm and desired $\alpha = 0.005$.

5. Conclusions

In this chapter, the term high performance product is coined for products with built-in redundancies and a screening scheme for these products is presented. A modified binomial distribution is used in describing the two subpopulations of the product. The scheme introduced here focuses on detecting the failure-prone ones within the minor population of nondefect-free (NDF) product. The NDF product with unacceptable failure rate can be detected effectively by implementing the scheme in the inspection procedure. A numerical example is given and it shows that the scheme is effective in detecting failure-prone items.

For future research, the frequency of observing a NDF product should be considered in the scheme, as producing too many NDF products will also affect the overall quality level of the product. In addition, planning of the corresponding stress test and the optimality of the decision variables of the screening test (k, and α) can also be investigated.

References

1. E. A. Elsayed, *Reliability Engineering* (Addison Wesley Longman, Massachusetts, 1996).
2. P. A. Tobias and D. C. Trindade, *Applied Reliability*, 2nd edn. (Van Nostrand Reinhold, New York, 1995).
3. N. Thorsen, *Fiber Optics and the Telecommunications Explosion* (Prentice Hall PTR, Upper Saddle River, NJ, 1998).
4. G. F. Hughes, J. F. Murray, K. Kreutz-Delgado and C. Elkan, Improved disk-drive failure warnings, *IEEE Trans. Reliab.* **5** (2002) 350–357.
5. N. L. Johnson, S. Kotz and A. W. Kemp, *Univariate Discrete Distributions*, 2nd edn. (Wiley, New York, 1992).
6. D. C. Montgomery, *Introduction to Statistical Quality Control*, 4th edn. (Wiley, New York, 2001).
7. D. J. Wheeler, *Advanced Topics in Statistical Process Control: The Power of Shewhart's Charts* (SPC Press, Inc., Tennessee, 1995).
8. T. N. Goh and M. Xie, Statistical control of a six sigma process, *Quality Engineering* **15** (2003) 587–592.

CHAPTER 20

Optimal Inspection Policies for a Self-Diagnosis System with Two Types of Inspections

Satoshi Mizutani and Toshio Nakagawa

Department of Industrial Engineering, Aichi Institute of Technology, 1247 Yachigusa, Yagusa-cho, Toyota 470-0392, Japan

Kodo Ito

Technology Training Center, Technical Headquarters, Mitsubishi Heavy Industries, Ltd., 1-50 Daikouminami 1-chome, Higashi-ku, Nagoya 461-0047, Japan

1. Introduction

In recent years, systems such as electronic control devices have greatly developed and become widely used. Therefore, the improvement of their reliability has become necessary and important. For instance, some failures of a system might incur great losses, and sometimes, might cause social confusion. To detect failures while a system is in service, it has to be checked periodically at suitable intervals.[1,2]

A typical example of such an inspection policy in a real system is electronic control devices which are periodically checked by the self-diagnosis program. The self-diagnosis function of systems is embedded in its electric circuits, and checks it periodically. On the other hand, the complexity of systems has dramatically increased, and as a result, it has been difficult to design the self-diagnosis program which can detect all possible failures. Moreover, the cost performance of the self-diagnosis increases as the coverage to detect failures increases.[1,3,4] Therefore, inspections should be classified into two types of high-cost inspections and low-cost self-diagnosis, where intervals of high-cost inspection would be larger than those of the self-diagnosis.

Barlow and Proschan[5] summarized the optimal inspection policies which minimize the expected cost until the detection of failure. Ito, Nakagawa and Nishi[6] considered two types of inspection policies for a system in storage.

In this chapter, we consider a system which is checked periodically by type-1 inspection or type-2 inspection: suppose that type-1 inspection checks a system more frequently than type-2 inspection, however, the cost of type-1 inspection is smaller than that of type-2 inspection. On the other hand, type-2 inspection can detect any failure, even those which cannot be detected by type-1 inspection (Fig. 1). When failures of a system are detected by some periodic inspection, it is maintained and is as good as new.

The inspection policy in reliability theory is applied to the above model[5,7,8]: type-1 inspection checks a system at periodic times jT ($j = 1, 2, \ldots$), and type-2 inspection checks a system at periodic time knT ($k = 1, 2, \ldots$). Consider the time from the beginning of

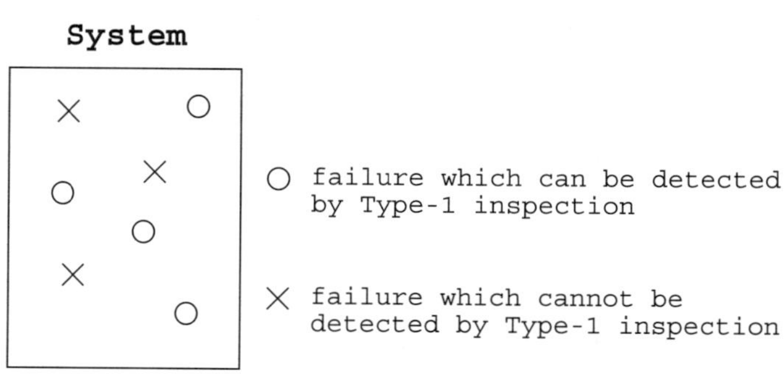

Fig. 1. System with two types of inspections.

system operation to the detection of failure as one cycle, and further, introduce a loss cost for the time elapsed between a failure and its detection. Then, the mean time of one cycle, the total expected cost for one cycle, and the expected cost per unit of time are obtained. The optimal numbers n^* which minimize the expected costs are analytically derived. Finally, numerical examples are given when the failure time distribution is exponential.

2. Expected Costs

Suppose that a system has a general failure distribution $F(t)$ $(t \geq 0)$ with finite mean $1/\lambda \equiv \int_0^\infty \bar{F}(t)dt < \infty$, where $\bar{F}(t) \equiv 1 - F(t)$. Then, a system is periodically checked by two types of inspections; type-1 inspection is performed at periodic times jT $(j = 1, 2, \ldots)$ and type-2 inspection is performed at periodic times knT $(k = 1, 2, \ldots)$ for some specified T and n $(n = 1, 2, \ldots)$, i.e., type-2 inspection is done at every n times of type-1 inspection.

When a system fails, its failure is detected in the following way: the failure can be detected by type-1 inspection with probability p $(0 < p < 1)$. On the other hand, the failure cannot be detected by type-1 inspection with probability $1 - p$ and can be always detected by type-2 inspection. If the failure is detected then a system is maintained and is as good as new. It is assumed that while the failure is detected, other failures do not occur. Further, let c_{i1} be the cost of type-1 inspection, $c_{i2} + c_{i1}$ be the cost of type-2 inspection, and c_d be the cost rate for the time elapsed between a failure and its detection.

Figure 2 shows the processes of system with two types of inspections: the horizontal axis represents the process of time. The upper side shows that when a sytem fails at time t $(knT + jT < t \leq knT + (j + 1)T)$, its failure is detected by type-1 inspection at time $knT + (j + 1)T$ with probability p, and the lower side shows that its failure is detected by type-2 inspection at time $(k + 1)nT$ with probability $1 - p$.

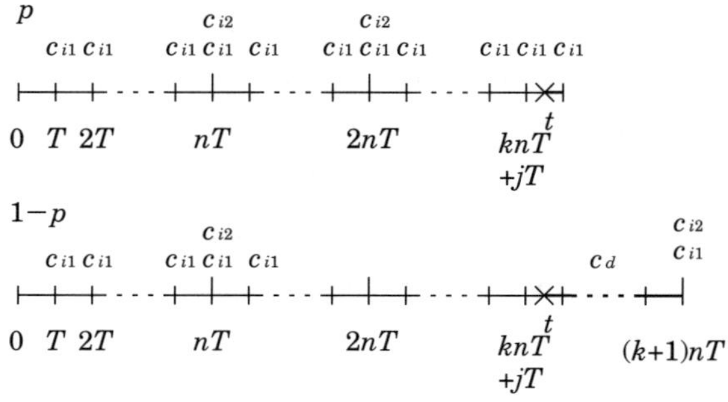

Fig. 2. Two types of inspections.

Then, the mean time of one cycle from system operation to the detection of failure is easily given from Fig. 2:

$$A(n;T) = p\sum_{k=0}^{\infty}\sum_{j=0}^{n-1}\int_{knT+jT}^{knT+(j+1)T}[knT+(j+1)T]dF(t) + (1-p)\sum_{k=0}^{\infty}\int_{knT}^{(k+1)nT}(k+1)nT\,dF(t)$$

$$= pT\sum_{k=0}^{\infty}\bar{F}(kT) + (1-p)nT\sum_{k=0}^{\infty}\bar{F}(knT) \quad (n=1,2,\dots). \tag{1}$$

Further, the total expected cost for one cycle is

$$B(n;T) = p\sum_{k=0}^{\infty}\sum_{j=0}^{n-1}\int_{knT+jT}^{knT+(j+1)T}\{c_{i1}(kn+j+1) + c_{i2}k + c_d[knT+(j+1)T-t]\}dF(t) + (1-p)\sum_{k=0}^{\infty}\int_{knT}^{(k+1)nT}\{(c_{i1}n+c_{i2})(k+1) + c_d[(k+1)nT-t]\}dF(t)$$

$$= (c_{i1} + c_d T)\left[p\sum_{k=0}^{\infty}\bar{F}(kT) + (1-p)n\sum_{k=0}^{\infty}\bar{F}(knT)\right]$$

$$+ c_{i2}\sum_{k=0}^{\infty}\bar{F}(knT) - pc_{i2} - \frac{c_d}{\lambda} \quad (n = 1, 2, \ldots). \qquad (2)$$

When $p = 1$ and $c_{i2} = 0$, this corresponds to the usual periodic inspection model.[8]

The expected cost $C(n; T)$ per unit of time is, from Eqs. (1) and (2),

$$C(n; T) \equiv \frac{B(n; T)}{A(n; T)}$$

$$= \frac{\begin{bmatrix} c_{i1}\left[p\sum_{k=0}^{\infty}\bar{F}(kT) + (1-p)n\sum_{k=0}^{\infty}\bar{F}(knT)\right] \\ + c_{i2}\left[\sum_{k=0}^{\infty}\bar{F}(knT) - p\right] - c_d/\lambda \end{bmatrix}}{\left[p\sum_{k=0}^{\infty}\bar{F}(kT) + (1-p)n\sum_{k=0}^{\infty}\bar{F}(knT)\right]T} + c_d$$

$$(n = 1, 2, \ldots). \qquad (3)$$

Assume that the failure distribution is exponential, i.e., $F(t) = 1 - e^{-\lambda t}$. Then, the total expected cost $B(n; T)$ in Eq. (2) can be rewritten as:

$$B(n; T) = (c_{i1} + c_d T)\left[\frac{p}{1 - e^{-\lambda T}} + \frac{(1-p)n}{1 - e^{-\lambda nT}}\right] + \frac{c_{i2}}{1 - e^{-\lambda nT}}$$

$$- pc_{i2} - \frac{c_d}{\lambda} \quad (n = 1, 2, \ldots), \qquad (4)$$

and the expected cost $C(n; T)$ in Eq. (3) is

$$C(n; T) = c_d + \frac{c_{i1}}{T} + \frac{c_{i2} - \left(\frac{1}{\lambda}c_d + pc_{i2}\right)(1 - e^{-\lambda nT})}{(1-p)n(1 - e^{-\lambda T}) + p(1 - e^{-\lambda nT})}$$

$$\times \left(\frac{1 - e^{-\lambda T}}{T}\right) \quad (n = 1, 2, \ldots). \qquad (5)$$

3. Optimal Policy 1

We seek an optimal number n_1^* of type-2 inspection which minimizes the total expected cost $B(n; T)$ in Eq. (4) for a fixed $T > 0$. Letting

$B(n+1;\,T) \geq B(n;\,T)$, we have

$$\sum_{k=1}^{n}(e^{\lambda kT}-1) \geq \frac{c_{i2}}{(1-p)(c_{i1}+c_d T)}\,. \tag{6}$$

It is easily seen that the left-hand side of Eq. (6) is strictly increasing in n from $e^{\lambda T}-1$ to ∞. Thus, there exists a finite and unique minimum n_1^* $(1 \leq n_1^* < \infty)$ which satisfies Eq. (6).

In particular, since $e^{\lambda kT}-1 > \lambda kT$, if there exists a minimum solution $\bar{n}_1$ to satisfy the inequality,

$$\sum_{k=1}^{n} k = \frac{n(n+1)}{2} \geq \frac{c_{i2}}{\lambda T(1-p)(c_{i1}+c_d T)}\,, \tag{7}$$

then $n_1^* \leq \bar{n}_1$. It is further noted from Eq. (6) that optimal n_1^* is decreasing in both $1-p$ and T, and $n_1^* \to \infty$ as $p \to 1$.

4. Optimal Policy 2

It is assumed that $c_d/\lambda > c_{i2}$, i.e., the downtime cost for the mean failure time is greater than the additional cost of one time of type-2 inspection. Then, we seek an optimal number n_2^* which minimizes the total expected cost $C(n;\,T)$ in Eq. (5). Letting $C(n+1;\,T) \geq C(n;\,T)$, we have

$$\frac{\sum_{k=1}^{n}(e^{\lambda kT}-1)}{n(1-p)+\frac{1}{1-e^{-\lambda T}}} \geq \frac{c_{i2}}{(1-p)\left[\frac{1}{\lambda}c_d-(1-p)c_{i2}\right]}\,. \tag{8}$$

Denoting the left-side hand of Eq. (8) by $L(n)$,

$$L(1) = \frac{e^{\lambda T}-1}{1-p+\frac{1}{1-e^{-\lambda T}}}\,,$$

$$L(\infty) = \lim_{n\to\infty} L(n) = \infty\,,$$

$$L(n+1)-L(n) = \frac{(1-p)\sum_{k=1}^{n}(e^{\lambda(n+1)T}-e^{\lambda kT})+\frac{e^{\lambda(n+1)T}-1}{1-e^{-\lambda T}}}{\left[n(1-p)+\frac{1}{1-e^{-\lambda T}}\right]\left[(n+1)(1-p)+\frac{1}{1-e^{-\lambda T}}\right]} > 0\,.$$

Thus, $L(n)$ is strictly increasing from $L(1)$ to ∞, and hence, there exists a finite and unique minimun n_2^* $(1 \leq n_2^* < \infty)$ which satisfies Eq. (8).

Since $e^{\lambda kT} - 1 > \lambda kT$, if there exists a minimum solution $\bar{n}_2$ to satisfy the inequality

$$\frac{\sum_{k=1}^{n} k}{n(1-p) + \frac{1}{\lambda T}} \geq \frac{c_{i2}}{\lambda T(1-p)\left[\frac{1}{\lambda}c_d - (1-p)c_{i2}\right]}, \tag{9}$$

then, $n_2^* \leq \bar{n}_2$. It is further noted that optimal n_2^* has no relation with c_{i1}, and is decreasing in both $1-p$ and T, and $n_2^* \to \infty$ as $p \to 1$.

5. Numerical Examples

We compute numerically optimal inspection numbers n_1^* and n_2^* which minimize the expected costs $B(n; T)$ and $C(n; T)$ when $F(t) = 1 - e^{-\lambda t}$, respectively, and compare n_1^* with $\bar{n}_1$ and n_2^* with $\bar{n}_2$. All costs are normalized to c_{i1} as a unit cost, i.e., they are divided by c_{i1}.

Table 1 presents the optimal n_1^* which minimizes $B(n; T)$ and its upper bound $\bar{n}_1$ for $1/(\lambda T) = 300, 600$, $c_d T/c_{i1} = 100, 1000$ and $c_{i2}/c_{i1} = 1, 2, 5, 10, 15, 20, 25, 30$ when $p = 0.9$. This indicates that n_1^* tends to increase as c_{i2}/c_{i1} or $1/(\lambda T)$ increases, and as $c_d T/c_{i1}$ decreases. For example, when the interval of type-1 inspection is $T = 1$ day, $1/\lambda = 300$, $c_d/c_{i1} = 100$ and $p = 0.9$, type-2 inspection should be performed almost every month for $c_{i2}/c_{i1} = 15$.

Table 2 shows the optimal n_1^* which minimizes $B(n; T)$ and $\bar{n}_1$ for $1/(\lambda T) = 300, 600$, $c_d T/c_{i1} = 100, 1000$ and $p = 0.1, 0.3, 0.5, 0.7, 0.8, 0.9, 0.95$ when $c_{i2}/c_{i1} = 10$. This indicates that the optimal n_1^* decreases with $1-p$. Thus, if $1-p$ is large, it would be better to perform type-2 inspection early.

Table 3 gives the optimal n_2^* which minimizes $C(n; T)$ and its upper bound $\bar{n}_2$ for $1/(\lambda T) = 300, 600$, $c_d T/c_{i1} = 100, 1000$ and $c_{i2}/c_{i1} = 1, 2, 5, 10, 15, 20, 25, 30$ when $p = 0.9$. This indicates that n_2^* is a little larger than n_1^* in Table 1.

Table 1. Optimal number n_1^* and its upper bound $\bar{n}_1$ for $1/(\lambda T)$, c_{i2}/c_{i1} and $c_d T/c_{i1}$ when $p = 0.9$.

	$1/(\lambda T) = 300$				$1/(\lambda T) = 600$			
	$c_d T/c_{i1}$				$c_d T/c_{i1}$			
	100		1000		100		1000	
$\frac{c_{i2}}{c_{i1}}$	n_1^*	$\bar{n}_1$	n_1^*	$\bar{n}_1$	n_1^*	$\bar{n}_1$	n_1^*	$\bar{n}_1$
1	8	8	2	2	11	11	3	3
2	11	11	3	3	15	15	5	5
5	17	17	5	5	24	24	8	8
10	24	24	8	8	34	34	11	11
15	29	30	9	9	42	42	13	13
20	34	34	11	11	48	49	15	15
30	41	42	13	13	59	60	19	19

Table 2. Optimal number n_1^* and its upper bound $\bar{n}_1$ for $1/(\lambda T)$, p and $c_d T/c_{i1}$ when $c_{i2}/c_{i1} = 10$.

	$1/(\lambda T) = 300$				$1/(\lambda T) = 600$			
	$c_d T/c_{i1}$				$c_d T/c_{i1}$			
	100		1000		100		1000	
p	n_1^*	$\bar{n}_1$	n_1^*	$\bar{n}_1$	n_1^*	$\bar{n}_1$	n_1^*	$\bar{n}_1$
0.1	8	8	3	3	11	12	4	4
0.3	9	9	3	3	13	13	4	4
0.5	11	11	3	3	15	15	5	5
0.7	14	14	4	4	20	20	6	6
0.8	17	17	5	5	24	24	8	8
0.9	24	24	8	8	34	34	11	11
0.95	34	34	11	11	48	49	15	15

Table 3. Optimal number n_2^* and its upper approximation $\bar{n}_2$ for $1/(\lambda T)$, c_{i2}/c_{i1} and c_dT/c_{i1} when $p = 0.9$.

	$1/(\lambda T) = 300$				$1/(\lambda T) = 600$			
	c_dT/c_{i1}				c_dT/c_{i1}			
	100		1000		100		1000	
$\frac{c_{i2}}{c_{i1}}$	n_2^*	$\bar{n}_2$	n_2^*	$\bar{n}_2$	n_2^*	$\bar{n}_2$	n_2^*	$\bar{n}_2$
1	8	8	2	3	11	11	3	4
2	11	11	3	4	15	16	5	5
5	17	17	5	6	24	25	8	8
10	24	24	8	8	34	35	11	11
15	30	30	9	10	42	43	13	13
20	34	35	11	11	49	49	15	16
30	42	43	13	13	59	60	19	19

Table 4 presents the optimal n_2^* which minimizes $C(n; T)$ and $\bar{n}_2$ for $1/(\lambda T) = 300, 600$, $c_dT/c_{i1} = 100, 1000$ and $p = 0.1, 0.3, 0.5, 0.7, 0.8, 0.9, 0.95$ when $c_{i2}/c_{i1} = 10$. For example, when the interval of type-1 inspection is $T = 1$ day, $1/\lambda = 600$, $c_d/c_{i1} = 100$ and $c_{i2}/c_{i1} = 10$, type-2 inspection should be performed every 34 days for $p = 0.9$ and every 16 days for $p = 0.5$. It is of interest that the upper bounds $\bar{n}_i$ $(i = 1, 2)$ give close approximations to optimal numbers in all tables.

Figure 3 shows the total expected cost $B(n; T)$ for $p = 0.8, 0.9$ when $1/(\lambda T) = 300$, $c_{i2}/c_{i1} = 10$, $c_dT/c_{i1} = 100$. For example, when $p = 0.9$, the optimal number is $n^* = 24$ and $B(n^*; T) = 589.3$. This indicates evidently that $B(n; T)$ decreases with p, that is, to decrease the expected life-cycle cost, we have to decrease the rate of failures which are detected only by type-2 inspection.

Figure 4 shows the expected cost $C(n; T)$ for $p = 0.8, 0.9$ when $1/(\lambda T) = 300$, $c_{i2}/c_{i1} = 10$, $c_dT/c_{i1} = 100$. For example, when

Table 4. Optimal number n_2^* and its upper approximation $\bar{n}_2$ for $1/(\lambda T)$, p and $c_d T/c_{i1}$ when $c_{i2}/c_{i1} = 10$.

	$1/(\lambda T) = 300$				$1/(\lambda T) = 600$			
	$c_d T/c_{i1}$				$c_d T/c_{i1}$			
	100		1000		100		1000	
p	n_2^*	$\bar{n}_2$	n_2^*	$\bar{n}_2$	n_2^*	$\bar{n}_2$	n_2^*	$\bar{n}_2$
0.1	8	8	3	3	12	12	4	4
0.3	9	9	3	3	13	13	4	4
0.5	11	11	4	4	16	16	5	5
0.7	14	14	5	5	20	20	6	6
0.8	17	17	5	6	24	25	8	8
0.9	24	25	8	8	34	35	11	11
0.95	34	35	11	11	48	49	15	16

$p = 0.9$, the optimal number is $n^* = 24$ and $C(n^*; T) = 1.95$. This also shows the same tendency as Fig. 3.

6. Conclusions

We have proposed optimal inspection policies for a system with two types of inspections. There might exist some failures in many practical systems which cannot be detected by type-1 inspection and can be done only through type-2 inspection. This assumption would be realistic, and the model is also simple. Further, it is easy to understand the results obtained and the techniques used in this chapter.

Using the inspection policy in reliability theory, we have derived the mean time and the total expected cost until the detection of failure, and the expected cost per unit of time. We have discussed analytically the optimal inspection policies which minimize the expected costs. We have given numerical examples when the failure time distribution

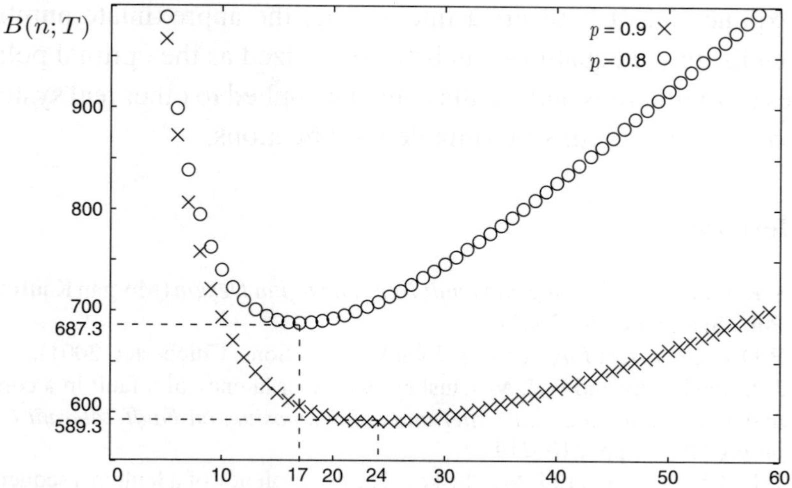

Fig. 3. Total expected cost $B(n; T)$ for $p = 0.8, 0.9$ when $1/(\lambda T) = 300$, $c_{i2}/c_{i1} = 10$ and $c_d T/c_{i1} = 100$.

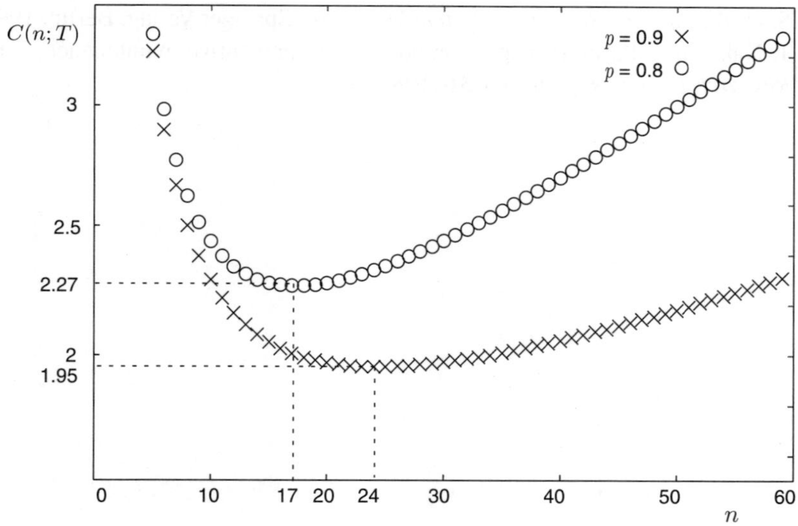

Fig. 4. Expected cost $C(n; T)$ for $p = 0.8, 0.9$ when $1/(\lambda T) = 300$, $c_{i2}/c_{i1} = 10$ and $c_d T/c_{i1} = 100$.

is exponential. It is of great interest that the approximate numbers given in simple equations can be fully utilized as the optimal policy. These formulations and results could be applied to other real systems such as digital circuits by suitable modifications.

References

1. P. K. Lala, *Self-Checking and Fault Tolerant Digital Desgin* (Morgan Kaufmann Pub., San Francisco, 2001).
2. P. O'Connor, *Test Engineering* (John Wiley & Sons, Chichester, 2001).
3. J. J. Shedletsky and E. J. McCluskey, The error latency of a fault in a combinational digital circuit, *5th International Symposium on Fault-Tolerant Computing* (1975), pp. 210–214.
4. J. J. Shedletsky and E. J. McCluskey, The error latency of a fault in a sequential digital circuit, *IEEE Trans. Computers* **C-24** (1975) 655–659.
5. R. E. Barlow and F. Proschan, *Mathematical Theory of Reliability* (John Wiley & Sons, New York, 1965).
6. K. Ito, T. Nakagawa and K. Nishi, Extended optimal inspection policies for a system in storage, *Mathematical and Computer Modeling* **22** (1995) 83–87.
7. S. Osaki, *Applied Stochastic System Modeling* (Springer Verlag, Berlin, 1992).
8. T. Nakagawa, Periodic inspection policy with preventive maintenance, *Naval Research Logistics Quarterly* **31** (1984) 33–40.

CHAPTER 21

Maintenance of a Cumulative Damage Model and Its Application to Gas Turbine Engine of Co-Generation System

Kodo Ito

Technology Training Center, Technical Headquarters,
Mitsubishi Heavy Industries, Ltd.,
1-ban-50, Daikouminami 1-chome, Higashi-ku, Nagoya 461-0047, Japan

Toshio Nakagawa

Department of Marketing and Information Systems,
Aichi Institute of Technology,
1247 Yachigusa, Yagusa-cho, Toyota 470-0392, Japan

1. Introduction

A co-generation system produces coincidentally both electric power and process heat in a single integrated system, and today, is exploited as the distributed power plant.[1] Various kinds of generators, such as steam turbine, gas turbine engine, gas engine, and diesel engine are adopted as the power sources of co-generation systems. A gas turbine engine has some attractive advantages as compared with other power sources, because its size is the smallest, its exhaust gas emission is the cleanest, and both its noise and its vibration level are the lowest

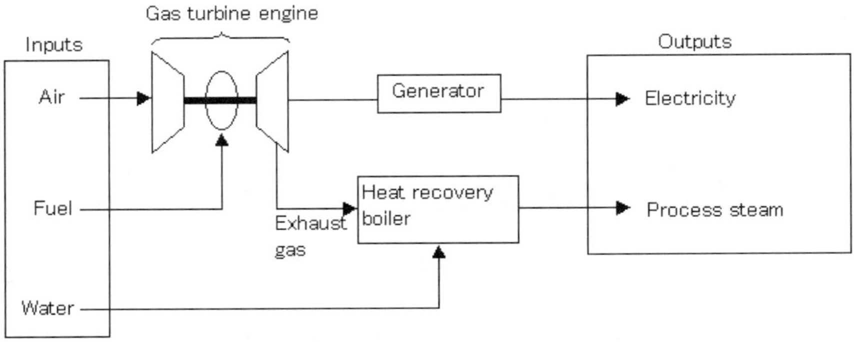

Fig. 1. Schematic diagram of gas turbine co-generation system.

in all power sources of the same power output. So, gas turbine co-generation systems are now widely utilized in factories, hospitals, and intelligent buildings to reduce costs of fuel and electricity. A schematic diagram of gas turbine engine co-generation system is shown in Fig. 1.

Maintenance is essential to uphold system availability, however, its cost may oppress customers financially. System suppliers should propose the effective maintenance plan to minimize the financial load on customers. Because the maintenance cost of gas turbine engine dominates mostly the maintenance costs of a whole system, an efficient maintenance policy should be established.

Cumulative damage models have been proposed by many authors.[2–13] In this chapter, we discuss the maintenance plan of gas turbine engine utilizing cumulative damage models. The engine is overhauled when its cumulative damage exceeds a managerial damage level. The expected cost per unit time is obtained and an optimal damage level which minimizes it is derived. Numerical examples are given to illustrate the results.

2. Model and Assumptions

Customers have to operate their co-generation system based on their respective operation plans. A gas turbine engine suffers the mechanical damage when it is turned on and operated, and it is assured to hold its required performance in a prespecified number of cumulative turning on and a certain cumulative operating period. So, the engine has to be overhauled before it exceeds the number of cumulative turning on or the cumulative operating period, whichever occurs first. When a co-generation system is continuously operated throughout the year, the occasion to perform the overhaul is restricted strictly, such as Christmas vacation period, because the overhaul needs a definite period and customers want to avoid the loss of unoperation.

We consider the following assumptions' policies:

(1) The jth turning on and operation of the engine arises an amount W_j of damage, where random variables W_j have an identical probability distribution $G(x)$ with finite mean, independent of the number of operation, where $\bar{G}(x) \equiv 1 - G(x)$. These damages are assumed to be accumulated to the current damage level, and the cumulative damage $Z_j \equiv \sum_{i=1}^{j} W_i$ up to the jth turning on and operation has

$$\Pr\{Z_j \leq x\} = G^{(j)}(x) \quad (j = 0, 1, 2, \ldots), \tag{1}$$

where $Z_0 \equiv 0$, $G^{(0)}(x) \equiv 0$ for $x < 0$ and 1 for $x \geq 0$, and in general, $\Phi^{(j)}(x)$ is the j-fold Stieltjes convolution of $\Phi(x)$ with itself.

(2) When the cumulative damage exceeds a prespecified level K at which the engine vendor prescribes, the customer of co-generation system performs the engine overhaul immediately, because the assurance of engine performance expires. A cost c_K is needed for the sum of the overhaul cost and the intermittent loss of operation.

(3) The customer performs the massive system maintenance annually, and checks all major items of the system precisely in several weeks. When the cumulative damage at such maintenance exceeds a managerial level k $(0 \leq k < K)$ at which the customer prescribes, the customer performs the engine overhaul. A cost

$c(z)$ is needed for the overhaul cost at the cumulative damage $z\,(k \leq z < K)$. It is assumed that $c(0) > 0$ and $c(K) < c_K$, because it is not required to consider the loss of operational interruption.

3. Analysis

The probability that the cumulative damage is less than k at the jth turning on and operation, and between k and K at the $(j+1)$th is

$$\int_0^k \left[\int_{k-u}^{K-u} dG(x) \right] dG^{(j)}(u)\,. \tag{2}$$

The probability that the cumulative damage is less than k at the jth turning on and operation, and more than K at the $(j+1)$th is

$$\int_0^k \bar{G}(K-u)dG^{(j)}(u)\,. \tag{3}$$

It is evident that Eq. (2)+ Eq. (3) $= G^{(j)}(k) - G^{(j+1)}(k)$.

When the cumulative damage is between k and K, the expected maintenance cost is, from Eq. (2),

$$\sum_{j=0}^{\infty} \int_0^k \left[\int_{k-u}^{K-u} c(x+u)dG(x) \right] dG^{(j)}(u)$$
$$= \int_0^k \left[\int_{k-u}^{K-u} c(x+u)dG(x) \right] dM(u)\,, \tag{4}$$

where $M(x) \equiv \sum_{j=0}^{\infty} G^{(j)}(x)$. Similarly, when the cumulative damage is more than K, the expected maintenance cost is, from Eq. (3),

$$c_K \int_0^k \bar{G}(K-u)dM(u)\,. \tag{5}$$

Next, we define a random variable X_j as the time interval from the $(j-1)$th to the jth turning on and operation, and its distribution as $\Pr\{X_j \leq t\} \equiv F(t)\,(j = 1, 2, \ldots)$ with finite mean

$1/\lambda \equiv \int_0^\infty [1 - F(t)]dt$. Then, the probability that the jth turning on and operation occurs until time t is

$$\Pr\left\{\sum_{i=1}^{j} X_i \le t\right\} = F^{(j)}(t) . \tag{6}$$

From Eq. (6), the mean time that the cumulative damage exceeds k at the jth turning on and operation, is

$$\sum_{j=1}^{\infty} \int_0^\infty t[G^{(j-1)}(k) - G^{(j)}(k)]dF^{(j)}(t) = \frac{M(k)}{\lambda} . \tag{7}$$

Therefore, the expected cost $C(k)$ per unit time is, from Ref. 14,

$$\frac{C(k)}{\lambda} = \frac{\int_0^k \left[\int_{k-u}^{K-u} c(x+u)dG(x)\right]dM(u) + c_K \int_0^k \bar{G}(K-u)dM(u)}{M(k)} , \tag{8}$$

especially, the expected costs at $k = 0$ and $k = K$ are, respectively,

$$\frac{C(0)}{\lambda} = \int_0^K c(x)dG(x) + c_K\bar{G}(K) , \tag{9}$$

$$\frac{C(K)}{\lambda} = \frac{c_K}{M(K)} . \tag{10}$$

4. Optimal Policy

We find an optimal damage level k^* which minimizes the expected cost $C(k)$ in Eq. (8). Differentiating $C(k)$ with respect to k and setting it equal to zero, we have

$$[c_K - c(K)] \int_{K-k}^{K} M(K-x)g(x)dx + \int_0^k \left[\int_k^K g(x-u)dc(x)\right] M(u)du - c(k) = 0 , \tag{11}$$

where $g(x) \equiv dG(x)/dx$ which is a density function of $G(x)$. When we denote the left-hand side of Eq. (11) as $Q(k)$, we easily have

$$Q(0) = -c(0) < 0\,, \tag{12}$$

$$Q(K) = [c_K - c(K)]M(K) - c_K\,. \tag{13}$$

Thus, if $Q(K) > 0$, i.e., $M(K) > c_K/[c_K - c(K)]$, then there exists a finite k^* $(0 < k^* < K)$ which minimizes $C(k)$, and the resulting cost is

$$\frac{C(k^*)}{\lambda} = \int_0^{K-k^*} [c(k^* + x) - c(k^*)]dG(x) + [c_K - c(k^*)]\bar{G}(K - k^*)\,. \tag{14}$$

When $c(z) = c_1 z + c_0$ $(k \le z < K)$ where $c_1 K + c_0 < c_K$, Eqs. (8) and (9) are rearranged as, respectively,

$$\frac{C(k)}{\lambda} = \frac{\int_0^k \left[\int_{k-u}^{K-u} (c_1(u+x) + c_0)dG(x)\right]dM(u) + c_K \int_0^k \bar{G}(K-u)dM(u)}{M(k)}\,, \tag{15}$$

$$\frac{C(0)}{\lambda} = \int_0^K (c_1 x + c_0)dG(x) + c_K \bar{G}(K)\,, \tag{16}$$

and $C(K)/\lambda$ is equal to Eq. (10).

Differentiating $C(k)$ in Eq. (15) with respect to k and putting it to zero, we have

$$(c_K - c_1 K - c_0)\int_{K-k}^{K} M(K-u)dG(u) - c_1 \int_{K-k}^{K} M(K-u)\bar{G}(u)du = c_0\,. \tag{17}$$

Letting denote the left-hand side of Eq. (17) by $T(k)$, we have

$$T(0) = 0\,, \tag{18}$$

$$T(K) = (c_K - c_1 K - c_0)[M(K) - 1] - c_1 K\,. \tag{19}$$

Thus, if $T(K) > c_0$, i.e., $M(K) > c_K/(c_K - c_1 K - c_0)$, then there exists a finite k^* $(0 < k^* < K)$ which minimizes $C(k)$.

Next, suppose that $G(x) = 1 - \exp(-\mu x)$, i.e., $M(x) = \mu x + 1$. Then, if $\mu K + 1 > c_K/(c_K - c_1 K - c_0)$, i.e., $\mu > (c_1 + c_0/K)/(c_K - c_1 K - c_0)$, then there exists a finite k^* $(0 < k^* < K)$. Further, differentiating $T(k)$ with respect to k, we have

$$T'(k) = (\mu k + 1)e^{-\mu(K-k)}(c_K - c_1 K - c_0) \times \left(\mu - \frac{c_1}{c_K - c_1 K - c_0}\right) > 0\,, \tag{20}$$

since $(c_1 + c_0/K)/(c_K - c_1 K - c_0) > c_1/(c_K - c_1 K - c_0)$.

Therefore, we have the following optimal policy:

(i) If $\mu K > (c_1 K + c_0)/(c_K - c_1 K - c_0)$ then there exists a finite and unique k^* $(0 < k^* < K)$ which satisfies

$$ke^{-\mu(K-k)} = \frac{c_0}{\mu(c_K - c_1 K - c_0) - c_1}\,, \tag{21}$$

and the resulting cost is

$$\frac{C(k^*)}{\lambda} = \frac{c_1}{\mu}(1 - e^{-\mu(K-k^*)}) + (c_K - c_1 K - c_0)e^{-\mu(K-k^*)}\,. \tag{22}$$

(ii) If $\mu K \le (c_1 K + c_0)/(c_K - c_1 K - c_0)$ then $k^* = K$ and $C(K)/\lambda = c_K/(\mu K + 1)$.

5. Numerical Illustration

Suppose that $G(x) = 1 - \exp(-\mu x)$ and $c(z) = c_1 z + c_0$ $(k \le z < K)$. Then, the expected cost is, from Eq. (15),

$$\frac{C(k)}{\lambda} = \frac{\frac{c_1}{\mu}[1 - e^{-\mu(K-k)}] + c_1 k + c_0 + (c_K - c_1 K - c_0)e^{-\mu(K-k)}}{\mu k + 1}\,, \tag{23}$$

and the optimal policy is given in (i) and (ii).

Table 1. Optimal managerial level k^* and expected cost $C(k^*)/\lambda$.

c_1	c_0	c_K	μ	K	k^*	$C(k^*)/\lambda$	$C(K)/\lambda$
1	1	200	1	50	41.3	1.02	3.92
0.1	1	200	1	50	41.0	0.12	3.92
1	10	200	1	50	43.6	1.23	3.92
1	1	1000	1	50	39.5	1.03	19.61
1	1	200	0.5	50	34.3	2.06	7.69
1	1	200	1	25	17.0	1.06	7.69

Table 1 gives the optimal managerial level k^* and its minimum cost $C(k^*)/\lambda$ when $c_1 = 0.1, 1$, $c_0 = 1, 10$, $c_K = 200, 1000$, $\mu = 0.5, 1$, and $K = 25, 50$. $C(k^*)$s are smaller than $C(K)$s and $C(k^*)/C(K)$ changes from 0.05 to 0.31 in this case. It is natural that k^* decreases when c_1, c_0, and $1/c_K$ decrease. The reduction of c_1 and c_0 ought to be equal to the increase of c_K. So, it is of interest in this illustration that $C(k^*)/\lambda$ decreases when c_1 and c_0 decrease, and $C(k^*)/\lambda$ slightly increases when c_K gains. It is obvious that k^* decreases and $C(k^*)/\lambda$ increases when K decreases. In this illustration, k^* decreases and $C(k^*)/\lambda$ increases when μ decreases.

The maintenance plan is settled at the beginning of co-generation system operation and the optimal managerial level k^* is calculated. The system is continuously operated and the cumulative damage is monitored. The system maintenance is performed annually and the customer decides whether the overhaul of gas turbine engine should be performed or not by comparing the monitored cumulative damage and k^*.

6. Conclusions

We have considered the optimal maintenance policy for gas turbine engine of a co-generation system. When the cumulative damage of gas turbine engine, which is caused by every turning on and operation, exceeds a managerial level k, the engine is overhauled. The expected cost per unit time has been derived and an optimal policy which minimizes it has been analytically discussed, employing the cumulative damage model. We have exhibited that a finite and unique k^* exists when the cumulative damage has an exponential distribution. Finally, a numerical illustration has been given, and characteristics of k^* and the minimum expected cost have been revealed.

We have discussed the optimal policy in only the case of $c(z) = c_1 z + c_0$ for $k \leq z < K$. We could consider easily several cost structures according to those of actual systems. For example, when the maintenance cost increases discretely with every step of the amount of cumulative damage, i.e., $c(z) = c_j$ for $k_{j-1} \leq z < k_j$ $(j = 1, 2, \ldots, n)$ and c_K for $z \geq K$ where $k_0 \equiv k$, $k_n \equiv K$, $c_{n+1} \equiv c_K$ and $c_j < c_{j+1}$ $(j = 1, 2, \ldots, n)$, the expected cost is, from Eq. (8),

$$\frac{C(k)}{\lambda} = \frac{c_1 + \sum_{j=1}^{n}(c_{j+1} - c_j)\int_0^k \bar{G}(k_j - u)dM(u)}{M(k)}. \tag{24}$$

In particular, when $n = 1$,

$$\frac{C(k)}{\lambda} = \frac{c_1 + (c_K - c_1)\int_0^k \bar{G}(K - u)dM(u)}{M(k)}. \tag{25}$$

References

1. L. C. Witte, P. S. Schmidt and D. R. Brown, *Industrial Energy Management and Utilization* (Hemisphere Publishing Corporation, New York, 1988).
2. P. J. Boland and F. Proschan, Optimal replacement of a system subject to shocks, *Operations Research* **31** (1983) 697–704.
3. D. R. Cox, *Renewal Theory* (Methuen, London, 1962).
4. J. D. Esary, A. W. Marshall and F. Proschan, Shock models and wear processes, *The Annals of Probability* **1** (1973) 627–649.
5. R. M. Feldman, Optimal replacement with semi-Markov shock models, *Journal of Applied Probability* **13** (1976) 108–117.
6. R. M. Feldman, Optimal replacement with semi-Markov shock models using discounted costs, *Math. Operations Research* **2** (1977) 78–90.
7. M. S. A. Hameed and F. Proschan, Nonstationary shock models, *Stochastic Processes and Their Applications* **1** (1973) 383–404.

8. M. S. A. Hameed and I. N. Shimi, Optimal replacement of damaged devices, *Journal of Applied Probability* **15** (1978) 153–161.
9. M. J. M. Posner and D. Zuckerman, A replacement model for an additive damage model with restoration, *Operations Research Letters* **3** (1984) 141–148.
10. P. S. Puri and H. Singh, Optimum replacement of a system subject to shocks: A mathematical lemma, *Operations Research* **34** (1986) 782–789.
11. H. M. Taylor, Optimal replacement under additive damage and other failure models, *Naval Research Logistic Quart.* **22** (1975) 1–18.
12. D. Zuckerman, Replacement models under additive damage, *Naval Research Logistic Quart.* **24** (1977) 549–558.
13. D. Zuckerman, A note on the optimal replacement time of damaged devices, *Naval Research Logistic Quart.* **27** (1980) 521–524.
14. S. M. Ross, *Stochastic Processes* (John Wiley & Sons, New York, 1983).

CHAPTER 22

An Inspection-Maintenance Model for Degraded Repairable Systems

Wenjian Li and Hoang Pham

Department of Industrial and Systems Engineering, Rutgers University, Piscataway, NJ 08854, USA

1. Introduction

Maintenance has evolved from simple model that deals with machinery breakdowns, to time-based preventive maintenance (PM), to today's condition based maintenance (CBM). It is of importance to avoid the failure of a system during its actual operating. CBM has the potential to greatly reduce costs by avoiding the occurrence of failures. In this research, we adopt a CBM strategy to develop an inspection-maintenance model for periodically inspected degraded systems subject to a continuous and increasing degradation measured by a process $(Y(t))_{t \geq 0}$ and random shocks measured by a compound Poisson process $(D(t))_{t \geq 0}$. The condition of the system at time t is described by $(Y(t), D(t))_{t \geq 0}$.

Lam[1] considered the geometric process replacement model when the repair times are increasing stochastic sequence. Sheu[2] studied a generalized replacement model where a deteriorating system has two types of failures. The repair time is not negligible and time-to-repair sequence is an increasing randomly geometric process. Grall *et al.*[4]

studied the inspection-maintenance strategy for a single unit deteriorating system. The joint influence of the preventive maintenance threshold and inspection dates function based on the average long-run cost rate was developed. Inspection dates are determined based on an inspection scheduling function.

Pham *et al.*[3] presented a model for predicting the reliability of k-out-*of*-n systems assuming that components are subject to several stages of degradation as well as catastrophic failures. Markov approach is used to obtain state probability. Klutke and Yang[5] investigated the availability of a degraded system subject to a graceful degradation and random shocks. Li and Pham[6] recently developed a reliability model for multi-state degraded systems subject to multiple competing processes. However, they did not consider the time to repair and failures can only detect by the inspection.

It is assumed that the state of the system is found only through an inspection. The failure is self-announcing which is detected immediately without inspection. The purpose of the inspection is to identify degradation of systems during their service and to provide an early warning so that remedial action can be taken before failure occurs. Although continuous monitoring and inspection is possible, in this research, discrete inspection is taken due to the cost and other practical constrains.

Two maintenance actions are considered in this paper: PM and corrective maintenance (CM). The system state is divided into three regions: maintenance-free, PM and CM zone. When the system state falls into PM zone, a PM action is taken. When the system fails, a CM action is performed. An advantage of PM action is that it can be planned and hence the total system cost might be cheaper. Requiring a random time to perform either PM or CM action is assumed in this chapter.

Markov process,[7–9] semi-Markov process,[10] and the stationary degradation process[4] have been commonly used to develop models for the systems subject to degradation. In this paper, the two continuous and increasing degradation functions are considered as follows:

(1) $Y(t) = A + Bg(t)$ where $A > 0$ and $B > 0$ are independent and $g(t)$ is an increasing function of time t.
(2) The function $Y(t) = We^{Bt}/(A + e^{Bt})$ is called the randomized logistic degradation path function, where A and B are independent non-negative random variables, and W is a constant. The random variable A represents an initial threshold degradation level and B describes the rate at which degradation accumulates.

For the shock process, it is modeled by $D(t) = \sum_{i=0}^{N(t)} X_i$ where X_i's are independent and identically distributed (i.i.d.) with $X_i > 0$ and $N(t)$ is a random variable that follows a Poisson process.

It is assumed that the state of the system is only revealed upon an inspection (except for the failure). The inspection is scheduled as $\{I, 2I, \ldots, nI, \ldots\}$. In this chapter, we develop a condition-based maintenance model for determining the optimal inspection time I and PM threshold L that minimizes the average system cost.

2. Assumptions and Model Description

2.1. *Notation*

Cc:	Cost per CM action.
Cp:	Cost per PM action.
C_i:	Cost per inspection.
$Y(t)$:	Degradation process.
$D(t)$:	Cumulative shock damage value up to time t.
G:	Threshold value for degradation process.
S:	Threshold value for shock damage.
L:	PM threshold value for degradation process.
$C(t)$:	Cumulative maintenance cost up to time t.
$E[C_1]$:	Average total maintenance cost during a cycle.
$E[W_1]$:	Mean renewal cycle length.
$E[N_I]$:	Expected number of inspections during a renewal cycle.

I: Inspection time-interval.
R_1: Random time to perform a PM action.
R_2: Random time to perform a CM action.
T: Time to failure.
P_i: Probability that there are total i inspections in a renewal cycle.
P_p: Probability that a renewal cycle ends by a PM action.
P_c: Probability that a renewal cycle ends by a CM action.
$EC(L, I)$: Expected long-run cost rate function.

2.2. *Assumptions*

The system starts at a new condition. The assumptions are as follows:

(1) The system is not continuously monitored, its state can be detected only by inspection. But the system failure is self-announcing without inspection.
(2) After a PM or CM action, the system will store it back to as-good-as-new state.
(3) A CM action is more costly than a PM and a PM costs much more than an inspection. That implies $C_c > C_p > C_i$.
(4) $Y(t)$ and $D(t)$ are independent.
(5) Repair time is not negligible.

Although continuous monitoring process to some systems is feasible, however the cost to monitor the process and the labor extensive would not make it realistic in practices. Therefore, it will make sense to consider criteria that will improve the system performance by performing periodic inspections with a maintenance action and minimizing the average total system maintenance cost. Since the system due to deterioration while running that leads to system failure, it proves to be better to assume that, as in this paper, the degradation paths are continuous and increasing functions. Consequently, the degradation processes, methods and the criteria studied in our paper differ from those of Refs. 3, 5 and 8–10.

2.3. *Inspection-maintenance policy*

In this research, the system is proposed to be periodically inspected at times $\{I, 2I, \ldots, nI, \ldots\}$. We assume that the degradation $\{Y(t)\}_{t\geq 0}$ and random shock $\{D(t)\}_{t\geq 0}$ are independent. Let T denote the time-to-failure and is defined as $T = \inf\{t > 0 : Y(t) > G \text{ or } D(t) > S\}$, where G is the critical value for $\{Y(t)\}_{t\geq 0}$ and S is the threshold level for $\{D(t)\}_{t\geq 0}$. L and G (G is fixed) effectively divides the system state into three regions as illustrated in Fig. 1, which are doing nothing zone, PM zone and CM zone.

Maintenance action will be performed when either of the following situations occurs:

(1) The current inspection reveals that the system condition falls into PM zone, however this state is not found at previous inspection. At the inspection time iI, the system falls into PM zone which

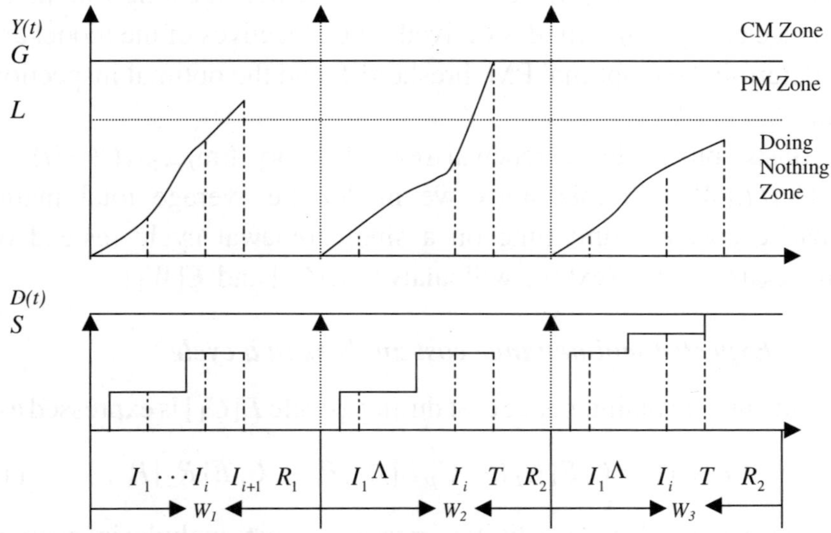

Fig. 1. The evolution of the system.

means $\{Y((i-1)I) \leq L, D((i-1)I) \leq S\} \cap \{L < Y(iI) \leq G, D(iI) \leq S\}$. Then PM action is performed and it takes a random time R_1.

(2) When the system fails at T, a CM action is taken immediately and takes time R_2.

It is assumed that both PM and CM actions are considered to be perfect. Even though both PM and CM actions bring the system back to as good as new state, physically, they are not necessarily same, since a CM has to be performed on a worse system. Hence, CM is likely to be more complex and expensive. Therefore, it is realistic to assume that the repair time is not negligible in which we consider in this paper that the PM action will take R_1 random amount of times and CM action will take R_2 random time. After a PM or a CM action is performed, the system is renewed. A new sequence of inspection begins which is defined in the same way.

3. Maintenance Cost Modeling

In this section, an explicit expression for the average long-run maintenance cost per unit time is derived. The objectives of the model are to determine the optimal PM threshold L and the optimal inspection time I.

Based on the basic renewal reward theory $\lim_{t\to\infty}(C(t)/t) = E[C_1]/E[W_1]$; in this study we model the average total maintenance cost per unit time on a single renewal cycle instead of $\lim_{t\to\infty}(C(t)/t)$. Next we will analyze $E[C_1]$ and $E[W_1]$.

3.1. *Expected maintenance cost analysis in a cycle*

The mean total maintenance cost during a cycle $E[C_1]$ is expressed as:

$$E[C_1] = C_i E[N_I] + C_p E[R_1] P_p + C_c E[R_2] P_c . \qquad (1)$$

During a renewal cycle, activities in terms of costs include: inspection cost, time-to-repair, PM or CM action. Renewal cycle will end by either a PM or a CM action. With a probability of P_p, the cycle will end by a PM action and it will take on the average $E[R_1]$ amount of times to complete a PM action with a corresponding cost $C_p E[R_1] P_p$.

Similarly, if a cycle ends by a CM action with probability P_c, it will take on the average $E[R_2]$ amount of times to complete a CM action with corresponding cost $C_c E[R_2] P_c$.

In the following, we will perform the analytical analysis of $E[C_1]$.

(1) **Calculate** $E[N_I]$. Let $E[N_I]$ denote the expected number of inspections during a cycle. $E[N_I]$ can be obtained as:

$$E[N_I] = \sum_{i=1}^{\infty} (i) P\{N_I = i\}\,. \tag{2}$$

It is obvious that $\sum_{i=1}^{\infty} P\{N_I = i\} = 1$. There will be a total of i inspections during a cycle if the first time to trigger a PM within the time interval $((i-1)I, iI]$ or the system condition is in the doing nothing zone before the time iI while the system fails during the interval $(iI, (i+1)I]$. In other words, the inspection will stop when the ith inspection finds that a PM condition satisfied while this situation is not revealed in the previous inspection or the system fails during the interval $iI < T \le (i+1)I$ while the system is in the doing nothing zone before iI.

Let $P\{N_I = i\}$ denote the probability that there are a total of i inspections occurred in a renewal cycle. Then

$$\begin{aligned} P\{N_I = i\} = {} & P\{Y((i-1)I) \le L, D((i-1)I) \le S\} \\ & \times P\{L < Y(iI) \le G, D(iI) \le S\} \\ & + P\{Y(iI) \le L, D(iI) \le S\} P\{iI < T \le (i+1)I\}\,, \end{aligned} \tag{3}$$

$$\begin{aligned} E[N_I] = \sum_{i=1}^{\infty} & i\{P\{Y((i-1)I) \le L, D((i-1)I) \le S\} \\ & \times P\{L < Y(iI) \le G, D(iI) \le S\} \\ & + P\{Y(iI) \le L, D(iI) \le S\} P\{iI < T \le (i+1)I\}\}\,. \end{aligned} \tag{4}$$

We now discuss in details Eq. (3). First, we calculate the term $P\{Y((i-1)I) \le L, D((i-1)I) \le S\}$ and $P\{L < Y(iI) \le G, D(iI) \le S\}$ with the following two different expressions for $Y(t)$.

(A) Assume $Y(t) = A + Bg(t)$, where $A \sim N\big(\mu_A, \sigma_A^2\big)$, $B \sim N\big(\mu_B, \sigma_B^2\big)$, and A and B are independent. Given $g(t) = t$. $D(t) = \sum_{i=0}^{N(t)} X_i$ where X_i's are i.i.d. and $N(t) \sim \mathit{Possion}(\lambda)$. Then

$$\begin{aligned} &P\{Y((i-1)I) \le L, D((i-1)I) \le S\} \\ &\quad = P\{A + B(i-1)I \le L\} P\left\{D((i-1)I) = \sum_{i=0}^{N((i-1)I)} X_i \le S\right\} \\ &\quad = \Phi\left(\frac{L - (\mu_A + \mu_B(i-1)I)}{\sqrt{\sigma_A^2 + \sigma_B^2((i-1)I)^2}}\right) e^{-\lambda(i-1)I} \\ &\qquad \times \sum_{j=0}^{\infty} \frac{(\lambda(i-1)I)^j}{j!} F_X^{(j)}(S)\,. \end{aligned} \tag{5}$$

$$\begin{aligned} &P\{L < Y(iI) \le G, D(iI) \le S\} \\ &= \left\{\Phi\left(\frac{G - (\mu_A + \mu_B iI)}{\sqrt{\sigma_A^2 + \sigma_B^2 (iI)^2}}\right) - \Phi\left(\frac{L - (\mu_A + \mu_B iI)}{\sqrt{\sigma_A^2 + \sigma_B^2 (iI)^2}}\right)\right\} e^{-\lambda iI} \\ &\quad \times \sum_{j=0}^{\infty} \frac{(\lambda iI)^j}{j!} F_X^{(j)}(S)\,. \end{aligned} \tag{6}$$

(B) Assume $Y(t) = W(e^{Bt}/(A + e^{Bt}))$, where W is a constant, $A \sim U[0, a]$, $a > 0$; $B \sim \exp(\beta)$, $\beta > 0$, A and B are independent. $D(t) = \sum_{i=0}^{N(t)} X_i$ where X_i's are i.i.d. and $N(t) \sim \mathit{Possion}(\lambda)$. Then

$$\begin{aligned} &P\{Y((i-1)I) \le L, D((i-1)I) \le S\} \\ &= \left\{1 - \frac{1}{a}\left(\frac{1-u_1}{u_1}\right)^{\beta/I_{i-1}} \left(\frac{(i-1)I}{(i-1)I - \beta}\right)\right. \\ &\quad \left. \times \left(a^{1-(\beta/(i-1)I_1)} - 1\right)\right\} e^{-\lambda(i-1)I} \sum_{j=0}^{\infty} \frac{[\lambda(i-1)I]^j}{j!} F_X^{(j)}(S)\,, \end{aligned} \tag{7}$$

where $u_1 = L/W$. Similarly,

$$\begin{aligned}
&P\{L < Y(iI) \le G, D(iI) \le S\} \\
&= \left\{\frac{1}{a}\left(\frac{iI}{iI-\beta}\right)\left(a^{1-(\beta/iI)}\right)\left[\left(\frac{1-u_3}{u_3}\right)^{\beta/iI}\right.\right. \\
&\quad \left.\left. -\left(\frac{1-u_2}{u_2}\right)^{\beta/iI}\right]\right\} e^{-\lambda iI}\sum_{j=0}^{\infty}\frac{(\lambda iI)^j}{j!}F_X^{(j)}(S)\,, \qquad (8)
\end{aligned}$$

where $u_2 = G/W$, $u_3 = L/W$.

(2) Next, we discuss the calculation of $P\{iI < T \le (i+1)I\}$ in Eq. (3). The definition of T is $T = \inf\{t > 0 : Y(t) > G \text{ or } D(t) > S\}$. According to the definition, we derive the expression in the following:

$$\begin{aligned}
&P\{iI < T \le (i+1)I\} \\
&\quad = P\{Y(iI) \le L, Y((i+1)I) > G\}P\{D((i+1)I) \le S\} \\
&\qquad + P\{Y((i+1)I) \le L\}P\{D(iI) \le S, D((i+1)I) > S\}\,. \qquad (9)
\end{aligned}$$

In Eq. (9), since $Y(iI)$, $Y((i+1)I)$ are not independent, we could obtain the joint p.d.f $f_{Y(iI),Y((i+1)I)}(y_1, y_2)$ in order to compute $P\{Y(iI) \le L, Y((i+1)I) > G\}$. We consider two different expressions for $Y(t)$ as follows:

(A) Assume $Y(t) = A + Bg(t)$, where $A > 0$ and $B > 0$ are independent random variables, and $g(t)$ is an increasing function of time t. Assume that $A \sim f_A(a)$, $B \sim f_B(b)$.

$$\text{Let}\begin{cases} y_1 = a + bg(iI)\,, \\ y_2 = a + bg((i+1)I)\,. \end{cases}$$

After simultaneously solving the above equations in terms of y_1 and y_2, we obtain:

$$a = \frac{y_1 g((i+1)I) - y_2 g(iI)}{g((i+1)I) - g(iI)} = h_1(y_1, y_2)\,,$$

$$b = \frac{y_2 - y_1}{g((i+1)I) - g(iI)} = h_2(y_1, y_2)\,.$$

The random vector $(Y(iI), Y((i+1)I))$ has a joint continuous p.d.f. as follows:

$$f_{Y(iI),Y((i+1)I)}(y_1, y_2) = |J| f_A(h_1(y_1, y_2)) f_B(h_2(y_1, y_2))\,, \quad (10)$$

where J, the *Jacobian*, is given by:

$$J = \begin{vmatrix} \frac{\partial h_1}{\partial y_1} & \frac{\partial h_1}{\partial y_2} \\ \frac{\partial h_2}{\partial y_1} & \frac{\partial h_2}{\partial y_2} \end{vmatrix} = \left| \frac{1}{g(iI) - g((i+1)I)} \right|.$$

(B) Assume $Y(t) = We^{At}/(B + e^{At})$, where $A > 0$ and $B > 0$ are independent. Assume $A \sim f_A(a)$, $B \sim f_B(b)$.

$$\text{Let} \begin{cases} y_1 = \dfrac{We^{aiI}}{b + e^{aiI}}\,, \\ y_2 = \dfrac{We^{a(i+1)I}}{b + e^{a(i+1)I}}\,. \end{cases}$$

The solutions for a and b can be easily solved from the above equations in terms of y_1 and y_2 as follows:

$$\begin{cases} a = \dfrac{\ln\left(\dfrac{y_2(y_1 - W)}{y_1(y_2 - W)}\right)}{I} = h_1(y_1, y_2)\,, \\ b = -e^{\left(\ln\left(\frac{y_2(y_1-W)}{y_1(y_2-W)}\right)(i+1)I\right)/I}(y_2 - W)y_2 = h_2(y_1, y_2)\,. \end{cases}$$

Similarly, the random vector $(Y(iI), Y((i+1)I))$ has a joint density function as given in Eq. (10).

As for the term $P\{D(iI) \le S, D((i+1)I) > S\}$ in Eq. (9), since $D(t) = \sum_{i=0}^{N(t)} X_i$ is a compound Poisson process, the compound Poisson process has stationary independent increment property. Therefore, random variables $D(iI)$ and $D((i+1)I) - D(iI)$ are independent. Using the *Jacobian* transformation, random vector $(D(iI), D((i+1)I) - D(iI))$ is distributed same as vector $(D(iI), D((i+1)I))$. Note that $D(iI)$ and $D(I_{i+1})$ are independent. Therefore,

$$\begin{aligned} &P\{D(iI) \le S, D((i+1)I) > S\} \\ &\quad = P\{D(iI) \le S\}P\{D((i+1)I) > S\}\,. \end{aligned} \tag{11}$$

(3) **Calculate P_p.** It should be noted that either a PM or CM action will end a renewal cycle. In other words, PM and CM, these two events, are mutually exclusive at renewal time point. As a consequence, $P_p + P_c = 1$. The probability P_p can be obtained as follows:

$$\begin{aligned} P_p &= P\{\text{PM ending a cycle}\} \\ &= \sum_{i=1}^{\infty} P\{Y(i-1)I) \le L, L < Y(iI) \le G\}P\{D(iI) \le S\}\,. \end{aligned} \tag{12}$$

3.2. *Expected cycle length analysis*

Since the renewal cycle ends either by a PM action with probability P_p or a CM action with probability P_c, the mean cycle length $E[W_1]$ is calculated as follows:

$$\begin{aligned} E[W_1] &= \sum_{i=1}^{\infty} E[(iI + R_1)I_{PM \text{ occur in } ((i-1)I, iI]}] \\ &\quad + E[(T + R_2)1_{CM \text{ occur}}] \\ &= \left\{ \sum_{i=1}^{\infty} iIP\{Y((i-1)I) \le L, D((i-1)I) \le S\} \right. \\ &\quad \left. \times P\{L < Y(iI) \le G, D(iI) \le S \right\} \\ &\quad + E[R_1]P_p + (E[T] + E[R_2])P_c\,, \end{aligned} \tag{13}$$

where $I_{PM \text{ occurs in } ((i-1)I, iI]}$ and $I_{CM \text{ occurs}}$ are indicator functions.

The mean time to failure, $E[T]$, is calculated as follows:

$$E[T] = \int_0^{\infty} P\{T > t\}dt$$

$$= \int_0^{\infty} P\{Y(t) \leq G, D(t) \leq S\}dt$$

$$= \int_0^{\infty} P\{Y(t) \leq G\} \sum_{j=0}^{\infty} \frac{(\lambda_2 t)^j e^{-\lambda_2 t}}{j!} F_X^{(j)}(S)dt\,,$$

or, equivalently, that:

$$E[T] = \sum_{j=0}^{\infty} \frac{F_X^{(j)}(S)}{j!} \int_0^{\infty} P\{Y(t) \leq G\}(\lambda_2 t)^j e^{-\lambda_2 t}dt\,. \qquad (14)$$

4. Maintenance Cost Optimization

We determine the optimal inspection time I and PM threshold L such that the long-run average maintenance cost rate $EC(L, I)$ is minimized. Mathematically, it is to minimize the following objective function:

$$EC(L, I)$$

$$= \frac{\sum_{i=1}^{\infty} iP\{Y(I_{i-1}) \leq L, D(I_{i-1}) \leq S\} \times P\{L < Y(I_i) \leq G, D(I_i) \leq S\}}{\left\{\sum_{i=1}^{\infty} I_i P\{Y(I_{i-1}) \leq L, D(I_{i-1}) \leq S\} P\{L < Y(I_i) \leq G, D(I_i) \leq S\right\} + E[R_1]P_p + E[R_2]P_c}$$

$$+ \frac{\sum_{i=1}^{\infty} iV_i \left\{P\{Y(I_i) \leq L,\, Y(I_{i+1}) > G\} P\{D(I_{i+1}) \leq S\} + P\{Y(I_{i+1}) \leq L\} P\{D(I_i) \leq S, D(I_{i+1}) > S\}\right\}}{\left\{\sum_{i=1}^{\infty} I_i P\{Y(I_{i-1}) \leq L, D(I_{i-1}) \leq S\} P\{L < Y(I_i) \leq G, D(I_i) \leq S\right\} + E[R_1]P_p + E[R_2]P_c}$$

$$
+\frac{C_pE[R_1]\sum_{i=1}^{\infty}P\{Y(I_{i-1})\le L,D(I_{i-1})\le S\}\times P\{L<Y(I_i)\le G,D(I_i)\le S\}}{\left\{\sum_{i=1}^{\infty}I_iP\{Y(I_{i-1})\le L,D(I_{i-1})\le S\}P\{L<Y(I_i)\le G,D(I_i)\le S\right\}+E[R_1]P_p+E[R_2]P_c}
$$

$$
+\frac{C_cE[R_2]\left\{1-\sum_{i=1}^{\infty}P\{Y(I_{i-1})\le L,\ D(I_{i-1})\le S\}\times P\{L<Y(I_i)\le G,D(I_i)\le S\}\right\}}{\left\{\sum_{i=1}^{\infty}I_iP\{Y(I_{i-1})\le L,D(I_{i-1})\le S\}P\{L<Y(I_i)\le G,D(I_i)\le S\right\}+E[R_1]P_p+E[R_2]P_c}, \quad (15)
$$

where $I_{i-1}=(i-1)I$, $I_i=iI$, $I_{i+1}=(i+1)I$ and $V_i=P\{Y(iI)\le L,\ D(iI)\le S\}$.

The above complex objective function is a nonlinear optimization problem and it is hard to obtain closed-form optimal solutions for L and I. Our proposed step-by-step algorithm below based on Nelder–Mead[11] downhill simplex method, which does not require the calculation of derivatives, is given as follows:

Step 1. Choose $(n+1)$ distinct vertices as an initial set $\{Z^{(1)},\ldots,Z^{(n+1)}\}$. Then calculate the function value $f(Z)$ for $i=1,2,\ldots,(n+1)$ where $f(Z)=EC(I,\ L)$. Putting the values $f(Z)$ in an increasing order where $f(Z^{(1)})=\min\{EC(I,\ L)\}$ and $f(Z^{(n+1)})=\max\{EC(I,\ L)\}$. Set $k=0$.

Step 2. Compute the best-n centroid $X^{(k)}=\frac{1}{n}\sum_{i=1}^{n}Z^{(i)}$.

Step 3. Use the centroid $X^{(k)}$ in Step 2 to compute away-from-worst move direction $\Delta X^{(k+1)}=X^{(k)}-Z^{(n+1)}$.

Step 4. Set $\lambda=1$ and compute $f(X^{(k)}+\lambda\Delta X^{(k+1)})$.
If $f(X^{(k)}+\lambda\Delta X^{(k+1)})\le f(Z^{(1)})$ then go to Step 5.
Otherwise, if $f(X^{(k)}+\lambda\Delta X^{(k+1)})\ge f(Z^{(n)})$ then go to Step 6.
Otherwise, fix $\lambda=1$ and go to Step 8.

Step 5. Set $\lambda = 2$ and compute $f(X^{(k)} + 2\Delta X^{(k+1)})$.
If $f(X^{(k)} + 2\Delta X^{(k+1)}) \leq f(X^{(k)} + \Delta X^{(k+1)})$ then set $\lambda = 2$.
Otherwise set $\lambda = 1$.
Then go to Step 8.

Step 6. If $f(X^{(k)} + \lambda\Delta X^{(k+1)}) \leq f(Z^{(n+1)})$ then set $\lambda = 1/2$. Compute $f\left(X^{(k)} + \frac{1}{2}\Delta X^{(k+1)}\right)$.
If $f\left(X^{(k)} + \frac{1}{2}\Delta X^{(k+1)}\right) \leq f(Z^{(n+1)})$ then set $\lambda = 1/2$ and go to Step 8.
Otherwise, set $\lambda = -1/2$ and if $f\left(X^{(k)} - \frac{1}{2}\Delta X^{(k+1)}\right) \leq f(Z^{(n+1)})$ then set $\lambda = -1/2$ and go to Step 8.
Otherwise, go to Step 7.

Step 7. Shrinking the current solution set toward best $Z^{(1)}$ by $Z^{(i)} = \frac{1}{2}(Z^{(1)} + Z^{(i)})$, $i = 2, \ldots, n+1$. Compute the new $f(Z^{(2)}), \ldots, f(Z^{(n+1)})$, let $k = k+1$, and return to Step 2.

Step 8. Replace the worst $Z^{(n+1)}$ by $X^{(k)} + \lambda\Delta X^{(k+1)}$. If

$$\sqrt{\frac{1}{n+1}\sum_{i=1}^{n+1}[f(Z^{(i)} - \bar{f}]^2} < 0.5\,,$$

where $\bar{f}$ is an average value, then STOP. Otherwise, let $k = k+1$ and return to Step 2.

It should be noted that the criterion in Step 8 is not unique but will depend on how you would like the algorithm to stop when the vertices function values are close. Here, we consider the difference between the maximum and the minimum values of f to be less than 0.5.

5. Numerical Examples

Assuming that the degradation process is described by $Y(t) = A + Bg(t)$, where A and B are independent and follow the uniform distribution with parameter interval [0, 4] and exponential distribution with parameter 0.3, i.e., $B \sim \exp(-0.3t)$, respectively, and $g(t) = \sqrt{t}e^{0.005t}$.

Assume that the random shock damage is represented by $D(t) = \sum_{i=1}^{N(t)} X_i$ where X_i follows the exponential distribution, i.e., $X_i \sim \exp(-0.04t)$ and $N(t) \sim Poisson(0.1)$. Given $G = 50$, $S = 100$ and the cost parameters are: $C_i = 900/\text{inspection}$, $C_c = 5600/CM$, $C_p = 3000/PM$. $R_1 \sim \exp(-0.1t)$, $R_2 \sim \exp(-0.04t)$.

We now determine both the values of I and L so that the average total cost per unit time *EC(I, L)* is minimized.

Following are step-by-step using our proposed procedure:

Step 1. Since there are two decision variables I and L, we need $(n + 1) = 3$ initial distinct vertices, which are $Z^{(1)} = (25, 20)$, $Z^{(2)} = (20, 18)$, and $Z^{(3)} = (15, 10)$. Set $k = 0$.

We calculate the value of $f(Z^{(\cdot)})$ corresponding to each vertices and sort them in an increasing order in terms of $EC(I, L)$.

Step 2. Calculate the centroid: $X^{(0)} = (Z^{(1)} + Z^{(2)})/2 = (22.5, 19)$.

Step 3. Generate the searching direction: $\Delta X = X^{(0)} - Z^{(3)} = (7.5, 9)$.

Step 4. Set $\lambda = 1$, it will produce a new minimal $EC(30, 28) = 501.76$ which leads to try an expansion with $\lambda = 2$ that is $(37.5, 38)$.

Step 5. Set $\lambda = 2$. Similarly, calculate $f(Z)$ that leads to $EC(37.5, 38) = 440.7$. Go to Step 8. This result turns out to be a better solution, hence (15, 10) is replaced by (37.5, 38).

The iteration continues and stops at $k = 6$ (see Table 1) since

$$\sqrt{\frac{1}{3}\sum_{i=1}^{3}\left[EC(Z^{(i)}) - \overline{EC(I,\ L)}\right]^2} < 0.5\,,$$

where $\overline{EC(I,\ L)}$ is the average value.

Table 1. Nelder–Mead algorithm result.

k	$Z^{(1)}$	$Z^{(2)}$	$Z^{(3)}$	Search result
0	(25, 20) $EC(I, L) = 564.3$	(20, 18) $EC(I, L) = 631.1$	(15, 10) $EC(I, L) = 773.6$	(37.5, 38) $EC(I, L) = 440.7$
1	(37.5, 38) $EC(I, L) = 440.7$	(25, 20) $EC(I, L) = 564.3$	(20, 18) $EC(I, L) = 631.1$	(42.5, 40) $EC(I, L) = 481.2$
2	(37.5, 38) $EC(I, L) = 440.7$	(42.5, 40) $EC(I, L) = 481.2$	(25, 20) $EC(I, L) = 564.3$	(32.5, 29) $EC(I, L) = 482.2$
3	(37.5, 38) $EC(I, L) = 440.7$	(42.5, 40) $EC(I, L) = 481.2$	(32.5, 29) $EC(I, L) = 482.2$	(32.5, 33.5) $EC(I, L) = 448.9$
4	(37.5, 38) $EC(I, L) = 440.7$	(32.5, 33.5) $EC(I. L) = 448.9$	(42.5, 40) $EC(I, L) = 481.2$	(38.75, 37.125) $EC(I, L) = 441.0$
5	(37.5, 38) $EC(I, L) = 440.7$	(38.75, 37.125) $EC(I, L) = 441.0$	(32.5, 33.5) $EC(I, L) = 448.9$	(35.3125, 35.25) $EC(I, L) = 441.1$
6	(37.5, 38) $EC(I^*, L^*) = 440.7$	(38.75, 37.125) $EC(I, L) = 441.0$	(35.3125, 35.25) $EC(I, L) = 441.4$	Stop

Table 1 illustrates the process of Nelder–Mead algorithm. In the table, $Z^{(\cdot)} = (I, L)$.

From Table 1, we observe that the optimal value is $I^* = 37.5$, $L^* = 37$ and $EC^*(I, L) = 440.7$.

Table 2 illustrates the various values of L on P_c for given $I = 37.5$. From Table 2, we observe that the probability P_c increases as L increases. In other words, a larger the value L will put the system in a high failure risk.

Figure 2 pictures the relationship between L and P_c for different I values such as $I = 35$, $I = 37.5$, and $I = 40$. From Fig. 2, we observe that P_c is an increasing function on L, that means a higher preventive maintenance threshold is more likely to result in a failure.

Figure 3 depicts the effect of the first inspection time on P_p for various given L values such as $L = 33$, $L = 35$, $L = 37$ and $L = 39$. Smaller the inspection time will cause more frequent inspection, as the results, it will result in a PM action with a larger probability. From Fig. 3, we also observe that, for the smaller L value ($L = 33$ and $L = 35$), the curve slightly decreases as I increases; while the bigger

Table 2. The effect of L on P_c for $I = 37.5$.

L	P_c
33	0.465
35	0.505
37	0.654
39	0.759

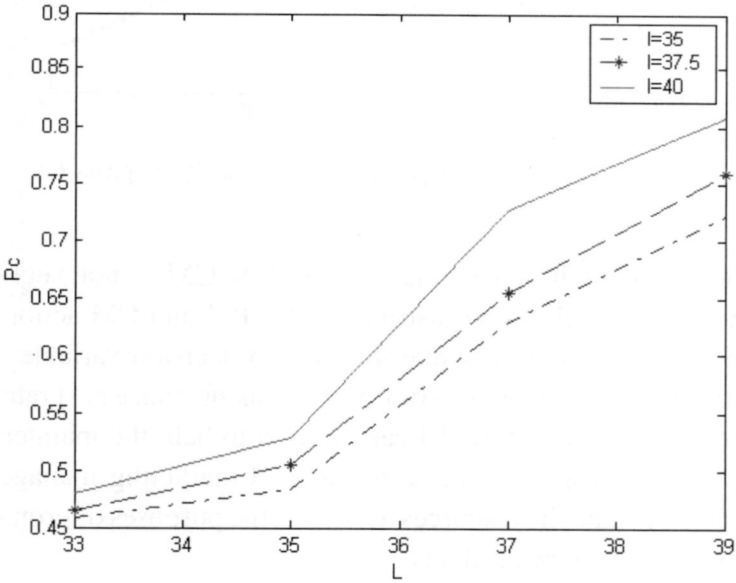

Fig. 2. P_c versus L.

the value L, such as $L = 37$ and $L = 39$, the curve has relatively bigger decrease as I increases. We also observe that the curve is more sensitive to the value L, especially when L is large.

6. Remarks

The state of working condition of the system discussed in this chapter is revealed by inspection, except that failure is self-announcing. The

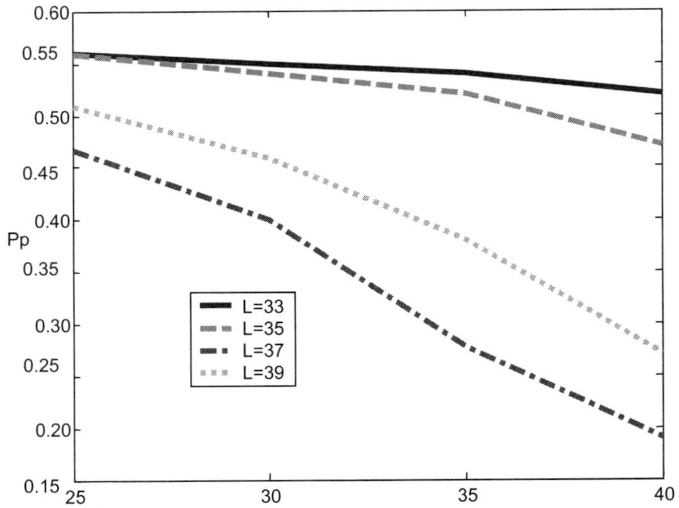

Fig. 3. The effect of inspection sequence on P_p for given L.

time to repair with consideration of PM or CM is not negligible and would take a different distribution for PM and CM action. PM threshold and inspection sequence are two decision variables. The objective is to minimize long-run average maintenance cost rate. The results of the proposed model can be used to help the maintenance managers and inspectors in particular and marketing managers in general to allocate the resources as well as the purposes of promotion strategies for the new products.

References

1. Y. Lam, A note on the optimal replacement problem, *Advanced Applied Probability* **20** (1983) 851–859.
2. S. H. Sheu, Extended optimal replacement model for deteriorating systems, *European Journal of Operational Research* **112** (1999) 503–516.
3. H. Pham, A. Suprasad and R. B. Misa, Reliability and MTTF prediction of k-out-of-n complex systems with components subjected to multiple stages of degradation, *International Journal of Systems Science* **27** (1996) 995–1000.

4. A. Grall, L. Dieulle, C. Berenguer and M. Roussignol, Continuous-time predictive-maintenance scheduling for a deteriorating system, *IEEE Trans. on Reliability* **51** (2002) 141–150.
5. G. A. Klutke and Y. J. Yang, The availability of inspected systems subjected to shocks and graceful degradation, *IEEE Trans. on Reliability* **44** (2002) 371–374.
6. W. Li and H. Pham, An inspection-maintenance model for systems with multiple competing processes, *IEEE Trans. on Reliability* **54** (2005) 318–327.
7. S. Bloch-Mercier, A preventive maintenance policy with sequential checking procedure for a Markov deteriorating system, *European Journal of Operational Research* **147** (2002) 548–576.
8. C. T. Lam and R. H. Yeh, Optimal maintenance-policies for deteriorating systems under various maintenance strategies, *IEEE Trans. on Reliability* **43** (1994) 423–430.
9. M. J. Zuo, B. Liu and D. N. P. Murthy, Replacement-repair policy for multi-state deteriorating products under warranty, *European Journal of Operational Research* **123** (2000) 519–530.
10. R. M. Feldman, Optimal replacement with semi-Markov shock models, *Journal of Applied Probability* **13** (1976) 108–117.
11. R. L. Rardin, *Optimization in Operations Research* (Prentice Hall, 1998).

CHAPTER 23

Age-Dependent Failure Interaction

Qing Zhao, Takashi Satow and Hajime Kawai

Department of Social Systems Engineering,
Faculty of Engineering,
Tottori University,
4-101, Koyama-Minami, Tottori, 680-8552, Japan

1. Introduction

In many practical maintenance situations, failure interaction has a tremendous impact on multi-component system reliability. For the repairable multi-component systems, it has been shown that to repair or replace the failed components only is more economic than to replace the whole system, and to replace components jointly is more economic than do separately. However, optimal maintenance policies for multi-component systems become more complex due to the aspects of structural, economic or stochastic dependences between components. For the multi-component systems that do not have simple structure, several author have investigated the maintenance policies. Cho and Parlar[1] surveyed the work done before 1991, they divided the maintenance models into two prime categories, preventive and preparedness. The article covers a wide variety of literature in multi-unit systems. The most famous economic dependence models are opportunistic maintenance models. Several maintenance models (see Refs. 2–6) show that the opportunistic maintenance policies are more economic than traditional maintenance policies in some cases. In a multi-component system, the failure or degradation of one component may affect the failure of the other components. Such kind of dependence, which occurs commonly in practical situations, is called stochastic dependence. Stochastic dependences are usually considered to be difficult to model due to the uncertainness of dependence between components. This chapter provides maintenance policies that are useful for maintaining large complex systems with interaction failure. The basic concepts, practical considerations and numerical observations of an age-dependent failure interaction model for a system with three components are investigated. These are designed to make the considerations of age-dependent models, which are suitable for practical situations to be understood easily. Several traditional maintenance models are also provided. The cost criterion composed of preventive replacement, corrective replacement and minimal repair costs is designed. The conditions which guarantee that there exists an unique solution to the optimality equation are also discussed. The objective of this chapter is to provide readers with some background on formulating the cost function and conducting the optimal results to the optimality equation of a multi-component system with failure interaction.

The outline of this chapter is as follows. The next section introduces the basic concepts and terms relating to age-dependent maintenances such as perfect repair and minimal repair. In Sec. 3, first, the background of failure interaction with a brief review of interactions among components in multi-component systems is introduced. Then the details of an age-dependent failure interaction model for a system with three components are illustrated. The optimal policies defined as that minimize the long-run average cost per unit time for both the scheduled time T and the Nth minor failure are conducted. Finally, the unique solution to the optimality equation and its existing conditions are also surveyed. Numerical examples designed to help readers to understand how to formulate a stochastic model and enable the model to be implemented easily in practical situations are provided in Sec. 4.

2. Age-Dependent Maintenance

The maintenance approaches concerning stochastic behaviors of repairable system are usually classified into two categories, preventive maintenance and corrective maintenance. For a complex system, the former carried out all necessary activities (cleaning, adjusting, changing, repairing/replacing, etc.) to keep the system running smoothly in the normal status, or being capable of performing system functions. Usually it consists of routine maintenance and scheduled maintenance. The latter deals with any maintenance (repair or replacement) when failure occurs. In most cases, when a system fails, considerable expense is required to allocate manpower on an emergency basis, repair/replacement parts, and lost revenues due to nonproduction can mount rapidly depending upon the manufacturing process or product. In addition, an unexpected failure can be dangerous to personnel and facilities. Accounting into the consideration of cost, lifetime, criteria level, multivariate, etc., many maintenance models propose two types of repair, perfect repair and imperfect repair. The former restores a system operating state to as good as new, while the latter makes a system back to some state that is well enough to operate. Furthermore, the imperfect repair, which returns a system to its functioning condition just prior to failure, is called minimal repair. The most significant characteristics of minimal repair and perfect repair are that the failure rate is not changed after a minimal repair and perfect repair regenerates a system to a new state, see Ref. 7. In many theoretical analysis situations, it is convenient to assume the probability of failure or repair/replacement to be a constant. However, when a component fails at age t, what level should it be restored to or whether to carry out a perfect repair or a minimal repair is commonly related to its age t, which is to say, it seems more realistic if the probability of maintenance activity is considered as an age-dependent variable in many practical situations.

Assume that a system starts at $t = 0$. Define the time to the first system failure as a random variable Y which has a distribution function $F(t)$ that is a continuous function of t. The hazard rate of the system is defined as $r(t) = f(t)/\bar{F}(t)$, where $f(t) = \partial F(t)/\partial t$ is the density function of $F(t)$, and $\bar{F}(t)$ $(= 1 - F(t))$ is the survival function of the new system. If the system fails at age t, it is either completely repaired with probability $p(t)$ or undergoes a minimal repair with probability $q(t) = 1 - p(t)$ which are both age-dependent variables. Special cases are showed as follows.

Case 1. $p(t) = 1, q(t) = 0$. In this case, the system undergoes perfect repair. As each repair renews the system, the survival function can be given by $\bar{G}(t) = \bar{F}(t) = \exp\{-\Lambda(t)\}$, see Ref. 8.

Case 2. $0 < p(t) < 1, 0 < q(t) < 1$. The cumulative hazard function defined above can be expressed as:

$$\Lambda(t) = \int_0^t p(x)r(x)dx\,. \tag{1}$$

Hence, we have

$$\bar{G}(t) = \Pr\{Y > t\} = \exp\left\{-\int_0^t p(x)r(x)dx\right\}. \tag{2}$$

For the detailed proof of Eq. (2), see Ref. 9.

Case 3. $p(t) = 0$, $q(t) = 1$. In this case, the system failures occur according to a non-homogeneous Poisson process with cumulative hazard function $\Lambda(t)$. Assume that the survival probability of the system at time t is $\bar{G}(t)$ (which means that no system renewal occurs for t unit of time). Then we have $\bar{G}(t) = 1$.

2.1. *Traditional age replacement model*

Suppose that the costs of preventive maintenance and corrective maintenance are c_p and c_f respectively. Following directly from the well-known renewal reward theory, which gives the ratio of the expected time between successive system renewals and the expected cost of one cycle of the system. Assume that the time of replacement can be negligible. Define the long-run average cost per unit time as $C(T)$, the traditional age replacement model is defined as follows.

A new system starts to operate at time $t = 0$. The system is replaced with an identical new one at scheduled time T or at failure whichever occurs first. The long-run average cost per unit time can be given by:

$$C(T) = \frac{c_f \Pr\{Y \le T\} + c_p \Pr\{Y > T\}}{\int_0^T \bar{G}(t)dt} = \frac{c_f + (c_p - c_f)\bar{G}(t)}{\int_0^T \bar{G}(t)dt}, \tag{3}$$

where $\bar{G}(t) = \Pr\{Y > t\}$.

2.2. *Typical minimal repair model*

Suppose that a system fails at $S_0, S_1, \ldots$, the system is replaced at the Nth failure, and between replacements it undergoes minimal repair. Define $\Pr\{S_j \le t\} = G_j(t)$, the long-run average cost per unit time is given by:

$$C(N) = \frac{c_m(N-1) + c_f}{\int_0^\infty G_N(t)dt}, \tag{4}$$

where c_m is the cost of minimal repair. The time for minimal repair is negligible, especially, when the system up times defined as $X_1, X_2, \ldots$, where $X_i = S_i - S_{i-1}$ $(i = 1, 2, \ldots)$ are supposed to be exponentially distributed with rate λ, the long-run average cost per unit time can be rewritten as:

$$C(N) = \frac{\lambda}{N}\left[c_m(N-1) + c_f\right]. \tag{5}$$

2.3. *Age replacement with minimal repair model*

Brown and Proschan[10] considered the imperfect repair model where a component is renewed with probability $p (0 < p < 1)$, and minimally repaired with probability $q \, (= 1 - p)$. Suppose that the system is replaced preventively at scheduled time T. If it fails before T, it is replaced with probability p or undergoes minimal repair with probability q. Hence, we have

$$C(T) = \frac{c_f + (c_p - c_f)\bar{G}(T) + c_m q \int_0^T r(t)\bar{G}(t)dt}{\int_0^\infty \bar{G}(t)dt}. \quad (6)$$

Block *et al.*[8] extended the above model to an age-dependent minimal repair model that the probability of minimal repair p is dependent on the system age t. Thus, the model can be redefined as; whenever the system fails, it is completely replaced with probability $p(t)$, and minimally repaired with probability $q(t)$ $(= 1 - p(t))$, where t is its age at failure.

We can also consider the cost of minimal repair is dependent on age t. In this case, the total expected cost of minimal repairs during the interval $(0, T]$ can be expressed as $\int_0^T c_m(t)q(t)r(t)\bar{G}(t)dt$. For more details, see Ref. 11.

3. Failure Interaction Analysis

In a multi-unit system, the complex structure would be divided into components. If components constituting the complex system are interconnected in such a way that they can be considered as structurally, economically or stochastically independent, the optimal maintenance policy for thus complex system reduced to that of each single component, see Ref. 12. The interactions between components are usually classified into three types, structural, economic and stochastic dependence. At the design stages, a fault-tolerant, easy-maintenance, reliability-based structural design are desired, and a computer simulation method is considered as very effective to be applied to reliability structure designs and predictions. The maintenances of multi-component systems whose components are economically dependent are considered as that the cost can be saved when several components are jointly maintained instead of separately, especially, when the set-up cost and the downtime cost are considered as large, it is more economic to replace more than one component. Opportunity maintenances are considered to be effective to the systems with economic dependence, see Ref. 13.

In the following sections, we focus on the stochastic dependence.

3.1. *Failure interaction*

All systems deteriorate with age and are subject to stochastic failure. A failure is caused not only by a single factor but by a combination of factors. Usually we can consider the failure mechanism as the combination of wear out failure and random failure. In these cases, it may not be economical to replace the whole system at failure but only repair or replace the failed component. If the replaced component is a minor part of the system, then when the system is returned to an operating state, its age would be the same as if no failure had occurred, which leads to the concept of minimal repair that restores a component to its functioning condition just prior to failure. A component failure can occur by failure in the component itself, or by failure of an adjacent component. These types of interactions between components have been termed as failure interaction by Murthy and Nguyen.[14]

Murthy and Nguyen considered a two-unit system model of interaction failure. Whenever unit 1(2) fails, it induces a failure of unit 2(1) with probability $p(q)(0 \leq p, q \leq 1)$, and has no effect on unit 2(1) with probability $1-p(1-q)$. Failed unit is replaced immediately with new one. If no induced failure occurs, only one unit is replaced at failure and with induced failure both the units are replaced. They gave the expressions for the expected cost of operating the system for both finite and infinite time horizons. Future, Murthy and Casey[15] considered preventive maintenances of a two-unit system with shock damage interaction. However, when a component fails at age t, what level it is restored to or whether it undergoes a perfect repair or a minimal repair is related to its age t, which is to say, it seems more realistic if the probability p and q are considered as age dependent variables in many practical situations.

3.2. *Problem definition and research methodology*

In the following subsections, we give a detailed analysis of failure interaction with a three-component system. Each component of the system fails alone or induce a failure of the other one or two components. A system failure occurs when all the components failed simultaneously. The probability of each kind of failure is age dependent. The system undergoes minimal repair at component level failures, and replaced at age T or at the Nth system minor failure or at a system failure whichever occurs first.

Introduction to Terminology. The interaction failure process discussed in this section is as follows. A system with three new components (which do not need to be identical) starts at time $t = 0$. Whenever component i ($i = 1, 2, 3$) fails at age t, it fails alone (Natural failure) with probability $p_i(t)$ or induces a failure of one of the others with probability $q_{il}(t)(l \neq i;\ l = 1, 2, 3)$ (Induced failure I) or induces a system failure with probability $\gamma_i(t)$ (Induced failure II) (see Fig. 1). At system failures, each of the three components is replaced with an identical new one, and the other failures (called minor failure of the system including Natural failure and Induced failure I) are corrected with minimal repairs. It is easy to see that

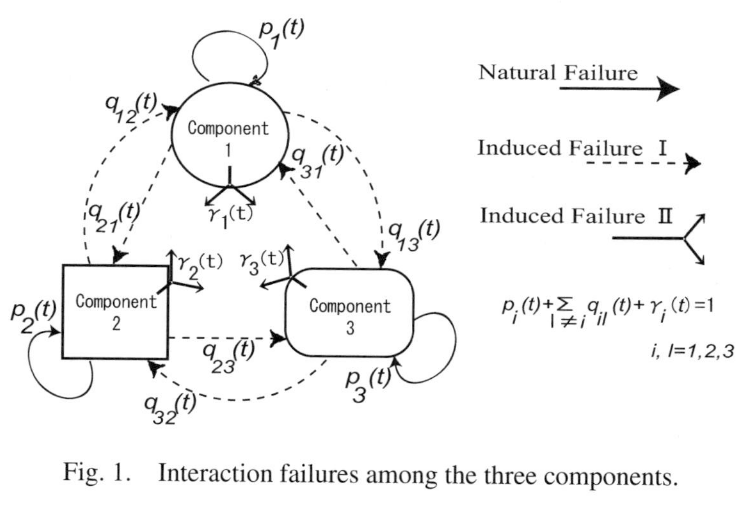

Fig. 1. Interaction failures among the three components.

$p_i(t) + \sum_{l \neq i} q_{il}(t) + \gamma_i(t) = 1$. The probability $p_i(t)$, $q_{il}(t)$ and $\gamma_i(t)$ are dependent on the age of component i. The system is minimally repaired with cost c_m and preventively replaced at age T or at the Nth minor failure with cost c_p or correctively replaced at an induced failure II with cost $c_f (c_f > c_p)$ whichever occurs first. Thus, the occurrence of Induced failure II corresponds to a renewal point of the system.

Notation and Assumptions.

Notation

Y: Time to the first system failure (r.v.).
$G(\cdot)$: The failure distribution function of the system.
$\bar{G}(\cdot)$: The survival function of the system, $\bar{G}(\cdot) = 1 - G(\cdot)$.
$g(\cdot)$: The probability density function of the system.
$r_s(\cdot)$: The hazard rate function of $G(\cdot)$.
Y_i: Time to the first induced failure II caused by component i $(i = 1, 2, 3)$ (r.v.).
$F_i(\cdot)$: The failure distribution function of component i.
$\bar{F}_i(\cdot)$: The survival function of component i, $\bar{F}_i(\cdot) = 1 - F_i(\cdot)$.
$f_i(\cdot)$: The probability density function of component i.
$r_i(\cdot)$: The hazard rate function of $F_i(\cdot)$.
$p_i(t)$: The probability of natural failure of component i at age t.
$q_{il}(t)$: The probability of induced failure I of component l $(l = 1, 2, 3; l \neq i)$ caused by component i at age t.
$\gamma_i(t)$: The probability of induced failure II caused by component i at age t.
c_p: The cost of a system preventive replacement.
c_f: The cost of a system corrective replacement $(c_f > c_p)$.
c_m: The cost of a minimal repair.
T: The age replacement period.
T^*: The optimal value of T.
N: The number of system minor failures.
N^*: The optimal value of N.

$N(t)$: The total number of minor failures occurred up to time t.
S_j: The arrival time of the jth minor failure ($j = 1, 2, \ldots$).
$H_N(t)$: The probability of the Nth system minor failure occurred up to time t.

To simplify the analysis, the following assumptions are made here:

- Minimal repair and replacement time are negligible.
- Three components do not fail at the same time except both induced failures occur.
- The state of replaced components are considered to be as good as new.

3.3. *Long-run average cost per unit time*

Let the random variable Y be the time to the first system failure with the distribution function $G(t)$ and failure rate $r_s(t)$ respectively, $r_s(t) = g(t)/\bar{G}(t)$ ($g(t) = \partial G(t)/\partial t$). Then we have the survival function $\bar{G}(t) = \Pr\{Y > t\}$. As Y_i ($i = 1, 2, 3$) is the time to the first induced failure II caused by component i, the survival function of component i can be given by:

$$\bar{G}_i(t) = \exp\left\{-\int_0^t \gamma_i(x) r_i(x) dx\right\}, \tag{7}$$

see Ref. 8, where $r_i(t) = f_i(t)/\bar{F}_i(t)$ ($f_i(t) = \partial F_i(t)/\partial t$) is the failure rate of component i at age t. It is easy to be seen that $Y = \min\{Y_1, Y_2, Y_3\}$, that is, the system fails whenever component i caused the induced failure II occur. Thus, we have

$$\bar{G}(t) = \prod_{i=1}^{3} \bar{G}_i(t) = \exp\left\{-\int_0^t \sum_{i=1}^{3} \gamma_i(x) r_i(x) dx\right\}. \tag{8}$$

Consider a counting process of the total minor failure of the system at time t. Let $\{N(t), t \geq 0\}$ be a nonhomogeneous Poisson process whose mean value function is given by $\Lambda(t)$. Let $S_1, S_2, \ldots$ denote the successive arrival times of natural failure and induced failure I, so that the probability of the kth system minor failure occurred up to time t is $\Pr\{S_k \le t\}$. Define $H_k(t) = \Pr\{S_k \le t\}$, hence we have

$$H_k(t) = \sum_{j=k}^{\infty} \Pr\{N(t) = k\} = \sum_{j=k}^{\infty} \frac{[\Lambda(t)]^j}{j!} e^{-\Lambda(t)}, \tag{9}$$

where $\Lambda(t) = \sum_{i=1}^{3} \Lambda_i(t)$, and $\Lambda_i(t)$ is the mean number of minor failures of component i occurred up to time t, which can be expressed as:

$$\Lambda_i(t) = \int_0^t \left\{ p_i(x) r_i(x) + \sum_{l \ne i} q_{il}(x) r_i(x) \right\} dx . \tag{10}$$

Let $C(T, N)$ be the long-run average cost per unit time. It is to be a critical function of this model. With the renewal reward theory, see Ref. 16, we have

$$C(T, N) = \frac{\alpha(T, N)}{\beta(T, N)}, \tag{11}$$

which means that the long-run average cost per unit time equals the expected cost during a cycle which is expressed by $\alpha(T, N)$ divided by the expected time of a cycle which is defined as $\beta(T, N)$. To simplify the expression, let's rewrite S_N as Z. Consequently, the probability that the system replaced at age T is

$$\begin{aligned} \Pr\{T < \min(Y, Z)\} &= \int_T^{\infty} [1 - H_N(t)] dG(t) \\ &\quad + \int_T^{\infty} [H_N(t) - H_N(T)] dG(t) \\ &= \bar{H}_N(T) \bar{G}(T) . \end{aligned} \tag{12}$$

The probability that the system replaced at the Nth minor failure is

$$\Pr\{Z < \min(Y, T)\} = \int_0^T H_N(t)dG(t) + \bar{G}(T)H_N(T)\,. \tag{13}$$

The probability that the system undergoes corrective replacement is

$$\begin{aligned}\Pr\{Y \le \min(T, Z)\} &= G(T)[1 - H_N(T)] + \int_0^T [H_N(T) \\ &\quad - H_N(t)]\, dG(t) \\ &= \int_0^T \bar{H}_N(t)dG(t)\,. \end{aligned} \tag{14}$$

Here we have Eq. (12) + Eq. (13) + Eq. (14) = 1. With Eqs. (12), (13) and (14), we have

$$\begin{aligned}\alpha(T, N) &= c_p[\Pr\{T < \min(Y, Z)\} + \Pr\{Z < \min(Y, T)\}] \\ &\quad + c_f \Pr\{Y \le \min(Z, T)\} + c_m M(T, N) \\ &= c_p + (c_f - c_p)\int_0^T \bar{H}_N(t)dG(t) + c_m M(T, N)\,, \end{aligned} \tag{15}$$

where $M(T, N)$ is the expected number of minor failures over a cycle. Under the (T, N) policy, it is given by:

$$\begin{aligned}M(T, N) &= \Lambda(T)\bar{G}(T)\bar{H}_N(T) + N\int_0^T \bar{G}(t)dH_N(t) \\ &\quad + \sum_{j=0}^{N-1} j\int_0^T h_j(t)dG(t)\,, \end{aligned} \tag{16}$$

where $h_j(t) = \Pr\{N(t) = j\} = H_j(t) - H_{j+1}(t) = ([\Lambda(t)]^j/j!)\times \exp\{-\Lambda(t)\}$. As the system is replaced at age T or at the Nth minor failure or at an Induced failure II whichever occurs first, we have that $\beta(T, N) = E[\min(T, Z, Y)]$. In view of Eqs. (12), (13) and (14), the expected cycle length $\beta(T, N)$ is given by:

$$\begin{aligned}\beta(T, N) &= T\bar{H}_N(T)\bar{G}(T) + \int_0^T \int_0^t x dH_N(x) dG(t) \\ &\quad + \bar{G}(T) \int_0^T t dH_N(t) + \bar{H}_N(T) \int_0^T t dG(t) \\ &\quad + \int_0^T t[H_N(T) - H_N(t)] dG(t) \\ &= \int_0^T \bar{H}_N(t)\bar{G}(t) dt\,. \end{aligned} \tag{17}$$

Thus, we obtain the expression of long-run average cost per unit time as:

$$\begin{aligned} C(T, N) &= \frac{\alpha(T, N)}{\beta(T, N)} \\ &= \frac{c_p + (c_f - c_p)\int_0^T \bar{H}_N(t) dG(t) + c_m M(T, N)}{\int_0^T \bar{H}_N(t)\bar{G}(t) dt}\,. \end{aligned} \tag{18}$$

3.4. *The optimal T^* policy*

In this subsection, we wish to determine the optimal period T^* which minimizes $C(T, \infty)$, the long-run average cost per unit time, as $N \to \infty$, that is, the system is only preventively replaced at age T. According to Eq. (18) we have

$$C(T, \infty) = \frac{c_p + (c_f - c_p)G(T) + c_m M(T, \infty)}{\int_0^T \bar{G}(t) dt}\,. \tag{19}$$

From Eq. (16), the expected number of minor failures can be rewritten as:

$$\begin{aligned} M(T, \infty) &= \lim_{N\to\infty} M(T, N) = \int_0^T \bar{G}(t) d\Lambda(t) \\ &= \int_0^T \bar{G}(t) \sum_{i=1}^{3} \left\{ p_i(t) r_i(t) + \sum_{l=1, l\neq i}^{3} q_{il}(t) r_i(t) \right\} dt\,. \end{aligned} \tag{20}$$

Hence, let $M_1(T) = M(T, \infty)$, we rewrite $C(T, \infty)$ as:

$$C(T, \infty) = \frac{c_p + (c_f - c_p)G(T) + c_m M_1(T)}{\int_0^T \bar{G}(t)dt}. \tag{21}$$

Differentiate Eq. (21) with respect to T and set it equal to zero. As $\partial C(T, \infty)/\partial t = 0$, if and only if

$$\frac{\alpha'(T, \infty)}{\beta'(T, \infty)} = \frac{\alpha(T, \infty)}{\beta(T, \infty)}, \tag{22}$$

where $\alpha'(T, \infty) = \partial\alpha(T, \infty)/\partial T$, $\beta'(T, \infty) = \partial\beta(T, \infty)/\partial T$. From Eq. (22), we obtain

$$\begin{aligned} r_s(T) \int_0^T \bar{G}(t)dt + \bar{G}(T) + \frac{c_m}{c_f - c_p} \\ \times \left\{ \frac{M_1'(T)}{\bar{G}(T)} \int_0^T \bar{G}(t)dt - M_1(T) \right\} = \frac{c_f}{c_f - c_p}. \end{aligned} \tag{23}$$

Define the left-hand side of Eq. (23) as $\mu(T)$, then

$$\begin{aligned} \mu(T) = r_s(T) \int_0^T \bar{G}(t)dt + \bar{G}(T) \\ + \frac{c_m}{c_f - c_p} \left\{ \frac{M_1'(T)}{\bar{G}(T)} \int_0^T \bar{G}(t)dt - M_1(T) \right\}. \end{aligned} \tag{24}$$

Differentiating $\mu(T)$ with respect to T, and we have

$$\frac{\partial\mu(T)}{\partial T} = \left[r_s'(T) + \frac{c_m}{c_f - c_p}\omega'(T) \right] \int_0^T \bar{G}(t)dt, \tag{25}$$

where $\omega(T) = m_1(T)/\bar{G}(T)$ and $m_1(T) = \partial M_1(T)/\partial T$.

If $r_s'(T)/\omega'(T) > c_m/(c_p - c_f)$ then $\partial\mu(T)/\partial T > 0$, thus, we have that $\mu(T)$ is increasing in T. As $\lim_{T\to 0} \mu(T) = 1$, denote $EY = \int_0^\infty \bar{G}(t)dt$, we have $\lim_{T\to\infty} \mu(T) = r_s(\infty)EY +$

$(c_m/(c_f - c_p))[\omega(\infty)EY - M_1(\infty)]$. If

$$r_s(\infty)EY + \frac{c_m}{c_f - c_p}[\omega(\infty)EY - M_1(\infty)] > \frac{c_f}{c_f - c_p}, \tag{26}$$

then $\mu(0) < c_f/(c_f - c_p) < \mu(\infty)$, hence there exists a unique and finite T^* to satisfy the optimality equation, and

$$C(T^*, \infty) = (c_f - c_p)r_s(T^*) + c_m\omega(T^*). \tag{27}$$

If

$$r_s(\infty)EY + \frac{c_m}{c_f - c_p}[\omega(\infty)EY - M_1(\infty)] \leq \frac{c_f}{c_f - c_p}, \tag{28}$$

then $C'(T, \infty) \leq 0$, T^* tends to be infinity, $C(T^*, \infty) = [c_f + c_m M_1(\infty)]/EY$, it implies that no preventive replacement is needed.

3.5. *The optimal N* policy*

For the infinite-horizon of T, we want to obtain the optimal N^* which minimizes $C(\infty, N)$, the long-run average cost per unit time, that is to say, the system is replaced only at the Nth minor failure or an Induced failure II whichever occurs first. According to Eq. (18), we have

$$C(\infty, N) = \frac{c_p + (c_f - c_p)\int_0^\infty \bar{H}_N(t)dG(t) + c_m M(\infty, N)}{\int_0^\infty \bar{H}_N(t)\bar{G}(t)dt}. \tag{29}$$

From Eq. (16), we can rewrite the expected number of minor failures as:

$$\begin{aligned} M(\infty, N) &= \lim_{T\to\infty} M(T, N) \\ &= N\int_0^\infty H_N(t)dG(t) + \sum_{j=0}^{N-1} j\int_0^\infty h_j(t)dG(t). \end{aligned} \tag{30}$$

Hence, let $M_2(N) = M(\infty, N)$, we can rewrite $C(\infty, N)$ as:

$$C(\infty, N) = \frac{c_p + (c_f - c_p)\int_0^\infty \bar{H}_N(t)dG(t) + c_m M_2(N)}{\int_0^\infty \bar{H}_N(t)\bar{G}(t)dt}. \tag{31}$$

We see that if $r_s(t) = g(t)/\bar{G}(t)$ is strictly increasing in t, then there exists a minimum N^* which satisfies the inequalities $C(\infty, N) < C(\infty, N-1)$ and $C(\infty, N+1) \geq C(\infty, N)$. Thus, we have:

$$Q(N^* - 1) < \frac{c_p}{(c_f - c_p)} \leq Q(N^*)\,, \tag{32}$$

where

$$Q(N) = \frac{V(N)W(N+1) - V(N+1)W(N)}{V(N+1) - V(N)} + \frac{c_m}{c_f - c_p}\frac{M_2(N+1)V(N) - M_2(N)V(N+1)}{V(N+1) - V(N)}\,, \tag{33}$$

and

$$V(N) = \int_0^\infty \bar{H}_N(t)\bar{G}(t)dt\,, \tag{34}$$

$$W(N) = \int_0^\infty \bar{H}_N(t)dG(t)\,. \tag{35}$$

Further,

$$Q(N+1) - Q(N) = \left\{\left[\frac{J(N+1)}{L(N+1)} - \frac{J(N)}{L(N)}\right] + \frac{c_m}{c_f - c_p} \times \left[\frac{K(N+2)}{L(N+1)} - \frac{K(N+1)}{L(N)}\right]\right\}V(N)\,, \tag{36}$$

where

$$J(N) = \int_0^\infty h_N(t)dG(t)\,, \tag{37}$$

$$K(N) = \int_0^\infty H_N(t)dG(t)\,, \tag{38}$$

$$L(N) = \int_0^\infty h_N(t)\bar{G}(t)dt\,. \tag{39}$$

Theorem 1. *If the failure rate $r_s(t)$ is continuous and strictly increasing in t, then $\int_0^t h_N(x)dG(x)/\int_0^t h_N(x)\bar{G}(x)dx$ is increasing in N.*

Proof. What we want to show is that for each fixed $t > 0$,

$$\frac{\int_0^t h_{N+1}(x)dG(x)}{\int_0^t h_{N+1}(x)\bar{G}(x)dx} - \frac{\int_0^t h_N(x)dG(x)}{\int_0^t h_N(x)\bar{G}(x)dx} > 0\,. \tag{40}$$

As $\int_0^t h_{N+1}(x)\bar{G}(x)dx \int_0^t h_N(x)\bar{G}(x)dx > 0$, let

$$\begin{aligned}\psi(t) = &\int_0^t h_{N+1}(x)dG(x)\int_0^t h_N(x)\bar{G}(x)dx \\ &- \int_0^t h_N(x)dG(x)\int_0^t h_{N+1}(x)\bar{G}(x)dx\,,\end{aligned} \tag{41}$$

and $\psi(0) = 0$. Differentiating $\psi(t)$ with respect to t, we have

$$\begin{aligned}\frac{\partial\psi(t)}{\partial t} = &\frac{[\Lambda(t)]^N}{(N+1)!N!}\exp\{-\Lambda(t)\}\bar{G}(t)\int_0^t [\Lambda(x)]^N \exp\{-\Lambda(x)\} \\ &\times \bar{G}(x)[r_s(t) - r_s(x)][\Lambda(t) - \Lambda(x)]dx \geq 0\,.\end{aligned} \tag{42}$$

It means that for any small enough $\varepsilon > 0$, $\forall x \in [t-\varepsilon, t+\varepsilon]$, if $r_s(x)$ is continuous and strictly increasing in x, then $\partial\psi(t)/\partial t > 0$, that is, $\int_0^t h_N(x)dG(x)/\int_0^t h_N(x)\bar{G}(x)dx$ is increasing in N. □

Theorem 2. *If $\lambda(t)(\lambda(t) = \partial\Lambda(t)/\partial t)$ is continuous and strictly increasing in t, then $\int_0^t H_{N+1}(x)dG(x)/\int_0^t h_N(x)\bar{G}(x)dx$ is increasing in N for any $t > 0$.*

Proof. Rewrite $\int_0^t H_{N+2}(x)dG(x)$ as $\int_0^t \lambda(x)h_{N+1}(x)\bar{G}(x)dx$. We want to show that for any fixed $t > 0$,

$$\frac{\int_0^t \lambda(x)h_{N+1}(x)\bar{G}(x)dx}{\int_0^t h_{N+1}(x)\bar{G}(x)dx} - \frac{\int_0^t \lambda(x)h_N(x)\bar{G}(x)dx}{\int_0^t h_N(x)\bar{G}(x)dx} > 0\,. \tag{43}$$

As $\int_0^t h_{N+1}(x)\bar{G}(x)dx \int_0^t h_N(x)\bar{G}(x)dx > 0$, same as Theorem 1, let

$$\phi(t) = \int_0^t \lambda(x)h_{N+1}(x)\bar{G}(x)dx \int_0^t h_N(x)\bar{G}(x)dx - \int_0^t \lambda(x)h_N(x)\bar{G}(x)dx \int_0^t h_{N+1}(x)\bar{G}(x)dx\,. \tag{44}$$

We have $\phi(0) = 0$. Differentiating $\phi(t)$ with respect to t, and we have

$$\frac{\partial\phi(t)}{\partial t} = \frac{[\Lambda(t)]^N}{(N+1)!N!}\exp\{-\Lambda(t)\}\bar{G}(t)\int_0^t [\Lambda(x)]^N \exp\{-\Lambda(x)\} \times \bar{G}(x)[\lambda(t)-\lambda(x)][\Lambda(t)-\Lambda(x)]dx > 0\,. \tag{45}$$

□

Thus, we have that $Q(N+1) - Q(N) \geq 0$, that is, $Q(N)$ is increasing in N. Further, if $Q(\infty) > c_p/(c_f - c_p)$, there exists a minimum N^* which satisfies Eq. (32). Conversely, if $Q(\infty) \leq c_p/(c_f - c_p)$, then $N^* = \infty$, $C(\infty,\infty) = [c_f + c_m M(\infty,\infty)]/EY$, that is to say, system is replaced at Induced failure II only.

4. Numerical Observations

To illustrate the results obtained in the previous sections, we give a numerical example in this section. We only show the T policy here as an example.

4.1. *Definitions*

Define the hazard rate of each component as $r_i(t) = 2t/i$ $(i = 1, 2, 3)$. Let the random variable Y be the time to the first system failure with a distribution function $G(t)$ and failure rate $r_s(t)$ respectively. Define Y_i as the time to the first Induced failure II caused by component i. Suppose that the probability of Induced failure II caused by each component has the Gamma characteristics which is defined as $\gamma_i(t) = t/2i(1+t)$. From Eq. (8), we have the survival function of each component as:

$$\bar{G}_i(t) = \exp\left\{-\frac{1}{i^2}\int_0^t \frac{x^2}{1+x}dx\right\}. \tag{46}$$

The graph characteristics of $\bar{G}_i(t)$ are shown in Fig. 2.

As the survival function of the system is defined as $\bar{G}(t) = \prod_{i=1}^{3} \bar{G}_i(t)$, and then the mean system failure time can be given by $EY = \int_0^\infty \bar{G}(t)dt$. Define the probabilities of Induced failure I caused by component i, $q_{il_1}(t)$ ($l_1 = 1$ to 3, $l_1 \neq i$) and $q_{il_2}(t)$ ($l_2 = 1$ to 3, $l_2 \neq (i, l_2)$), as $1 - \exp\{-it/20\}$ and $\exp\{-it\}$, respectively. Then the probability of Natural failure caused by each component itself $p_i(t)$ equals $1 - \gamma_i(t) - \sum_{l=1, l\neq i}^{3} q_{il}(t)$. The probability of Natural failure, the probability of Induced Failure I and II of component 2 are shown in Fig. 3. From Fig. 3, it can be seen that the probability of Induced failure II and one of the probability of Induced failure I are increasing in t.

The mean number of minor failure of each component in the time interval (0, t] are shown in Fig. 4.

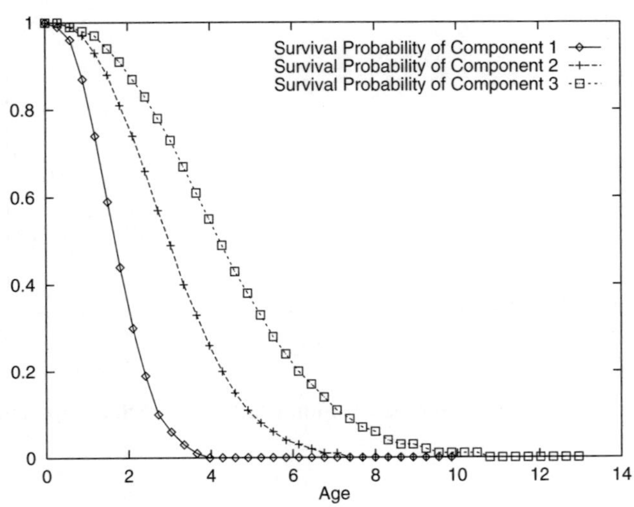

Fig. 2. The survival function of each component.

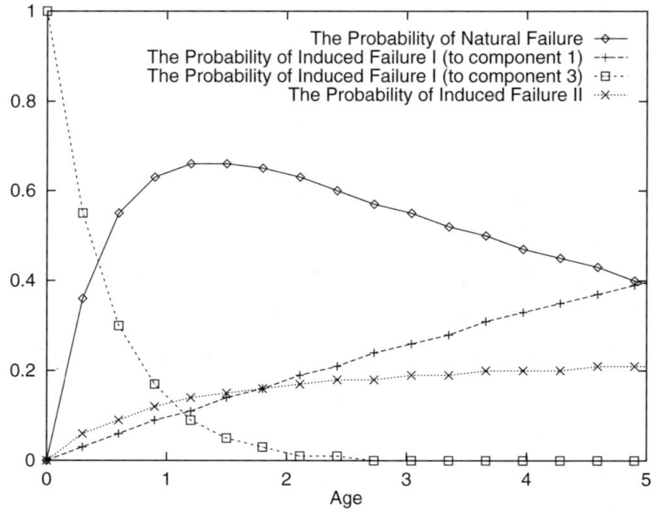

Fig. 3. The failure probabilities of component 2.

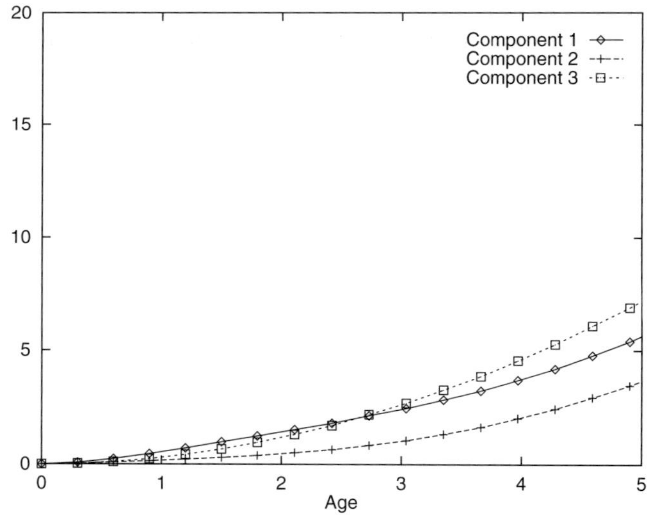

Fig. 4. The mean number of minor failure of each component.

4.2. *The optimal T* and C(T*)*

Define the preventive replacement and minimal repair costs as $c_p = 20$, $c_m = 6$, respectively, and range the cost of corrective replacement from 20_+ to 400. From Eq. (19), we can calculate the optimal T^* which minimizes the long-run average cost per unit time. The graph of $C(T^*)$ and the ratio of c_f to c_p, and the value of optimal T^* and the minimum $C(T^*)$ are shown in Fig. 5 and Table 1 respectively.

Time unit is not specified. It can be considered as measured in years. Figure 5 and Table 1 show that the optimal policy and the optimal cost are nonmonotonic. The time of minimum cost is influenced by the ratio of the corrective replacement cost to the preventive replacement cost (c_f/c_p). As c_f/c_p approaches 1, it implies that no preventive maintenance action is needed, and preventive replacements should be carried out before the system failure, while the mean time between renewals EY equals 1.55. Equation (46) shows that the expected life time of the system is only influenced by the probability of Induced failure II except the hazard rate (see Fig. 6). However, the minimum cost and optimal policy T^* are also affected by

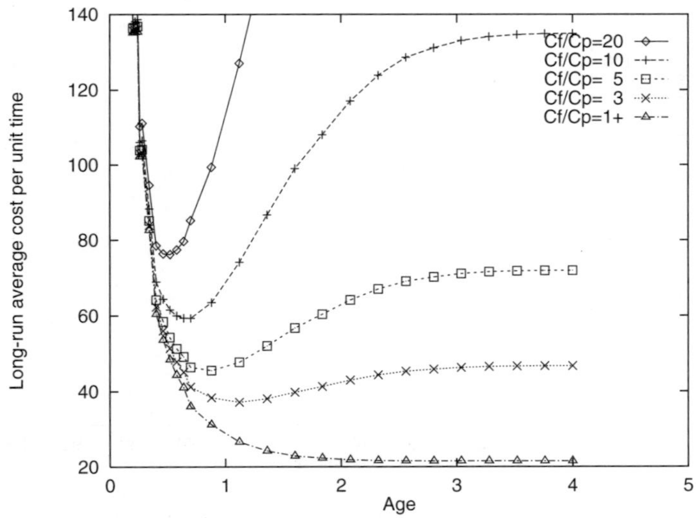

Fig. 5. Long-run average cost per unit time.

Table 1. Optimum T^* and the minimum cost $C(T^*)$.

$\frac{c_f}{c_p}$	T^*	$EY = 1.55$ $C(T^*)$
1_+	2.99	21.76
2	1.34	31.39
3	1.04	37.31
4	0.88	41.77
5	0.80	45.44
6	0.74	48.61
7	0.68	51.41
8	0.66	53.95
9	0.62	56.27
10	0.60	58.43
20	0.46	74.58
50	0.32	102.76
100	0.26	130.84

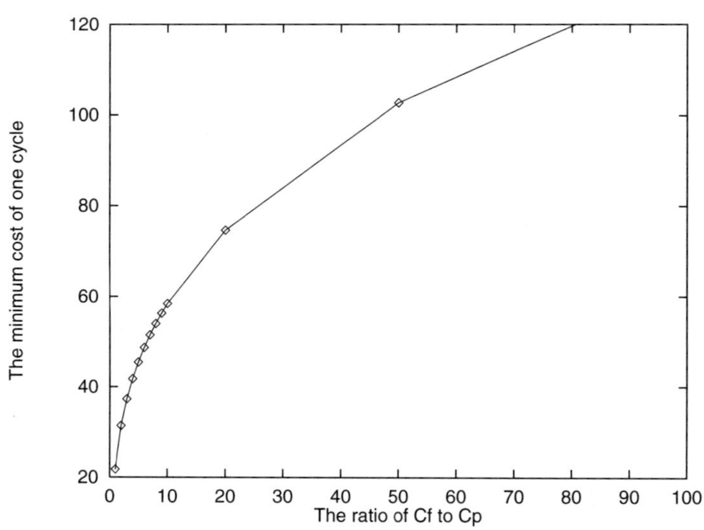

Fig. 6. The minimum costs $C(T^*)$ with respect to the ratio of c_f to c_p.

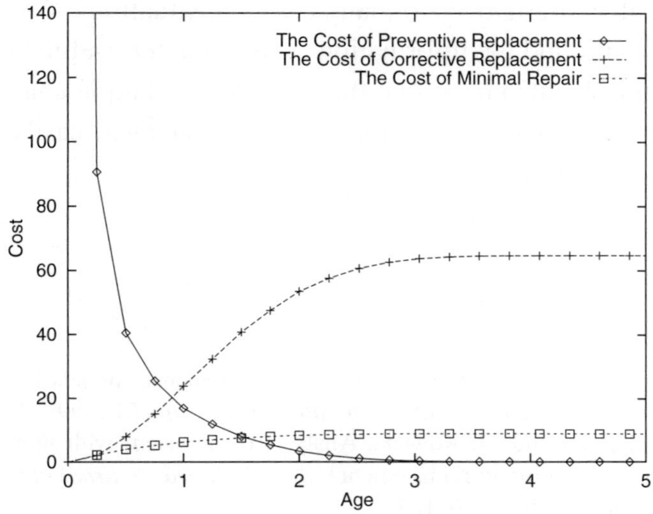

Fig. 7. The distribution of maintenance costs.

the other probabilities. The cost rate of preventive replacement, corrective replacement and minimal repair when c_f is 100 are shown in Fig. 7.

5. Conclusion

In this chapter, we provide a brief review of maintenance actions and failure interaction. We then give a detailed discussion on the cost function relating to the long-run average cost per unit time, and the optimal replacement policies which are defined as minimizing the cost function for both T and N for a three-component system with age dependent failure interaction are investigated. This is followed by an detailed illustration of numerical examples. However, it is difficult to gather and present practical data/information in this chapter. Actually, the numerical example we have given is just one of many interesting cases. The possibility that some specified cases such as one component just causes Induced failure I, and the other components may only fail at Induced failure II or fail alone etc. are considered to be feasible alternatives to many practical situations. We hope that readers would find the information in this chapter useful and easy to understand. We also hope that the numerical example analyzed step by step would enable the model to be implemented easily in many practical situations.

References

1. D. L. Cho and M. Parlar, A survey of maintenance models for multi-unit systems, *European Journal of Operational Research* **51** (1991) 1–23.
2. T. Nakagawa and M. Kowada, Analysis of a system with minimal repair and its application to replacement policy, *European Journal of Operational Research* **12** (1983) 176–183.
3. S. Özekici, Optimal period replacement of multicomponent reliability systems, *Operations Research* **36** (1988) 542–552.
4. Y. S. Sherif and M. L. Smith, Optimal maintenance models for systems subject to failure — A review, *Naval Research Logistics Quarterly* **21** (1981) 949–951.
5. H. Wang, A survey of maintenance policies of deteriorating systems, *European Journal of Operation Research* **139** (2002) 469–489.
6. Q. Zhao, T. Satow and H. Kawai, Optimal replacement policy for a three-component system with age-dependent failure interaction, *Proceedings of 9th ISSAT* (2003), pp. 55–59.
7. H. Pham and H. Wang, Imperfect maintenance, *European Journal Operational Research* **94** (1996) 425–438.
8. H. W. Block, W. S. Borges and T. H. Savits, Age-dependent minimal repair, *Journal of Applied Probability* **22** (1985) 370–385.
9. H. W. Block, W. S. Borges and T. H. Savits, A general age replacement model with minimal repair, *Naval Research Logistics* **35** (1988) 365–372.
10. M. Brown and F. Proschan, Imperfect repair, *Journal of Applied Probability* **20** (1983) 851–859.
11. M. Chen and R. M. Feldman, Optimal replacement policies with minimal repair and age-depend costs, *European Journal of Operational Research* **98** (1997) 75–84.
12. L. C. Thomas, A survey of maintenance and replacement models for maintainability and reliability of multi-item systems, *Reliability Engineering* **16** (1986) 297–309.
13. R. Dekker, Applications of maintenance optimization models: A review and analysis, *Reliability Engineering and System Safety* **51** (1996) 229–240.

14. D. N. P. Murthy and D. G. Nguyen, Study of two-component system with failure interaction, *Naval Research Logistics Quarterly* **32** (1985) 239–248.
15. D. N. P. Murthy and R. T. Casey, Optimal policy for a two component system with shock type failure interaction, *Proceedings of the 8th National Conference of Australian Operation Research Society* **8** (1987) 161–172.
16. S. M. Ross, *Applied Probability Models with Optimization Applications* (Holden-Day, San Francisco, 1991).

Index

accelerated life models, 107
accelerated life testing, 112
activity code, 228
age replacement model, 462
age-dependent failure, 459
alternating renewal process, 11
analysis of variance, 192
ARL, 48
Arrhenius model, 115
availability, 9
average fraction, 333
average run length, 48

Banach space, 4

capability maturity model, 273
central limit theorem, 306
certificate revocation list, 68
certification authority, 67
checkpointing interval, 29
CMM, 273
combined step stress test, 119
complexity concept, 156
conditional arrival distribution, 361
control charts, 47
convolution tools, 7
corrective maintenance, 440
cost model, 434
cost modeling, 444
cost of nonconformance, 237
counting process, 293, 359
CUSUM chart, 48

damage model, 429
decision rules, 409
decision trees, 216
delayed S-shaped, 357
delta CRL, 69
detecting change, 60
Dirac initial distribution, 6
disappointment probability, 373
discrete-time economic, 81
discriminant modeling, 136
double modular system, 31

effort code, 241
engineering effort code, 241
Erlang distribution, 50
error masking system, 34
error types, 322
expected cost, 72, 93, 397, 419
expected cost model, 444
exponential distribution, 47, 90, 307

factorial effects, 196
failure interaction, 464
failure rate, 90
failure-prone product, 406
failures, 291
fault-detection rate, 343
faults, 291
finite volume method, 9

gamma distribution, 20
gas turbine engine, 429
Gaussian white noise, 341
genetic programming, 203
geometric distribution, 90
geometric failure, 90
geometric repair, 90

HALT process, 110
HALT testing, 107
high performance products, 405
homogeneous Poisson process, 362
human errors, 29
human factors, 185

imperfect debugging, 289
inflection S-shaped model, 342
inspection policies, 417
inspection scheme, 408
inspection-maintenance, 439
inter-arrival time, 43
inverse power model, 115

Jelinska–Moranda model, 307

key process area, 275

Lebesgue measure, 5
log–normal distribution, 25
long-run average cost, 468

maintenance, 429, 460
maintenance cost, 451
maintenance plan, 436
marginal distribution, 3
Markov chain, 324
Markov chain process, 49
martingale approach, 289
maturity design testing, 109
maximum likelihood estimate, 296, 345
measuring effort, 225
measuring process quality, 234
minimal repair, 463
minor errors, 322
MTBF, 370
multi-objective optimization, 210

net present value, 84
NHPP, 340
normal distribution, 307

object-oriented software, 155
OC curve, 413
occurrence of defects, 407
optimal checking time, 399
optimal checkpointing, 37
optimal inducer condition, 196
optimal interval, 74
optimal policy, 422, 433, 471
optimal release policy, 350
optimization, 451
orthogonal array, 191

Pareto-optima, 210
partial random testing, 336
periodic inspection, 395
permanent faults, 30
Poisson process, 47
Poisson regression model, 131
preventive maintenance, 440
principal components, 140
process cost, 225
process deterioration, 52
process improvement, 55
production-inventory system, 95
public key infrastructure, 67

QCD, 276
quality engineering, 183
quality, cost, delivery, 276
queuing model, 363

random inspection, 394
rapid thermal stress test, 120
recovery scheme, 32
redundant system, 31
regression model, 134
renewal cycle, 449
renewal density, 8
renewal equation, 377
repairable component, 26
roll-forward recovery, 35
rollback recovery, 35

sampling plan, 315
screening scheme, 408
self-announcing, 456
self-diagnosis system, 417
self-exciting point processes, 290
semi-Markov, 1
software availability, 373
software process improvement, 273
software product measures, 138
software quality improvement, 201
software reliability, 289, 339
software reliability function, 369
software reliability growth model, 358
spatial complexity measures, 159
SRGM, 290
statistical control chart, 43
steady-state probability, 327
stochastic differential equation, 341
stochastic process, 376
stress test, 112
system maintenance, 431

task codes, 251
thermal step stress test, 111
tree-based software quality, 201
two-level testing plan, 319

voltage step stress test, 124

Weibull distribution, 13, 61, 399
Wiener process, 341
work breakdown structure, 226, 279

zero-modified distributions, 407